The Small Business Innovation Research Program: An Assessment of the Department of Defense Fast Track Initiative

Charles W. Wessner, Editor

Board on Science, Technology, and Economic Policy

Policy Division

National Research Council

NATIONAL ACADEMY PRESS
Washington, D.C.

NATIONAL ACADEMY PRESS • 2101 Constitution Avenue, N.W. • Washington, D.C. 20418

NOTICE: The project that is the subject of this report was approved by the Governing Board of the National Research Council, whose members are drawn from the councils of the National Academy of Sciences, the National Academy of Engineering, and the Institute of Medicine. The members of the committee responsible for the report were chosen for their special competences and with regard for appropriate balance.

This report has been reviewed by a group other than the authors according to procedures approved by a Report Review Committee consisting of members of the National Academy of Sciences, the National Academy of Engineering, and the Institute of Medicine. Any opinions, findings, conclusions, or recommendations expressed in this publication are those of the author(s) and do not necessarily reflect the views of the organizations or agencies that provided support for the project.

International Standard Book Number: 0-309-06929-7

Limited copies are available from:
Board on Science, Technology,
and Economic Policy
National Research Council
1055 Thomas Jefferson Street, N.W.
Washington, D.C. 20007
202-334-2200

Additional copies are available for sale from:
National Academy Press
Box 285
2101 Constitution Ave., N.W.
Washington, D.C. 20055
800-624-6242
202-334-3313 (in the Washington
Metropolitan Area)
Internet: http://www.nap.edu

Printed in the United States of America

THE NATIONAL ACADEMIES

National Academy of Sciences
National Academy of Engineering
Institute of Medicine
National Research Council

The **National Academy of Sciences** is a private, nonprofit, self-perpetuating society of distinguished scholars engaged in scientific and engineering research, dedicated to the furtherance of science and technology and to their use for the general welfare. Upon the authority of the charter granted to it by the Congress in 1863, the Academy has a mandate that requires it to advise the federal government on scientific and technical matters. Dr. Bruce M. Alberts is president of the National Academy of Sciences.

The **National Academy of Engineering** was established in 1964, under the charter of the National Academy of Sciences, as a parallel organization of outstanding engineers. It is autonomous in its administration and in the selection of its members, sharing with the National Academy of Sciences the responsibility for advising the federal government. The National Academy of Engineering also sponsors engineering programs aimed at meeting national needs, encourages education and research, and recognizes the superior achievements of engineers. Dr. William A. Wulf is president of the National Academy of Engineering.

The **Institute of Medicine** was established in 1970 by the National Academy of Sciences to secure the services of eminent members of appropriate professions in the examination of policy matters pertaining to the health of the public. The Institute acts under the responsibility given to the National Academy of Sciences by its congressional charter to be an adviser to the federal government and, upon its own initiative, to identify issues of medical care, research, and education. Dr. Kenneth I. Shine is president of the Institute of Medicine.

The **National Research Council** was organized by the National Academy of Sciences in 1916 to associate the broad community of science and technology with the Academy's purposes of furthering knowledge and advising the federal government. Functioning in accordance with general policies determined by the Academy, the Council has become the principal operating agency of both the National Academy of Sciences and the National Academy of Engineering in providing services to the government, the public, and the scientific and engineering communities. The Council is administered jointly by both Academies and the Institute of Medicine. Dr. Bruce M. Alberts and Dr. William A. Wulf are chairman and vice chairman, respectively, of the National Research Council.

For the National Research Council (NRC), this project was overseen by the Board on Science, Technology, and Economic Policy (STEP), a standing board of the NRC established by the National Academies of Sciences and Engineering and the Institute of Medicine in 1991. The mandate of the STEP Board is to integrate understanding of scientific, technological, and economic elements in the formulation of national policies to promote the economic well-being of the United States. A distinctive characteristic of STEP's approach is its frequent interactions with public- and private-sector decision makers. STEP bridges the disciplines of business management, engineering, economics, and the social sciences to bring diverse expertise to bear on pressing public policy questions. The members of the STEP Board* and the NRC staff are listed below:

*As of February 2000.

STEP Staff*

Stephen A. Merrill
Executive Director

John B. Horrigan
Staff Officer

Philip Aspden
Senior Program Officer

David E. Dierksheide
Program Associate

Charles W. Wessner
Program Director

Craig M. Schultz
Research Associate

McAlister T. Clabaugh
Program Associate

*As of February 2000.

Steering Committee for Government-Industry Partnerships for the Development of New Technologies*

Gordon Moore, *Chair*
Chairman Emeritus
Intel Corporation

M. Kathy Behrens
Managing Partner
Robertson, Stephens Investment Management
and STEP Board

Michael Borrus
Managing Director
Petkevich & Partners, LLP

Iain M. Cockburn
Professor of Finance and Economics
Boston University

Kenneth Flamm
Dean Rusk Chair in International Affairs
LBJ School of Public Affairs
University of Texas at Austin

James F. Gibbons
Professor of Engineering
Stanford University
and STEP Board

W. Clark McFadden
Partner
Dewey Ballantine

Burton J. McMurtry
General Partner
Technology Venture Investors

William J. Spencer, *Vice-Chair*
Chairman, SEMATECH
and STEP Board

Mark B. Myers
Senior Vice-President
Xerox Corporation
and STEP Board

Richard Nelson
George Blumenthal Professor of International and Public Affairs
Columbia University

Edward E. Penhoet
Dean, School of Public Health
University of California at Berkeley
and STEP Board

Charles Trimble
Vice-Chairman
Trimble Navigation

John P. Walker
Chairman and Chief Executive Officer
Axys Pharmaceuticals, Inc.

Patrick Windham
Adjunct Professor
Stanford University

*As of February 2000.

Project Staff*

Charles W. Wessner
Study Director

McAlister T. Clabaugh
Program Associate

John B. Horrigan
Staff Officer

David E. Dierksheide
Program Associate

*As of February 2000.

The Small Business Innovation Research Program: An Assessment of the Department of Defense Fast Track Initiative

Research Team

The National Research Council gratefully acknowledges the contributions of the members of the research team.*

Robert B. Archibald
The College of William and Mary

David B. Audretsch
Indiana University

Peter J. Cahill
BRTRC, Inc.

Reid Cramer
University of Texas at Austin

Maryann P. Feldman
Johns Hopkins University

David H. Finifter
The College of William and Mary

John B. Horrigan
National Research Council

Colin Kegler
Harvard University

Joshua Lerner
Harvard Business School

Albert N. Link
University of North Carolina at Greensboro

John T. Scott
Dartmouth College

Claudia Weigand
Indiana University

Juergen Weigand
Indiana University

Charles W. Wessner
National Research Council

Robert H. Wilson
University of Texas at Austin

* Biographies of the Research Team are included in Annex A.

National Research Council
Board on Science, Technology, and Economic Policy

Sponsors

The National Research Council gratefully acknowledges
the support of the following sponsors:

U.S. Department of Defense

National Science Foundation

U.S. Department of Energy

Office of Naval Research

National Institutes of Health

National Institute of Standards and Technology

National Aeronautics and Space Administration

Sandia National Laboratories

Electric Power Research Institute

International Business Machines

Kulicke and Soffa Industries

Merck and Company

Milliken Industries

Motorola

Nortel

Proctor and Gamble

Silicon Valley Group, Incorporated

Advanced Micro Devices

Contents

I

PREFACE

Preface

As we enter a new century full of technological promise, there is a renewed emphasis on the role of industrial entrepreneurs with new, innovative solutions to the problems and opportunities of twenty-first century America. Yet the reliance on industrial entrepreneurs and their symbiotic relationship with the federal government is not new. Driven by both the exigencies of national defense and the requirements of transportation and communication across the American continent, the federal government has played an instrumental role in fostering the development of new production techniques and technologies from the earliest years of the republic. To do so, government has often turned to individual entrepreneurs with innovative ideas. For example, in 1798 the federal government laid the foundation for the first machine tool industry with a contract to the inventor, Eli Whitney, for interchangeable musket parts.[1] A few decades later, in 1842, a hesitant Congress appropriated funds to demonstrate the feasibility of Samuel Morse's telegraph.[2] Both Whitney and Morse fostered significant inno-

[1]Whitney missed his first delivery date and encountered substantial cost overruns. However, his invention of interchangeable parts, and the machine tools to make them, was ultimately successful. The muskets were delivered and the foundation of a new industry was in place. As early as the 1850s, the United States had begun to export specialized machine tools to the Enfield Arsenal in Great Britain. The British described the large-scale production of firearms, made with interchangeable parts, as "the American system of manufacturers." See David C. Mowery and Nathan Rosenberg, *Paths of Innovation: Technological Change in 20th Century America*. New York: Cambridge University Press, 1998, p. 6.

[2]For a discussion of Samuel Morse's 1837 application for a grant and the congressional debate, see Irwin Lebow, *Information Highways and Byways*. New York: IEEE, 1995, pp. 9-12. For a more detailed account, see Robert Luther Thompson, *Wiring a Continent: The History of the Telegraph Industry in the United States 1823-1836*. Princeton, NJ: Princeton University Press, 1947.

vations which led to whole new industries. Indeed, Morse's innovation was the first step on the road toward today's networked planet.

Despite Whitney's ultimate success and the enormous consequences of Morse's ground-breaking innovation, the appropriate role of government in the economy has remained a source of debate and discussion in the United States to this day. Perhaps the earliest articulation of the government's nurturing role with regard to the composition of the economy was Alexander Hamilton's 1791 *Report on Manufactures,* in which he urged an activist approach by the federal government to the creation and nurturing of new industries. At the time, Hamilton's views were controversial, although subsequent U.S. policy has largely reflected his beliefs.

During both the nineteenth and twentieth centuries, the federal government has had an enormous impact on the structure and composition of the economy through infrastructure development, regulation, procurement, and a vast array of policies to support industrial and agricultural development.[3] Between World War I and World War II these policies included support for the development of key industries—which we would now call dual-use—such as radio and aircraft frames and engines. The requirements of World War II generated a huge increase in government procurement and support for high-technology industries.[4] Following that war, the federal government began to fund basic research at universities on a significant scale, first through the Office of Naval Research and later through the National Science Foundation.[5]

During the Cold War, the United States continued to emphasize technological superiority as a means of ensuring U.S. security. Government funds and costplus contracts helped to support systems and enabling technologies such as semiconductors and new materials, radar, jet engines, missiles, and computer

---

[3]Examples abound. The government played a key role in the development of the U.S. railway network, growth of agriculture through the Morrill Act (1862) and the agricultural extension service, and support of industry through the National Bureau of Standards (1901). See Richard Bingham, *Industrial Policy American Style: From Hamilton to HDTV*. New York: M.E. Sharpe, 1998 for a comprehensive review.

[4]David Mowery, "Collaborative R&D: how effective is it?" *Issues in Science and Technology*. 1998, p. 37.

[5]The National Science Foundation was initially seen as the agency that would fund basic scientific research at universities after World War II. However, disagreements over the degree of Executive Branch control over the NSF delayed passage of its authorizing legislation until 1950, even though the concept for the agency was first put forth in 1945 in Vannevar Bush's report, *Science: The Endless Frontier*. The Office of Naval Research bridged the gap in basic research funding during those years. For an account of the politics of the NSF's creation, see G. Paschal Zachary, *Endless Frontier: Vannevar Bush, Engineer of the American Century*. New York: The Free Press, 1997, pp. 231. See also Daniel Lee Kleinman, *Politics on the Endless Frontier: Postwar Research Policy in the United States*. Durham, NC: Duke University Press, 1995.

hardware and software.[6] For example, the government played a central role in the creation of the first electronic digital computer, the ENIAC.[7]

In the post-Cold War period, the evolution of the American economy continues to be profoundly marked by the interaction of government-funded research and innovative entrepreneurs. Government support in areas such as microelectronics, robotics, biotechnology, the human genome, and in the development of ARPANET (the forerunner of today's Internet) are providing the underpinnings of a new economy. Individual entrepreneurs and researchers often played leading roles in developing new approaches and new businesses to exploit these research investments.[8] The emergence of new Net-based companies and biotechnology firms, the latter increasingly focused on genomic-based research, has produced major innovations and promises to be a source of substantial growth.

Despite the important role the U.S. government has played in the development of the American economy, there is little consensus concerning the principle of government participation and there is often considerable debate about the appropriate mechanisms of participation. At the same time, in light of the rising costs, substantial risks, and the breadth of potential applications of new technologies, some believe that a supportive policy framework by the government is necessary for entrepreneurs and entrepreneurial firms in order to bring new, welfare-enhancing and wealth-generating technologies to the market. Partnerships among industry, government, and universities can be an important element in such a framework.

The STEP Mission

Since 1991 the National Research Council's (NRC) Board on Science, Technology, and Economic Policy (STEP) has undertaken a program of activities to improve policy makers' understanding of the interconnections of science, technology, and economic policy and their importance for the American economy and its international competitive position. The Board's activities have corresponded with increased policy recognition of the importance of technology to economic growth. The new economic growth theory emphasizes the role of technology creation, believed to be characterized by significant growth externalities.[9]

---

[6]For an excellent review of the role of government support in nurturing the computer industry, see National Research Council, *Funding a Revolution: Government Support for Computing Research.* Washington, D.C.: National Academy Press, 1999.

[7]Kenneth Flamm, *Creating the Computer.* Washington, D.C.: The Brookings Institution, 1988, chapters 1-3.

[8]David B. Audretsch and Roy Thurik, *Innovation, Industry, Evolution, and Employment.* Cambridge, UK: Cambridge University Press, 1999.

[9]Paul Romer, "Endogenous technological change," *Journal of Political Economy,* vol. 98, 1990, pp. 71-102. See also Gene Grossman and Elhanan Helpman, *Innovation and Growth in the Global Economy,* Cambridge, MA: MIT Press, 1993.

A consequence of the renewed appreciation of growth externalities is the growing focus on the economic geography of economic development. With growth externalities coming about in part from the exchanges of knowledge among innovators, certain regions become centers for particular types of high growth activities. Innovators are able to take advantage of the tacit knowledge available in such centers or clusters of activity to acquire relevant technological innovation and to rapidly address other business development issues.[10]

In addition, some economists have suggested limitations to traditional trade theory, particularly with respect to the reality of imperfect international competition.[11] Recent economic analysis suggests that high-technology is often characterized by increasing rather than decreasing returns, justifying to some the proposition that governments can capture permanent advantage in key industries by providing relatively small, but potentially decisive support to assist national industries up the learning curve and down the cost curve. In part, this is why the economic literature now recognizes the relationship between technology policy and trade policy.[12] Recognition of these linkages and the corresponding ability of governments to shift comparative advantage in favor of the national economy provides the intellectual underpinning for government support for high-technology industry.[13]

Another widely recognized rationale for government support of high-technology exists in cases where technology generates benefits beyond those which can be captured by innovating firms, often referred to as spillovers.[14] There are also cases where the cost of a given technology may be prohibitive for individual companies, even though expected benefits to society are substantial and widespread.[15] The increasing recognition of the dynamic element of technological

---

[10]Paul Krugman, *Geography and Trade,* Cambridge, MA: MIT Press, 1991, p. 23, points out how the British economist Alfred Marshall initially observed in his classic *Principles of Economics* how geographic clusters of specific economic activities arose from the exchange of "tacit" knowledge among business people. Annalee Saxenian's review of the growth of Silicon Valley provides a recent example of the cluster phenomenon. Annalee Saxenian, *Regional Advantage: Culture and Competition in Silicon Valley and Route 128*, Cambridge, MA: Harvard University Press, 1994.

[11]Paul Krugman, *Rethinking International Trade*, Cambridge, MA: MIT Press, 1990.

[12]See J.A. Brander and B.J. Spencer, "International R&D Rivalry and Industrial Strategy," *Review of Economic Studies*, vol. 50, 1983, pp. 707-722, and "Export Subsidies and International Market Share Rivalry," *Journal of International Economics*, vol. 16, 1985, pp. 83-100.

[13]For a discussion of governments' efforts to capture new technologies and the industries they spawn for their national economies, see National Research Council, *Conflict and Cooperation in National Competition for High Technology Industry*. Washington, D.C.: National Academy Press, 1996, p. 28-40. For a critique of these efforts, see P. Krugman, *Peddling Prosperity: Economic Sense and Nonsense in an Age of Diminished Expectations*. New York: W.W. Norton Press, 1994.

[14]See, for example, Martin N. Baily and A. Chakrabati, *Innovation and the Productivity Crisis*. Washington, D.C.: The Brookings Institution, 1998, and Zvi Griliches, *The Search for R&D Spillovers*, Cambridge, MA: Harvard University Press, 1990.

[15]See Ishaq Nadiri, *Innovations and Technological Spillovers*, NBER Working Paper No. 4423, 1993, and Edwin Mansfield, "Academic Research and Industrial Innovation," *Research Policy*, Feb-

innovation, in particular its cumulative nature, has provided intellectual underpinning and incentives for local, state, and national efforts to create competitive advantage for a region, country, or industry.[16]

Project Origins

The growth in government programs to support high-technology industry within national economies and their impact on international science and technology cooperation and on the multilateral trading system are of considerable interest worldwide. Accordingly, these topics were taken up by STEP in a study carried out in conjunction with the Hamburg Institute for Economic Research and the Institute for World Economics in Kiel, Germany, which produced the 1996 report, *Conflict and Cooperation in National Competition for High-Technology Industry.* One of the principal recommendations for further work emerging from that study was a call for an analysis of the principles of effective cooperation in technology development, to include lessons from national and international consortia, including eligibility standards and assessments of what new cooperative mechanisms might be developed to meet the challenges associated with the development of new high-technology products.[17]

In many high-technology industries, the burgeoning development costs for new technologies, the dispersal of technological expertise, and the growing importance of regulatory and environmental issues have provided powerful incentives for public/private cooperation. Notwithstanding the unsettled policy environment in Washington, D.C., collaborative programs have steadily expanded,[18] with perhaps as many as 70 federal cooperative technology programs currently under way.[19] During the Reagan administration, the program examined here—the Small Business Innovation Research program (SBIR)—was created as a way

---

ruary, 1991. Council of Economic Advisers, *Supporting Research and Development to Promote Economic Growth: The Federal Government's Role*. Washington, D.C., 1995.

[16]The dynamic nature of international competition in high-technology industries is discussed in National Research Council, 1996, *op. cit.*, pp. 28-40. For a critique of these efforts, see Krugman, 1995, *op. cit.*

[17]The summary report of the project (National Research Council, 1996, *op. cit.*) recommends further analytical work concerning principles for effective cooperation in technology development (see Recommendation 24, p. 8). More recently, David Mowery has noted the rapid expansion of collaborative activities and emphasized the need for comprehensive assessment. David Mowery, "Collaborative R&D: how effective is it?" *op. cit.*, p. 44.

[18]In addition to programs such as SBIR, SEMATECH, and ATP, other legislative initiatives sought to encourage cooperation and improve the payoff from federal R&D. Examples include the Stevenson-Wydler Technology Innovation Act (1980), the Bayh-Dole University and Small Business Patent Act (1980), the National Cooperative Research Act (1984), and the Federal Technology Transfer Act (1986). These are described in the Introduction.

[19]Dan Berglund and Christopher Coburn, *Partnerships: A Compendium of State and Federal Cooperative Technology Programs*. Columbus, OH: Battelle Press, 1995, p. 481.

to capitalize on the innovative capacity of small business and as a means of using federal R&D dollars more effectively. The creation of the SBIR program was not an isolated phenomenon. Concerns about U.S. competitiveness in the 1980s and early 1990s contributed to the creation of a number of cooperative programs. Some targeted particular sectors; others sought to capitalize on U.S. research. For example, to meet unprecedented challenges in the semiconductor industry, the SEMATECH consortium was established, although only after much debate.[20] In the Bush administration, Congress first funded the Advanced Technology Program (ATP) in the National Institute of Standards and Technology.[21] This program gives a significant percentage of its awards to small business but is not focused on small business per se, but rather on the development of new enabling technologies with a broad impact on the economy. The Clinton administration came to office with an emphasis on civilian technology programs, substantially expanding the ATP and creating the Technology Reinvestment Program (TRP) and the Partnership for the Next Generation Vehicle (PNGV).[22] The rapid expansion of these cooperative programs encountered significant opposition, rekindling the national debate on the appropriate role of the government in fostering new technologies. Indeed, broader philosophical questions about the appropriate role for government in collaborating with industry have tended to obscure the need for policy makers to draw lessons from current and previous collaborative efforts.[23]

Given the considerable change in federal research and development budgets since the end of the Cold War, and the reduced role of many centralized laboratories in the private sector, government-industry collaboration is of growing importance, yet it has seen remarkably little objective analysis.[24] At one level, analysis

[20]For a review of SEMATECH, see the National Research Council, 1996, *op. cit.,* pp. 141-151. For one of the most comprehensive assessments of SEMATECH, see John B. Horrigan, "Cooperating Competitors: A Comparison of MCC and SEMATECH," monograph, National Research Council, 1999.

[21]For a recent review of the ATP, see Charles W. Wessner, ed., *The Advanced Technology Program: Challenges and Opportunities.* Washington, D.C.: National Academy Press, 1999.

[22]For an analysis of ATP, see Christopher T. Hill, "The Advanced Technology Program: opportunities for enhancement," in Lewis Branscomb and James Keller, eds. *Investing in Innovation: Creating a Research and Innovation Policy.* Cambridge, MA: MIT Press, 1998, pp. 143-173. For an excellent analysis of the TRP, see Jay Stowsky, "Politics and Policy: The Technology Reinvestment Program and the Dilemmas of Dual Use," Mimeo, University of California, 1996. See also, Linda R. Cohen, "Dual-use and the Technology Reinvestment Project." in Branscomb and Keller, *op. cit.,* pp. 174-193. For PNGV, see National Research Council, *Review of the Research Program of the Partnership for a New Generation of Vehicles: Third Report.* Washington, D.C.: National Academy Press, 1997.

[23]Dan Berglund and Christopher Coburn, *Partnerships: A Compendium of State and Federal Cooperative Technology Programs.* Columbus, OH: Battelle Press, 1995, p. 481.

[24]Richard Rosenbloom and William Spencer, *Engines of Innovation: U.S. Industrial Research at the End of an Era.* Boston: Harvard Business Press, 1996.

may contribute to a better appreciation of the role of collaboration between government and industry in the development of the U.S. economy. Writing 20 years ago, one well-known American economist, Richard Nelson, observed that Americans are still remarkably uninformed about their long history of policies aimed at stimulating innovation.[25] Today, many Americans appreciate the contribution of technology to the current period of robust economic growth.[26] Yet there is little evidence that Americans are aware of the key contributions of federal support for technological innovation, from interchangeable musket parts to radio to the Internet.

Leaving aside the desirability of having a better understanding of the role of partnerships in fostering new technologies, one compelling argument for assessment is the simple fact that government intervention in the market is fraught with risk. There are cases of major success, such as federal support to the computer or semiconductor industries, where the Department of Defense served as a source of R&D and as a reliable, early buyer of products.[27] There are also cases of major frustration. Landmarks would include projects such as the Supersonic Transport and the Synfuels Corporation.[28] Regular assessment is vital to ensure continued technical viability, though cost-sharing requirements can be an effective safeguard. Assessment can also help avoid "political capture" of projects, especially large commercial demonstration efforts.[29] Even successful collaborations face the challenge of adapting programs to rapidly changing technologies.[30] Assess-

---

[25] Otis L. Graham, *Losing Time: The Industrial Policy Debate.* Cambridge, MA: Harvard University Press, 1992, p. 250. Graham cites Richard Nelson's observations at the end of the Carter Administration. The situation may not have improved. Writing in 1994, James Fallows makes a similar observation (see *Looking into the Sun: The Rise of the New East Asian Economic and Political System.* New York: Pantheon Books, 1994, p. 196). See also Thomas McCraw's "Mercantilism and the market: antecedents of American industrial policy," in *The Politics of Industrial Policy,* Claude E. Barfield and William A. Schambra, eds., Washington, D.C.: American Enterprise Institute for Public Policy Research, 1986, pp. 33-62.

[26] For a review of support for computing and the Internet, see National Research Council, *Funding a Revolution: Government Support for Computing Research.* Washington, D.C.: National Academy Press, 1999, *op. cit.* Chapter 7.

[27] *Ibid.* See also Graham, *op. cit.*, p. 2.

[28] See Linda R. Cohen and Roger G. Noll, *The Technology Pork Barrel.* Washington, D.C.: The Brookings Institution, 1991, pp. 97-148, 259-320. An interesting review of technology development programs, mainly from the 1970s, the analysis is less negative than the title suggests. Indeed, the volume identifies some successful R&D projects such as the photovoltaic electricity program, p. 363. The recent analyses by the Academies of government support for the computing industry underscore the importance of sustained government support.

[29] Cohen and Noll stress that political capture by distributive congressional politics and industrial interests are one of the principal risks for government-supported commercialization projects. In cases such as the Clinch River project, they extensively document the disconnect between declining technical feasibility and increasing political support (see *op. cit.*, p. vii and pp. 242-257).

[30] One of the strengths of SEMATECH was its ability to redefine goals in the face of changing conditions. See National Research Council, 1996, *op. cit.*, p. 148. See also Grindley, et al.,

ment thus becomes a means of keeping programs relevant. Assessment can also have the virtue of reminding policymakers of the need for humility before the "black box" of innovation. As one observer notes, "experience argues for hedged commitments, constant reappraisal, maintenance of options, and pluralism of advice and decision makers."[31]

From an international perspective, understanding the benefits and challenges of these programs is also important insofar as they have been, and remain, a central element in the national development strategies of both industrialized and industrializing countries. Governments have shown a great deal of imagination in their choice of mechanisms designed to support industry. They have adopted a wide range of policies from trade regulations designed to protect domestic products from foreign competition to tax rebates intended to stimulate the export of selected domestic products. Government R&D funding is provided for enterprises of particular interest, and governments sometimes give overt support through direct grants, loans, and equity investments or more opaque support through mechanisms such as tax deferral.[32] Data collected by the Paris-based Organization for Economic Cooperation and Development (OECD) suggest that worldwide government expenditures in support of high-technology industries involve significant resources and are increasingly focused on what policy makers consider to be strategic industries.[33]

The United States is an active, if unavowed, participant in this global competition, at both the state and federal levels. Indeed, the United States has a remarkably wide range of public/private partnerships in high-technology sectors.[34] In addition to the well-known cases mentioned above, there are public/private consortia of many types. They can be classified in a number of ways, such as by the economic objective of the partnership; that is, to leverage the social benefits associated with federal R&D activity, to enhance the position of a national industry, or to deploy industrial R&D to meet military or other government missions.[35]

"SEMATECH and collaborative research: Lessons in the Design of High-technology Consortia." *Journal of Policy Analysis and Management,* 1994, p. 724 and Peter Grindley and William Spencer, "SEMATECH after five years: high-tech consortia and U.S. competitiveness," *California Management Review*, vol. 35, no. 4, pp. 9-33 and Horrigan, "Cooperating Competitors," *op. cit.*

[31]Otis Graham, *op. cit.*, p. 251. Graham is referring to work by Richard R. Nelson in *Government and Technological Progress*. New York: Pergamon Press, 1982, p. 454-455.

[32]National Research Council, 1996, *op. cit.*, Box B., pp. 39-40.

[33]*Ibid.* Concerning support for small business, the OECD gives a positive review of U.S. programs. See OECD, *Technology, Productivity, and Job Creation: Best Policy Practices*. Paris: OECD, 1997, p. 21. Germany, for example, has undertaken a series of initiatives to support and encourage development of seed capital for promising firms and the venture capital industry. See Federal Ministry of Education and Research (BMBF), *Report of the Federal Government on Research: Facts and Figures 1998*. Bonn, December, 1998.

[34]See Chris Coburn and Dan Bergland, *op. cit.,* 1995.

[35]See Albert Link, "Public/Private Partnerships as a Tool to Support Industrial R&D: Experiences in the United States." Paper prepared for the working group on Innovation and Technology Policy of

The program taken up in this volume—the Small Business Innovation Research Program (SBIR) —falls under the latter category.

Project Steering Committee

The continual importance of government-industry collaboration underscores the need for better understanding of the opportunities and limitations of these programs and the conditions most likely to ensure success. Reflecting the interest of policy makers in this topic, the STEP Board initiated the project on *Government-Industry Partnerships for the Development of New Technologies,* which has benefited from broad support among federal agencies. These include: the U.S. Department of Defense; the U.S. Department of Energy; the National Science Foundation; the National Institutes of Health; the National Cancer Institute; the National Institute of General Medical Sciences; the Office of Naval Research; the National Aeronautics and Space Administration; and the National Institute of Standards and Technology; as well as a diverse group of private corporations, listed in the front of the report. To carry out this analysis, the STEP Board has assembled a distinguished multidisciplinary Steering Committee for government-industry partnerships, listed in the front of this report. The Committee's principal tasks are to provide overall direction and relevant expertise to assess the issues raised by the project. At the conclusion of the project, the Steering Committee will develop a consensus report outlining their findings and recommendations on the programs and issues reviewed by the project.

As a basis for the consensus report, the Steering Committee is commissioning research and convening a series of fact-finding meetings in the form of workshops, symposia, and conferences as a means of both informing its deliberations and addressing current policy issues affecting government-industry partnerships. As the project progresses, the Steering Committee is making recommendations and findings on major elements of its work, particularly in response to requests from participating agencies. This volume includes the first set of findings and recommendations based on the Steering Committee's fact-finding meetings and its commissioned research, in this case on the SBIR program.[36] The commis-

---

the OECD Committee for Science and Technology Policy, Paris, 1998, p. 20. Partnerships can also be differentiated by the nature of public support. Some partnerships involve a direct transfer of funds to an industry consortium. Others focus on the shared use of infrastructure, such as laboratory facilities.

[36]A companion volume on SBIR entitled *The Small Business Innovation Research Program: Challenges and Opportunities,* Washington, D.C.: National Academy Press, 1999, provides an overview of the SBIR Program. Other volumes in this series include *Industry-Laboratory Partnerships: A Review of the Sandia Science and Technology Park Initiative*, Washington, D.C.: National Academy Press, 1999; and *The Advanced Technology Program: Challenges and Opportunities*, Washington, D.C.: National Academy Press, 1999. The international component of the project was addressed with the conference and report on *New Vistas in Transatlantic Science and Technology Cooperation*, Washington, D.C.: National Academy Press, 1999.

sioned research provides empirical support for the recommendations and findings. However, as noted below, responsibility for the recommendations rests with the Steering Committee and not with the individual researchers.

A number of individuals deserve recognition for their contributions to the preparation of this report and for their willingness to serve as reviewers. On behalf of the STEP Board we would like to express special recognition to Jon Baron and Robert L. Neal Jr. of the Department of Defense. Their interest and commitment to an objective assessment of the Fast Track was crucial to the success of the project. Similarly, special recognition is due to Peter Cahill of BRTRC, Inc., and Albert N. Link of the University of North Carolina at Greensboro for their many valuable insights and contributions. Among the STEP staff, special recognition goes to John B. Horrigan for his contributions to the development of the research plan, his independent analysis of the project results, and preparation of this volume, both initial drafts and in review. Recognition and thanks are also due to Laura T. Holliday for her many contributions to the organization of the 5 May 1999 conference. Subsequently, McAlister T. Clabaugh and David E. Dierksheide played an instrumental role in editing and preparing the report for publication. Without their collective energy and commitment this project could not have been completed.

This report has been reviewed in draft form by individuals chosen for their diverse perspectives and technical expertise, in accordance with procedures approved by the NRC's Report Review Committee. The purpose of this independent review is to provide candid and critical comments that will assist the institution in making the published report as sound as possible and to ensure that the report meets institutional standards for objectivity, evidence, and responsiveness to the study charge. The review comments and draft manuscript remain confidential to protect the integrity of the deliberative process. We wish to thank the following individuals for their participation in the review process: Dr. James Adams, University of Florida; Dr. William L. Baldwin, Dartmouth College; Dr. Kathryn Combs, University of St. Thomas; Dr. Lance Davis, National Academy of Engineering; Dr. Gerald Dinneen, National Academy of Engineering, Dr. Jeffrey Hart, Indiana University; Dr. Michael Luger, University of North Carolina at Chapel Hill; Dr. Paula Stephan, Georgia State University; and Dr. Cherisa Yarkin, University of California at Berkeley. Although these individuals have provided constructive comments and suggestions, responsibility for the final content of this report and its recommendations and findings rests with the Steering Committee, listed above, and the National Research Council Board on Science, Technology, and Economic Policy.

Charles W. Wessner

II

INTRODUCTION

Introduction

The Small Business Innovation Research (SBIR) program is one of the largest government-industry partnerships in terms of its annual budget. It is also one of the most dispersed in terms of the agencies responsible for its implementation, the diversity of program goals, and the variety of award recipients. Ten agencies and departments grant SBIR awards totaling $1.2 billion annually to support a wide variety of federal missions.[1] Despite the size of the program, its 18-year existence, and anecdotal evidence of its success, relatively little independent research and analysis of the program has been conducted.[2] The papers presented here and the accompanying recommendations represent an important step toward a better understanding of this innovative program and, in particular, the recent Fast Track initiative at the Department of Defense (DoD).

---

[1]Currently, the agencies and departments which have SBIR programs are the Departments of Agriculture, Commerce, Defense, Education, Energy, Health and Human Services, and Transportation; the Environmental Protection Agency, the National Aeronautics and Space Administration, and the National Science Foundation. Until recent budget cuts moved it below the required threshold, the Nuclear Regulatory Commission also had an SBIR program.

[2]The U.S. General Accounting Office has conducted a number of valuable studies. See, for example, the recent GAO report, U.S. GAO, 1999, *Federal Research: Evaluation of Small Business Innovation Research Can Be Strengthened*, GAO/RCED-99-114, Washington, D.C.: U.S. GAO. The Small Business Administration also carries out an extensive research program. As Harvard Business School's Josh Lerner and Colin Kegler note in the literature review appearing in this volume, the academic literature is remarkably limited. Aside from the analysis here—and Lerner's earlier article, "Public Venture Capital: Rationales and Evaluation," in the National Research Council's first volume on the SBIR, C. Wessner, ed., *The Small Business Innovation Research Program: Challenges and Opportunities*. Washington, D.C.: National Academy Press, 1999—there has been little independent assessment of the program's economic impact.

The SBIR Program

SBIR was established in 1982 as a way to channel federal research and development funds to small businesses, while meeting agency mission needs through the use of research and development (R&D) expertise that is often unique to small businesses. Initially, the SBIR program required agencies with R&D budgets in excess of $100 million to set aside 0.2 percent of their funds for SBIR. This totaled $45 million in 1983, the program's first year of operation. Over the next six years, the set-aside percentage grew to 1.25 percent and, in 1992, Congress renewed the program and doubled the set-aside rate to 2.5 percent.[3] For fiscal year 1998 this resulted in a program budget of approximately $1.2 billion across all federal agencies, with the Department of Defense having the largest SBIR program at $540 million, followed by the National Institutes of Health (NIH) at $266 million.[4] Since its inception, the SBIR program has made over 45,000 awards totaling $8.4 billion in 1998 dollars.[5]

From the start, the SBIR grant-making process has had three phases. Phase I is essentially a feasibility study in which award winners undertake a limited amount of research aimed at establishing an idea's scientific and commercial promise. Today, Phase I grants can be as high as $100,000. Phase II grants are larger—normally $750,000—and fund more extensive R&D to further develop the scientific and technical merit and the feasibility of research ideas. Phase III normally does not involve SBIR funds, but is the stage at which grant recipients should be obtaining additional funds either from an interested agency, private investors, or the capital markets to move the technology to the prototype stage and into the marketplace.

Growing Emphasis on Commercialization

Over the SBIR program's 18-year lifespan, the American economy has undergone substantial structural change, yet the relevance of the program has if anything increased. There are several reasons for this. First, policymakers and economists have shown growing appreciation of the role of the small firm in economic development. Starting in the late 1970s and accelerating in the 1980s, a growing body of empirical evidence began to indicate an increasing role for small businesses in job creation and innovation.[6]

---

[3]The Small Business Research and Development Enhancement Act, P.L. 102-564, October 28, 1992.

[4]See http://www.acq.osd.mil/sadbu/sbir/overview.html for information on DoD's SBIR program. For information on NIH's SBIR program, see http://grants.nih.gov/grants/funding/sbir.htm#sbir.

[5]George Brown and James Turner, "The Federal Role in Small Business Research," *Issues in Science and Technology,* Summer, 1999, p. 52.

[6]For an account of the growing importance of the small firm in employment and innovation, see Zoltan J. Acs and David B. Audretsch, *Innovation and Small Business.* Cambridge, MA: MIT Press,

Second, the program grew and matured at a time of increasing concern over the ability of U.S. companies to commercialize R&D results. The early 1980s witnessed a severe recession and the entire decade experienced economic performance below post-World War II norms. The trade deficit rose sharply amidst widespread worry that this was driven by a Japanese economy that was outperforming American manufacturing in important industries, such as steel, autos, and semiconductors.

At the program's inception, a prominent element in the diagnosis of America's economic ills involved the country's failure to successfully commercialize new technologies developed by researchers. A recent report by the STEP Board recalls how the "gloomy picture of U.S. industrial competitiveness" in the 1980s was frequently cast in terms of American industry's failure "to translate its research prowess into commercial advantage."[7] One of the strategies adopted by the United States in response to its loss in competitiveness (at least in some sectors) was to encourage greater cooperation among industry and between industry and government.

Such collaboration was by no means novel in the U.S. economy. As noted above, government funds had supported the demonstration and development of the telegraph in the last century, and after World War I the federal government fostered an independent radio industry.[8] Later, the federal government also provided active support through a variety of mechanisms for military and civil avia-

---

1991, p. 4. For specifics on job growth, see Steven J. Davis, John Haltiwanger, and Scott Schuh, "Small Business and Job Creation: Dissecting the Myth and Reassessing the Facts," *Business Economics*, vol. 29, no. 3, 1994, pp. 113-22. More recently, a report by the Organization for Economic Cooperation and Development (OECD) notes that small and medium-sized enterprises are attracting the attention of policymakers, not least because they are seen as major sources of economic vitality, flexibility, and employment. Small business is especially important as a source of new employment, accounting for a disproportionate share of job creation. See OECD, *Small Business Job Creation and Growth: Facts, Obstacles, and Best Practices*. Paris, 1997.

[7]David C. Mowery, "America's Industrial Resurgence (?): An Overview," in David C. Mowery, ed., *U.S. Industry in 2000: Studies in Competitive Performance*. Washington, D.C.: National Academy Press, 1999, p. 1. This volume examines eleven economic sectors, contrasting the improved performance of many industries in the late 1990s with the apparent decline that was subject to much scrutiny in the 1980s. Among the studies highlighting poor economic performance in the 1980s include Dertouzos, et al., *Made in America: The MIT Commission on Industrial Productivity*, Cambridge, MA: The MIT Press, 1989 and Eckstein, et al., *DRI Report on U.S. Manufacturing Industries*, New York: McGraw Hill, 1984.

[8]Josephus Daniels, Secretary of the Navy during the Wilson administration, appeared to feel that monopoly was inherent to the wireless industry and, if that were the case, the monopoly should be American rather than British. Britain had dominated pre-war Atlantic wireless traffic as well as the undersea telegraph cable. By pooling patents, providing equity, and encouraging General Electric's participation, the Radio Corporation of America was created. See Irwin Lebow, *Information Highways and Byways: From the Telegraph to the 21st Century*. New York: IEEE Press, 1995, pp. 97-98 and chapter 12.

tion and the electronics industry.[9] Yet the 1980s and early 1990s saw a conscious effort to expand cooperation, in part by using federal R&D funding more effectively, to meet what were seen as unprecedented competitive challenges.

A series of public and private initiatives in the 1980s demonstrate the renewed emphasis on cooperation. The change in public policy is illustrated by the number of major legislative initiatives passed by Congress. These included: the Stevenson-Wydler Technology Innovation Act (1980), the Bayh-Dole University and Small Business Patent Act (1980), the Small Business Innovation Development Act (1982), the Federal Technology Transfer Act (1986), the Omnibus Trade and Competitiveness Act (1988), the National Competitiveness Technology Transfer Act (1989), and the Defense Conversion, Reinvestment, and Transition Assistance Act (1992). These individual acts are summarized in the below.

Principal Federal Legislation Related to Cooperative Technology Programs[10]

- **Stevenson-Wydler Technology Innovation Act (1980)** Required federal laboratories to facilitate the transfer of federally owned and originated technology to state and local governments and the private sector. The Act includes a requirement that each federal lab spend a specified percentage of its research and development budget on transfer activities and that an Office of Research and Technology Applications (ORTA) be established to facilitate such transfer.
- **Bayh-Dole University and Small Business Patent Act (1980)** Permitted government grantees and contractors to retain title to federally funded inventions and encouraged universities to license inventions to

continued

[9]David C. Mowery and Nathan Rosenberg, *Technology and the Pursuit of Economic Growth.* Cambridge, UK: Cambridge University Press, 1989. See chapter seven especially pp. 181-194. The authors note that the commercial aircraft industry is unique among manufacturing industries in that a federal research organization, the National Advisory Committee on Aeronautics (founded in 1915 and absorbed by NASA in 1958), conducted and funded research on airframe and propulsion technologies. Before World War II, NACA operated primarily as a test center for civilian and military users. NACA made a series of remarkable contributions with regard to engine nacelle locations and the NACA "cowl" for radial air cooled engines. These innovations, together with improvements in engine fillets based on discoveries at Caltech and the development of monocoque construction, had a revolutionary effect on commercial and military aviation. These inventions made the long-range bomber possible, forced the development of high-speed fighter aircraft, and vastly increased the appeal of commercial aviation. *Ibid.,* and personal communication with Albert Flax, National Academy of Engineering.

[10]Drawn, with NRC modifications, from Berglund and Coburn, *op. cit.,* p. 485.

continued from previous page

industry. The Act is designed to foster interaction between academia and the business community. This law provided, in part, for title to inventions made by contractors receiving federal R&D funds to be vested in the contractor if they are small businesses, universities, or not-for-profit institutions.

- **Small Business Innovation Development Act (1982)** Established the Small Business Innovation Research (SBIR) Program within the major federal R&D agencies to increase government funding of research with commercialization potential in the small high-technology company sector. Each federal agency with an R&D budget of $100 million or more is required to set aside a certain percentage of that amount to finance the SBIR effort.
- **National Cooperative Research Act (1984)** The National Cooperative Research Act of 1984 eased antitrust penalties on cooperative research by instituting single, as opposed to treble, damages for antitrust violations in joint research. The Act also mandated a "rule of reason" standard for assessing potential antitrust violations for cooperative research. This contrasted with the per se standard by which any R&D collusion is an automatic violation, regardless of a determination of economic damage.
- **Federal Technology Transfer Act (1986)** Amended the Stevenson-Wydler Technology Innovation Act to authorize cooperative research and development agreements (CRADAS) between federal laboratories and other entities, including state agencies.
- **Omnibus Trade and Competitiveness Act (1988)** In addition to establishing the Competitiveness Policy Council, designed to enhance U.S. industrial competitiveness, the Act created several new programs (e.g., the Advanced Technology Program and the Manufacturing Technology Centers) housed in the Department of Commerce's National Institute of Standards and Technology and intended to help commercialize promising new technologies and to improve manufacturing techniques of small and medium-sized manufacturers.
- **National Competitiveness Technology Transfer Act (1989)** Part of the Department of Defense authorization bill, this act amended the Stevenson-Wydler Act to allow government-owned, contractor-operated laboratories to enter into cooperative R&D agreements.
- **Defense Conversion, Reinvestment, and Transition Assistance Act (1992)** Initiated the Technology Reinvestment Project (TRP) to establish cooperative, interagency efforts that address the technology development, deployment, and education and training needs within both the commercial and defense communities.

Even with these policy innovations, the persistent perception that the U.S. continued to lag in its economic competitiveness fostered cooperative efforts to more fully capture the benefits of its research programs.[11] A National Academy of Sciences study released in 1992 found that "U.S. technological performance is challenged less in the creation of new technologies than in their commercialization and adoption."[12] While noting the difficulty in determining whether the United States's ability to commercialize technology was deteriorating relative to foreign competitors, the Academy found that "the United States can strengthen technology commercialization, at a stage prior to that at which private firms invest in commercialization activities, through federal activities to facilitate precommercial R&D."[13]

SBIR's Reauthorization in 1992

The SBIR program approached reauthorization in 1992 in the context of continued worries about the U.S. economy's capacity to commercialize inventions. As noted above, the 1992 SBIR reauthorization resulted in the set-aside being raised from 1.25 percent to 2.5 percent. This increase was consistent with a recommendation from the National Academy of Sciences to increase SBIR funding as a means to improve the U.S. economy's ability to adopt and commercialize new technologies.[14] By 1992, the SBIR program had also become politically popular with increasingly influential small business advocates. In conjunction with the emergence of innovative small start-ups in computing, biotechnology, and advanced materials, there was ample support for program expansion in 1992.[15]

The increase in the percentage of R&D funds allocated to the program was accompanied by a stronger emphasis on encouraging the commercialization of SBIR-funded technologies. As Robert Archibald and David Finifter describe in detail in this volume, legislative language added to the SBIR program's charge in 1992 explicitly highlighted commercial potential as a criterion for awarding SBIR grants.[16] For Phase I awards Congress directed program administrators to assess whether projects have "commercial potential" in addition to scientific and techni-

---

[11]Private sector cooperation was encouraged by the reduction in antitrust concerns. In 1984 Congress overwhelmingly passed the National Cooperative Research Act, which eased antitrust penalties for companies conducting joint research and development. Responding to this new environment, the private sector also undertook a series of innovative approaches to address its competitive failings, e.g., SEMATECH.

[12]Committee on Science, Engineering, and Public Policy, *The Government Role in Civilian Technology: Building a New Alliance.* Washington, D.C.: National Academy Press, 1992, p. 29.

[13]*Ibid.*, p. 42.

[14]*Ibid*, p. 2.

[15]Brown and Turner, *op. cit.,* p. 53. In addition to an account of SBIR's evolution, Brown and Turner offer constructive criticisms of SBIR and recommendations for improvement.

[16]Robert B. Archibald and David H. Finifter, "Evaluation of the Department of Defense Small Business Innovation Research Program and Fast Track Initiative: A Balanced Approach," in this volume.

cal merit when evaluating SBIR applications. With respect to Phase II, evaluation of a project's commercial potential would consider the existence of second-phase funding commitments from the private sector or from non-SBIR sources and the existence of third-phase, follow-on commitments, along with other indicators of commercial potential.

Moreover, the reauthorization directed that a small business's record of commercialization be taken into account when considering the Phase II application. To further reinforce the emphasis on commercialization, the 1992 reauthorization moved the goal "to increase private sector commercialization" from fourth to second in the list of SBIR program goals. The reauthorization did not provide specific guidelines as to how much weight should be given to commercialization as compared with the program's other goals, such as technological innovation or importance to the agency mission.[17]

Program Diversity and Innovation

As noted, the SBIR program applies to the ten federal agencies with annual extramural R&D above $100 million. While the Small Business Administration is tasked with the coordination of the SBIR program, the dispersal of the program across departments and agencies with very different missions and modes of operation results in considerable variation across the participating organizations. And, in the case of DoD and NIH, the missions the program supports also vary substantially. This has resulted in a very diverse response to the 1992 mandate to increase the focus on commercialization.[18] One of the most important responses has come from the Department of Defense, which has the largest SBIR program.

The Fast Track Initiative in the Defense Department

As early as 1992, DoD's Ballistic Missile Defense Organization (BMDO) began to reward applications whose technologies demonstrated commercial potential. This BMDO initiative called "co-investment" was effectively an informal "fast track" program. Under this approach, the evaluation process for Phase II proposals gave preference to applicants who could demonstrate that they would commit internal funding or that they had financial or in-kind commitments from third parties to bring the technology to market in Phase III. With that commitment, applicants received essentially continuous funding from Phase I to Phase II.[19]

---

[17]A recent GAO study found that agencies have not adopted a uniform method for weighing commercial potential in SBIR applications. See U.S. General Accounting Office, *Federal Research: Evaluations of Small Business Innovation Research Can Be Strengthened*, *op. cit.*, p. 2.

[18]*Ibid.*

[19]Additional funding was not "required" for award selection. In 1992 less than half of awardees had such funding commitments. By 1996 this had risen to over 90 percent. The BMDO program did not always entail an external commitment.

In October 1995, Defense launched a broader Fast Track initiative to attract new firms and encourage commercialization of SBIR funded technologies throughout the Department.[20] The principal ways in which Fast Track seeks to improve commercialization is through preferential evaluation and efforts to close the funding gap that can develop between Phase I and Phase II grants. The time-lag between the conclusion of Phase I and the receipt of Phase II funds can create cash-flow problems for small firms. The Fast Track pilot addresses the gap by providing expedited review and essentially continuous funding from Phase I to Phase II *as long as* applying firms can demonstrate that they have obtained third-party financing for their technology.[21] In this context, third-party financing means that another company or government agency has agreed to invest in or purchase the SBIR firm's technology; it can also mean that a venture capitalist has committed to invest in the firm or that other private capital is available. The expedited decision-making process for the Phase II award is justified from the agency's perspective because outside funding validates the commercial promise of the technology. More broadly, the Fast Track program addresses the need to shorten government decision cycles in order to interact more effectively with small firms focused on rapidly evolving technologies.

Assessing the Fast Track Initiative

Two years into the Fast Track initiative the Under Secretary of Defense asked the National Research Council's Board on Science, Technology, and Economic Policy (STEP) to assess Fast Track and related SBIR policy matters. The Under Secretary's request focused on three issues:

---

[20]The Fast Track program includes 6-9 percent of the SBIR awards. A description of the program is included in Annex D. Additional information about DoD's Fast Track can be found at the following Website: http://www.acq.osd.mil/sadbu/sbir/fsttrack.html. As of April 2000, 164 Phase I projects had qualified for Fast Track by attracting the required investment. 95 percent of these were selected for Phase II awards. By contrast, on average, only 40 percent of DoD Phase I projects are selected for Phase II. Because the survey of Fast Track companies reported in this volume focused on Fast Track firms that had advanced to Phase II, the survey was sent only to firms from the first Fast Track solicitation, FY 1996, a total of 48 firms. However, surveys were also sent to early BMDO awardees who employed a similar approach.

[21]Some states have developed innovative loan programs to address this gap. For example, the State of New Jersey has a Small Business Innovation Research Bridge Loan Program which is intended to provide access to working capital for small New Jersey technology companies who are between Phase I and Phase II of a federal Small Business Innovation Research development project. The Bridge Loan Program provides a loan guarantee to a private sector lender, which in turn issues a term loan to the SBIR recipient company. Eligibility for a loan guarantee is limited to New Jersey companies which have received a federal Phase I SBIR award, have completed Phase I activity, and have submitted a follow-on Phase II application to the federal agency. See http://www.state.nj.us/scitech/sbirinfo.html for a full description of the New Jersey program.

1. Whether Fast Track projects are achieving, or appear likely to achieve, greater success in SBIR than comparable non-Fast Track projects;
2. Whether Fast Track projects progress at different rates than non-Fast Track projects;
3. What companies perceive as advantages and disadvantages of Fast Track participation.

The request also permitted the Board to review other issues relevant to the operation and performance of the SBIR program. The STEP Board accepted the Under Secretary's request, taking up the task under the auspices of its project on *Government-Industry Partnerships for the Development of New Technologies*, led by Gordon Moore, Chairman Emeritus, Intel, and Chairman of the distinguished Steering Committee listed in the front matter.

While the Department of Defense is to be commended for seeking an outside assessment of a major initiative for its SBIR program, the assessment of an initiative that has been in place for a relatively short period posed a special challenge. To meet that challenge, the Steering Committee decided to undertake a multifaceted approach to the research. The first phase was to bring together university researchers, award recipients, and DoD program managers to discuss the operational goals and practices of the program, the experiences of the awardees, and how the program might by evaluated.

Based on these initial meetings and the virtual absence of academic research on the SBIR program, the Committee decided to commission field research on the program with a special emphasis on the Fast Track initiative. To this end, a research team—whose members had not previously studied the SBIR program—was assembled.[22] The team examined the SBIR program awards and the Fast Track initiative from three different perspectives:

1. **Survey Research:** As a first step, the STEP research team developed a survey instrument and then commissioned an outside consulting firm experienced with the program to carry out a large-scale survey of DoD SBIR awardees, using a sample of firms that have participated in Fast Track and a control group. The roughly 300 firms[23] (294 firms doing 379 projects) queried constitute the largest survey to have focused on the Fast Track

[22]The contributions of the members of the research team are noted in the front matter. The researchers' papers, on which the recommendations and findings are largely based, are listed in the table of contents. The research team included Albert N. Link, University of North Carolina at Greensboro; John T. Scott, Dartmouth College; David B. Audretsch, Juergen Weigand, and Claudia Weigand, Indiana University; Reid Cramer and Robert Wilson, University of Texas at Austin; Robert B. Archibald and David H. Finifter, College of William and Mary; Maryann P. Feldman, Johns Hopkins University; and Colin Kegler and Josh Lerner, Harvard Business School. Peter Cahill, BRTRC, Inc., who has extensive expertise with the SBIR program, also served on the research team.

[23]Separate surveys for each of the 379 projects were sent to 294 firms.

pilot. Reflecting the diligence with which the survey was pursued, the survey response rate was high, with approximately 72 percent of the firms responding.

2. **Case Studies:** In parallel, members of the research team conducted a series of case studies of SBIR companies including, wherever possible, both Fast Track and non-Fast Track participants. The case studies looked at firms in New England, California, North Carolina, Indiana, Texas, New Mexico, Colorado, and the Washington, D.C., metropolitan area. A total of 55 case studies were conducted.
3. **Empirical Analysis:** Using survey results and case studies, STEP-commissioned researchers also examined whether SBIR-funded technologies would have been pursued without the SBIR award and what the social returns to SBIR-funded technologies were.

Overview of Papers and Recommendations in the Report

The recommendations and findings approved by the Committee appear in Chapter III of this volume. They are based on the commissioned research, described above, and on a series of fact-finding and review meetings. The largest of these review meetings was a symposium held on May 5, 1999, at the National Academies, where the papers on which the findings and recommendations are based were presented. Symposium attendees (listed in Annex B) included Department of Defense officials, academics, and staff members from Capitol Hill and the Executive Branch. The symposium permitted researchers to obtain feedback on their work, and researchers were able to subsequently revise their papers based on comments from discussants and symposium participants. The supporting analyses and papers are reproduced in Chapter IV. Additional references and background information are available in the annexes and the bibliography.

While it is important to keep in mind the limits of the research effort and the short history of the Fast Track initiative,[24] the findings of the researchers do reflect well on the program. Put simply, the case studies, surveys, and empirical research suggest that the Fast Track initiative is meeting its goals of encouraging commercialization and attracting new firms to the program. Consequently, the Committee recommends that Fast Track be continued and expanded where appropriate. The Committee does not recommend that Fast Track be applied to the entire SBIR program at DoD. To do so might put at risk other goals, such as research and concept development. It is also important to keep in mind the need for additional research to validate these results over time.

---

[24]These limitations and the measures taken to compensate for them are described in Chapter III below.

Research Papers

To further the reader's understanding of the program, this volume includes the individual research papers on which the recommendations and findings are largely based, as well as an overview of existing economic assessment. The Harvard Business School's Josh Lerner and Colin Kegler note in their review that despite the proliferation of public efforts, both in the U.S. and abroad, to finance small, high-technology firms, there has been relatively little assessment of the economic impacts of these programs. Their paper provides valuable background to the case studies in that it explores the challenges posed by the financing of new firms, the ways specialized financial intermediaries address them, and the rationales for and problems faced by programs designed to assist small firms. Lerner and Kegler also provide a valuable summary of previous evaluation efforts of such programs and their limitations. The dearth of research on these programs, noted above, and particularly the SBIR program, led the Committee to commission original field research in the form of case studies and a survey in order to better understand the operation of the SBIR at the Department of Defense, and in particular, its Fast Track initiative.

The case studies prepared by the research team are also included to provide additional texture and detail to the analysis. While the case studies were all based on a common template (included in Annex C), the researchers undertook their research independently and were encouraged to pursue fruitful lines of inquiry even if these diverged from the template. Consequently, while all the case studies address the issues requested by the Under Secretary of Defense, each paper reflects the individual perspective of the researchers and the differences in the questions posed. All the companies interviewed for the case studies also responded to the survey. In addition, several researchers pursued independent lines of inquiry. For example, the paper by David Audretsch, Jurgen Weigand, and Claudia Weigand (Indiana University) on SBIR recipients in Indiana looks closely at whether the program influences the behavior of individual researchers, especially those in university environments. Reid Cramer, with Robert H. Wilson (University of Texas, Austin), develops a classification of SBIR firm types from case studies of firms in the southwestern and mountain states. In looking at a small sample of SBIR firms in the southeastern states, Albert N. Link (University of North Carolina, Greensboro) discovers distinct differences between Fast Track and non-Fast Track firms in the area of funding gaps between Phase I and Phase II, commercial potential, cost of commercialization, and social returns. John T. Scott (Dartmouth College) conducted the largest set of case studies, focused on New England. He examined fourteen SBIR award recipients and found clear differences between Fast Track and non-Fast Track firms in the area of commercial potential and social rates of return.

The paper by Robert Archibald and David Finifter (College of William and Mary) and the paper by Maryann Feldman (Johns Hopkins University) also took different approaches to assessing SBIR and Fast Track. In addition to providing a detailed legislative history of SBIR, Archibald and Finifter explored whether a tradeoff exists between the program's goals to encourage scientific and technical excellence as well as commercial success. To explore this question, they developed an innovative, e-mail based survey of DoD technical monitors of SBIR awards. Contrary to their initial hypothesis, they found that the technical monitors rated the quality of SBIR-funded research as virtually equal to that of other research funded by the Defense Department. Maryann Feldman examined DoD SBIR funding of a particular technology area, namely the biosciences. Her work highlights the contributions of DoD funding of R&D in the biosciences, and how Defense SBIR awards have proven to be important catalysts to the commercial success of several growing biotechnology firms.

With respect to the empirical analysis of the survey of SBIR awardees, Peter Cahill (BRTRC, Inc.) reports that half had participated in the Fast Track or BMDO co-investment programs and half had not. The survey results provide indispensable baseline data for determining the differences (e.g., age of firm, the firm's past experience with SBIR, length of funding gap between Phase I and Phase II, and commercial potential of the technology) between Fast Track and non-Fast Track firms. Finally, using information gathered in case study research on firms in the Southeastern and New England states, Link and Scott develop a measure of the social returns to SBIR-funded technologies in excess of private returns. They find that social rates of return substantially exceed expected private returns, which they see as the "fundamental rationale" for the program. Link and Scott also find that the funded companies would not have undertaken the R&D without public support.[25]

In order to provide a basis for more systematic conclusions about differences in Fast Track and non-Fast Track projects, the Committee commissioned a statistical analysis of the influence of the Fast Track program. This analysis was a collaborative effort among Audretsch, Link, and Scott. The major conclusions of this analysis of the survey data were that Fast Track projects have greater expected sales, a shorter funding gap, and greater employment growth than non-Fast Track projects. These findings complemented the case-based analyses described above and were designed to correct for potential bias. The use of multiple research techniques—carried out by a heterogeneous team of researchers with inputs from both DoD managers and awardees—represents a cohesive effort to measure the program's impact. The fact that several overarching findings emerge from a variety of different approaches, applied by researchers working independently, suggests that the findings are reasonably robust.

---

[25]Their estimates are based on projected sales of SBIR awardees. They are consistent with other estimates of social return for R&D investment in the economic literature. See Mansfield, *op. cit.*, and Griliches, *op. cit.*

Findings and Recommendations

The findings and recommendations are grouped under general findings, specific research topics, and two sets of recommendations for the Fast Track program and for future research. This section on general findings addresses issues of relevance to the program as a whole and outlines the results of the survey and case studies specifically relevant to the Fast Track initiative. A second general grouping reviews the specific findings by research topic. These topics include the impact of SBIR awards on university researchers, the role of DoD awards in developing innovations in biotechnology, the regionally-based findings, and a series of findings and recommendations with regard to multiple award winners. There is also a section with specific findings and recommendations made by the Steering Committee for the DoD Fast Track program itself. Lastly, there is a series of recommendations for future research. The findings and recommendations appear in Chapter III.

Conclusion

Small business will continue to play a prominent role in the U.S. economy and innovation system. With the growth of the Internet and electronic commerce in the 1990s, along with the unfolding biotechnology revolution, entrepreneurs and small businesses will be a vibrant source of innovation and job growth. Programs such as SBIR can serve as a catalyst for encouraging innovation in this entrepreneurial climate. To remain responsive to small firms in rapidly changing markets, programs such as SBIR must engage in policy experiments. The Fast Track initiative represents an innovative way of matching program objectives to an economy in which the rewards to rapid innovation are growing. The analysis and recommendations in this volume underscore the positive contributions of the SBIR program to the Defense mission. Based on its evaluation of Fast Track, the Committee recommends that the program be continued and expanded where appropriate.[26] The recommendations also identify areas which would benefit from further research and analysis in order to ensure the Fast Track initiative and the SBIR program at the Department of Defense continue to improve.

Charles W. Wessner

[26]See, for example, the summary of the survey in Peter Cahill, "*Fast Track: Is It Speeding Commercialization of Department of Defense Small Business Innovation Research*?" in this volume.

III

RECOMMENDATIONS AND FINDINGS

Evaluation of the Fast Track Initiative for the Department of Defense SBIR Program

Summary of Findings

I. General Findings

As noted in the Introduction, the Findings and Recommendations summarized below have important attributes and equally important limitations. A major attribute of the study is that the Board commissioned a significant series of studies to better understand the SBIR program. The case studies, the survey, and the empirical analyses were conducted independently, generally by established researchers who had not previously studied the program. Secondly, the research findings were publicly presented and discussed in an open forum and subsequently subjected to internal review procedures. While the papers and their conclusions are the responsibility of the investigators, the Findings and Recommendations are the responsibility of the Committee. As such, they reflect the results of the field research, the tacit knowledge acquired about the SBIR program during this phase of the project, and of course the substantial and diverse expertise of the Committee itself.

However, as noted in the Introduction, and as the investigators themselves note, there are important caveats and limitations to this research. The first limitation concerns the relatively short time that the Fast Track Program has been in place. This necessarily limits our ability to assess the impact of the program. Secondly, although the case studies and surveys constitute what is clearly the largest independent assessment of the SBIR Fast Track Program at the Department of Defense, the study is nonetheless constrained by the limitations of the case-study approach and the size of the survey sample.[1] Research results are

[1]The relatively high rate of response to the survey is a positive feature of this research. Still, as with any research effort, there are a number of potential sources of bias. For example, the survey may be

necessarily preliminary. It is also important to keep in mind that this analysis of the SBIR—and the findings—are limited to the program of the Department of Defense. It is for this reason that the Committee has underscored (in Section IV below) the need for additional research. Additional assessment is required to confirm these results as technologies and firms continue to mature and as new firms increasingly take advantage of the SBIR Program. Moreover, regular assessment of the SBIR Program results and their comparison with the results of the Fast Track, both at Defense and other participating agencies, would provide a valuable means of understanding the operation of this important cooperative program, which currently operates at approximately $1.2 billion annually. As the resources allocated to federal research increase, under current law the SBIR program is destined to increase as well, underscoring the need for careful assessment of the program's efficiency and effectiveness.

Although this research represents a significant step in improving our understanding of the SBIR program, these findings should be appreciated for what they are; that is, a preliminary and limited effort by independent researchers and an informed Committee to understand the operation of an important government-industry partnership.

A. Program Wide Findings

1. The SBIR Program is contributing to the achievement of Department of Defense mission goals. Valuable innovative projects are being funded by the SBIR. It appears that a significant portion of these projects would not have been undertaken in the absence of the SBIR funding.[2]

---

subject to a form of response bias if subjects have an incentive to make favorable comments about the program (Lerner raises this point in his paper, but also notes other factors which may, in fact, limit the willingness of entrepreneurs to emphasize the impact of the program). Some point out that a self-selection bias can be present insofar as the better firms may "self-select" for the Fast Track Program, although this criticism has a circular element. And as noted above, there is a sampling bias in that all of the Fast Track award recipients had not completed their projects before being asked to evaluate the Fast Track Program. The researchers and the Committee were sensitive to these concerns from the outset. The Committee attempted to indemnify against interpretive implications of these biases from the beginning by supporting a portfolio of methodologies to evaluate Fast Track as well as a portfolio of researchers with heterogeneous skills. While this effort does not eliminate all potential sources of bias, where possible the researchers controlled statistically for these types of biases. Because this multi-faceted approach to the research provides a useful means to search for common themes, the researchers felt that on balance the gains to be derived from this approach outweighed the methodological limitations. The statistical analysis prepared by Professors Audretsch, Link, and Scott specifically addresses these potential limitations of the data. Their empirical results are both robust and complementary to the case-based analyses by other researchers. See "Statistical Analysis of the National Academy of Sciences Survey of SBIR Awardees: Analyzing the Influence of the Fast Track Program," especially pp. 298-304.

[2]See especially the papers in this volume of Robert Archibald and David Finifter, "Evaluation of the Department of Defense Small Business Innovation Research Program and the Fast Track Initia-

2. The Fast Track Program increases the effectiveness of the SBIR Program at the Department of Defense by encouraging the commercialization of new technologies.[3]
3. Case studies from various regions in the United States found that the Department of Defense SBIR Program facilitates the development and utilization of human capital and technological knowledge.[4] This holds true even for firms that exit the market. The experience and human capital generated by the program have economic value and can be applied by other firms.
4. The receipt of an SBIR award can serve a certification function with regard to the firm's technology and eventual market, thus encouraging private sector investment.[5]

B. Fast Track Findings: Survey and Case Study Results

1. The Fast Track Program, whose principal objective is to reduce the funding gap between Phase I and Phase II of the SBIR award process, works. Among the companies surveyed, over half the Fast Track firms experienced no funding gap and the average funding gap was 2.4 months. By contrast, over 80 percent of companies not participating in the Fast Track Program confronted a funding gap between the end of Phase I funding and the beginning of Phase II funding; the average gap was 4.7 months.[6] In particular,

   a. The average additional development funding received by Fast Track firms was five times greater than for a control group.[7] This suggests, as intended, that the Fast Track Program is both attracting firms with a greater potential for commercial success and, with its promise of virtu-

---

tive" and the paper of Albert N, Link and John T. Scott "Estimates of the Social Returns to Small Business Innovation Research Projects." This finding is consistent with the case studies conducted by Reid Cramer, David Audretsch, and Albert Link in their contributions to this volume.

[3]Peter Cahill, "Fast Track: Is It Speeding Commercialization of DOD's SBIR?" pp. 64-65, and Figures 7 and 8.

[4]The paper of Audretsch, Weigand, and Weigand, "Does the Small Business Innovation Research Program Foster Entrepreneurial Behavior? Evidence from Indiana," focuses particularly on how SBIR encourages the development of human capital.

[5]See the papers in this volume by Reid Cramer "Participation in the SBIR Program in the Southwestern and Mountain States," p. 152 and John Scott "Assessment of the Small Business Innovation Research Program in New England: Fast Track Compared with Non-Fast Track Projects," pp. 116-118. Harvard Business School's Josh Lerner initially identified this effect. See "Public Venture Capital": *The Small Business Innovation Research Program: Challenges and Opportunities,* Washington, D.C.: National Academy Press, 1999, pp. 120-122. See also Kegler and Lerner, "Evaluating the SBIR: A Literature Review," p. 321 in this volume.

[6]Peter Cahill, "Fast Track: Is It Speeding Commercialization of DOD's SBIR?" pp. 66-67.

[7]*Ibid.,* p. 44.

ally continual funding through Phase I and Phase II, is aiding in attracting capital to SBIR Fast Track firms.

b. Actual and expected commercialization is greater for Fast Track firms. Fast Track firms forecast significantly greater sales than do non-Fast Track firms.[8]

c. Two-thirds of the technical representatives of agencies within the Department of Defense evaluated Fast Track projects as *more* effective than other projects in advancing their research goals, with only 11.8 percent rating Fast Track as less effective[9]

2. The Fast Track Program appears useful in assisting companies in attracting outside investment.[10] However, individual companies have to weigh the benefits of Fast Track against its requirement for early outside investment.[11]
3. Neither the Fast Track Program, nor SBIR as a whole, appear to compensate for regional weaknesses in capital markets or university-industry interactions.[12]
4. The quality of research carried out under the Fast Track Program compares favorably with the research carried out under the regular SBIR award process.[13]
5. Fast Track attracts additional firms new to the program that are younger and smaller than non-Fast Track firms. Fifty-eight percent of Fast Track firms had no prior Phase II awards, compared with 30 percent with no prior Phase II awards for non-Fast Track firms.[14] Attracting new community-oriented firms for the SBIR program is one of the objectives of the Fast Track approach.

C. Case Studies Summaries

Case studies were undertaken relying on detailed interviews with the founders, owners, and employees of over fifty firms. All of the case study firms

---

[8]*Ibid.*

[9]Archibald and Finifter, *op. cit.* p. 243.

[10]Cahill, *op. cit.* pp. 68-70 and Reid Cramer, "Participation in the SBIR Program in the Southwestern and Mountain States," p. 152. Both authors add that Fast Track tends to attract firms with a commercial orientation and this is a factor beyond the Fast Track tool that facilitates attracting outside investment.

[11]See Cramer, *op. cit.* for a discussion of benefits and costs of Fast Track participation, pp. 149-150.

[12]For example, a recent GAO report notes that the "SBA has found that the distribution of SBIR awards generally resembles the distribution of non-SBIR expenditures for research and development, venture capital investments, and academic research funds." *Evaluation of Small Business Innovation Research Can Be Strengthened,* GAO (T-RCED-99-198), p. 4. See Kegler and Lerner, p. 313 in this volume. The authors underscore the role of clusters in encouraging geographic concentration of firms and therefore awards.

[13]Archibald and Finifter, *op. cit.*, p. 231.

[14]Cahill, *op. cit.*, p. 61.

had received SBIR assistance. They are dispersed across the United States and span a broad range of technologies, products, and industries. While some are new start-ups, others are established firms. These case studies examined the impact of the SBIR in a broad context. In particular, the case studies found that:

1. The benefits of the SBIR extend beyond the impact on the individual recipient firm. Analysis of 44 projects shows that the social rate of return from SBIR-funded research to be 84 percent, well in excess of the 25 percent average expected rate of return from the projects without SBIR funding.[15] This means that there are positive spillovers (i.e., net benefits to society) from SBIR-funded projects. The social rate of return, which incorporates this external positive impact, exceeds the private rate of return. The magnitude of the difference between social and private returns does not vary significantly, on average, between Fast Track and non-Fast Track projects.[16]
2. By definition, Fast Track firms attract additional financial capital. They also have higher rates of employment growth. Both suggest a greater likelihood of survival.[17]
3. Firms not participating in the Fast Track Program had nearly six times as many projects interrupt work due to a funding gap.[18]
4. There are three different types of firms involved with the SBIR. The first category is comprised of firms that are contractors, that is, firms whose mission is to conduct R&D on a contract basis for clients. The second category focuses on the development of a specific product or promising technology. The third category involves firms pursuing basic research outside of a university setting. The Fast Track Program seems to be the most effective with firms in the second category, which are involved in commercializing a new product.[19]

II. Specific Findings by Research Topic

A. Impact of SBIR Awards on University Researchers

1. There is evidence that SBIR induces scientists and engineers to change their career path and apply their technological knowledge to the development of a new firm.[20]

---

[15]Link and Scott, "Estimates of the Social Returns to Small Business Innovation Research Projects," p. 285.

[16]*Ibid.*, p. 275.

[17]Audretsch, Link, and Scott, "Statistical Analysis of the SBIR Awardees," pp. 293-305, especially p. 304 (Table 6).

[18]Cahill, *op. cit.*, p. 67.

[19]Cramer, *op. cit.* pp. 143-145.

[20]Audretsch, Weigand, and Weigand "Does the Small Business Innovation Research Program Fos-

2. The SBIR awards provide a source of funding for researchers to launch start-up firms that otherwise would not have had access to alternative sources of funding. [21]
3. SBIR awards can have a powerful demonstration effect. Scientists commercializing research results by starting companies induce colleagues to consider applications and the commercial potential of their research. The awards also encourage other scientists to submit their research to the award process for review.[22]

B. Biotechnology Awards

1. Department of Defense SBIR awards to biotechnology firms have made a significant contribution to biotechnology R&D. [23]
2. Many Department of Defense funded bioscience projects have dual civilian and military uses.[24]
3. Department of Defense and National Institutes of Health bioscience funding complement each other.[25]
4. The SBIR award can have a positive impact on innovation even when the firm fails. The knowledge created sometimes has value for other firms and in some cases leads to successful commercial products. From a public viewpoint, this has positive economic value.[26]

C. Regional Findings

1. Northeast[27]
   a. Innovative activity in the small business sector would have been lower in the absence of SBIR.

---

ter Entrepreneurial Behavior? Evidence from Indiana" elaborate on this; see especially the case of Star Enterprises, p. 166.

[21]*Ibid.*, pp. 176-178. Ten of 12 firms interviewed said that they would not have founded the firms without SBIR funds. See also particularly the discussion of Genetic Models. See also John Scott, "SBIR in New England," p. 122.

[22]*Ibid.*, p. 176; four of eight firms interviewed in Indiana said their SBIR experience encouraged colleagues to start a firm.

[23]See Maryann Feldman, "The Role of DoD in Building Biotech Expertise," in this volume. She reaches this conclusion based on her overall assessment of DoD's impact on the biotechnology industry; she also reports that DoD spent $241 million (1997 dollars) on biotech research in the SBIR program from 1983-1997; p. 255.

[24]*Ibid.*, pp. 268-271 in the discussion of HT Medical Systems; moreover, nearly all of the DoD SBIR-funded biotech companies sell to commercial markets as well as the military.

[25]*Ibid.* See especially the Martek case, pp. 266-268.

[26]See Feldman, *op. cit.*

[27]See John Scott's paper in this volume.

   b. Fast Track projects required less time for development subsequent to Phase II than did non-Fast Track Firms. This suggests that Fast Track firms are more oriented towards commercialization.
2. Southeast[28]
   a. Fast Track firms proceed to Phase II faster than non-Fast Track firms.
   b. Fast Track firms develop commercialization strategies sooner than non-Fast Track firms.
3. Mountain states and southwest[29]
   SBIR is perceived as being more useful to new firms oriented towards bringing new products to market than to other firms.

D. Multiple Award Winners

As noted above, firms approached the SBIR award process at different stages of development and with different objectives. Some firms are developing technology concepts; some firms see their vocation as contract research organizations; others actively seek to develop commercial products, either for public agencies or for the marketplace. Discussions with SBIR managers and awardees as well as field research suggest that in light of these different objectives:

1. Investigator-led firms, limited in size and focused on a single concept, may seek several awards as they advance the research.[30]
2. For firms that carry out research as a core activity, success is necessarily measured in multiple contract awards. Even with many awards, there is nothing intrinsically wrong with a process that provides high-quality research at a lower cost than might otherwise be available to the Department of Defense. Inexpensive exploration of new technological approaches can be valuable, particularly if they limit expenditure on technological dead ends. For research-oriented firms, the key issue is the quality of the research.
3. Those firms that seek to develop commercial products may, in an initial phase, seek multiple awards to rapidly develop a technology. Normally, this period is limited in time, before private investment becomes the principal source of funding.[31]
4. Responsibility for recognizing qualitative differences among multiple

[28]See Albert N. Link's paper in this volume.

[29]See Reid Cramer's paper in this volume.

[30]See Cramer, *op. cit.* p. 151, who makes the point about the incremental nature of technical advance, which sometimes necessitates several awards and Scott, *op. cit.,* in his discussion of Foster-Miller on p. 109.

[31]See Feldman, *op. cit.*, pp. 266-268, in her discussion of Martek as an example; see also Cramer, *op. cit.,* pp. 146-147, who discusses several firms that realized commercial success after several awards.

award winners and ensuring congruency of awards with agency interests and goals lies with the managers of SBIR programs. Senior management, in consultation with SBIR managers, should develop meaningful guidance to ensure maximum returns to the program.[32]

5. As noted below, additional research with respect to multiple award winners might well reveal differences by industry and by frequency of awards. Established companies that successfully seek large numbers of awards and fledgling companies that seek few awards clearly have different objectives. It is the responsibility of the agency management to determine if the mandated goals of the SBIR program and the needs of the agency are being achieved.

III. Recommendations for the Fast Track Program

A. The Fast Track Program initiative at the Department of Defense has proved effective at increasing commercialization. As intended, the program has attracted more commercially-oriented firms to the program, has substantially reduced the funding gap between Phase I and Phase II of SBIR, and has served as a tool to attract third party investors to SBIR awardees. On this basis, it should be continued.

B. This broadly positive assessment of Fast Track suggests that DoD should consider expanding the Fast Track Program within appropriate services, organizations, and agencies within the Department of Defense. This expansion should be undertaken with the recognition that the Fast Track process should *not* be applied to the entire award process. Other significant program goals, such as research and concept development, might otherwise be reduced or excluded.

C. In light of its recent implementation, continued research on the impact of the Fast Track Program is required, as noted below. Cross-agency comparisons of the impact of Fast Track could prove useful for the continued refinement of the Program.

D. Consideration should be given to allowing greater spread in the size of funding of awards, with larger awards reserved for commercially-oriented projects. The case studies and survey reveal that a variety of firm types participate in Fast Track, including start-up firms new to the program. This growing diversity suggests that greater flexibility in award size is warranted.

[32]For a discussion of DoD measures to ensure quality research, see *DOD's Small Business Innovation Research Program,* GAO (RCED-97-122), p. 3.

IV. Recommendations for Future Research

A. Regular assessment of SBIR program results and their comparison with Fast Track results would provide a valuable means of improving our understanding of program operations.

B. While this study represents the first systematic assessment of the Fast Track Program carried out to date, additional assessment is required to confirm these results as technologies and firms continue to mature.

C. The impact of SBIR awards in inducing researchers to undertake commercialization should be assessed through a larger study including more regions. Consideration should also be given to assessing this impact by technology.

D. An assessment of the costs and benefits of better integrating SBIR awards in the development of "clusters" around universities and technology parks should be undertaken.

E. An assessment of the costs and benefits of integrating particular SBIR awards to meet pressing agency needs through the development of focus programs, directed towards a particular technology or system, should be undertaken.

F. Further analysis is required concerning the need and effectiveness of outreach programs designed to publicize the opportunities of the SBIR program, to facilitate application to the program, and to reduce the application burden on first-time applicants.

The Steering Committee*

*For the Committee membership, see page *vii*.

IV

PAPERS

Fast Track: Is It Speeding Commercialization of the Department of Defense Small Business Innovation Research Projects?

Peter Cahill
BRTRC, Inc.

EXECUTIVE SUMMARY

Objective

This paper compares the commercialization potential and performance outcomes to date of research funded by the Department of Defense (DoD) Small Business Innovation Research (SBIR) program under standard program procedures and under Fast Track. It describes the operation of the SBIR program within DoD and discusses prior studies of SBIR commercialization. It further describes the methodology for establishing study and control group samples for the case studies and for surveys of both the contractors and government technical monitors of SBIR projects.

Methodology

A total of 379 projects were selected from among those that received Phase II awards during 1992–1996 for surveys of contractors and government technical monitors. The contractor questionnaire was derived from the one that the U.S. General Accounting Office (GAO) used in the 1991 seminal study of SBIR commercialization. The sample consisted of all 48 Fast Track winners in 1996, a control group representative of the 1996 DoD SBIR population (61 projects), 127 Ballistic Missile Defense Office (BMDO) co-investment projects awarded from 1992 to 1996, and a BMDO control group representative of the 1992–1996 DoD SBIR population (143 projects).

Results

Key findings of the contractor survey (70 percent responding) were:

- Sixty-one percent of Fast Track winners (compared to 32 percent of all other winners) had no prior Phase II awards.
- At time of application, Fast Track firms were, on average, five years younger (median founding year 1994) and smaller in annual revenue than firms in the other groups.
- The Fast Track funding gap between Phase I and Phase II was half that of control group. Five times as many control group projects had to stop work because of the gap.
- The average additional developmental funding received by Fast Track projects ($1,193,000) is almost five times that of the control group and double that of the more mature BMDO Co-investment and BMDO Control projects. Twenty percent of Fast Track firms have received venture capital compared to 3 percent of all SBIR firms.
- Although only 14 percent of Fast Track projects have completed Phase II, their average sales are already over $100,000.
- At $8,950,000, the average sales that Fast Track projects expect to achieve by the end of 2001 is over six times greater than expected by the 96 Control group and more than double that expected by the other two groups.
- Five times as many Fast Track firms have sold or are negotiating sale of partial ownership. (Such sales, probably linked to obtaining third-party funding, are viewed negatively by some SBIR participants and positively by others.)

Conclusions

Whether it is the validation by a third party to the commercial potential, the timing and magnitude of the additional funding, or merely the reduction in funding gap that contributes most to Fast Track, the program is working. By each primary measurement of commercialization success used in past SBIR studies (sales, additional developmental funding, and expected sales), Fast Track projects are clearly outperforming those in the control group.

INTRODUCTION

This paper documents a portion of the overall National Research Council evaluation of the Department of Defense (DoD) Small Business Innovation Research (SBIR) program and the Fast Track Initiative. It describes the operation of the SBIR program within DoD and discusses prior studies of SBIR commercialization. It further discusses the survey effort used to examine the program and the sampling methodology used to target the surveys of contractors. The samples used for the contractor survey provided the basis for the survey of government technical monitors of SBIR projects and the case studies, which were conducted by other members of the research team. The contractor surveys indicate that Fast Track is working. It is selecting projects that should succeed in commercialization and it is apparently contributing to their success.

SBIR PROGRAM WITHIN DoD

Background

Congress established the SBIR Program in 1982 to strengthen the research and development (R&D) role of small innovative companies. Ten federal agencies participate in the program in proportion to the size of their external R&D budget. As the agency with the largest R&D budget, DoD provides over half of the total federal SBIR funding. SBIR has become the primary vehicle through which DoD funds R&D projects at small technology companies. With funding of over a half billion dollars in 1998, the program offers DoD a unique opportunity to harness the talents of small technology companies—which studies show to be a potent source of innovation—for the benefit of DoD and the U.S. economy.

SBIR legislation describes three phases for SBIR projects. Phases I and II, funded by the SBIR Program, develop the innovative idea. Phase III involves follow-on non-SBIR government contracts for government application or use of nonfederal funds for commercial application of a technology. Commercial sales are a principal focus of Phase III.

Goals and Administration

The program is authorized by the Small Business Innovation Research Program Reauthorization Act of 1992. In the accompanying House Report, Congress noted that programs that effectively stimulate innovation and accelerate technological advance were key to national economic growth. The report stressed the concentration of scientific and technical talent in small companies and the ability of small businesses to transform R&D results into new products. The legislation addressed the congressional concern that although small businesses were the most productive source of significant innovation in the nation, their share of federally funded R&D was not commensurate with their abilities.

In authorizing the SBIR program, Congress designated four major goals:

1. to stimulate technological innovation,
2. to use small business to meet federal R&D needs,
3. to foster and encourage participation by minority and disadvantaged persons in technological innovation,
4. to increase private-sector commercialization of innovations derived from federal R&D.

In addition to establishing goals, the legislation determined agency participation and funding for the program. Agencies spending more than $100 million annually for external R&D were required to set aside a percentage of their total R&D funds for SBIR. This proportion has grown from 1.25 percent during 1987-1992 to 1.5 percent in 1993 and 1994 to 2.0 percent in 1995 and 1996. Since 1997, not less than 2.5 percent of external R&D funds have been set aside for SBIR.

Each agency with an SBIR program is unilaterally responsible for targeting research areas and administering its own SBIR funding agreements. SBIR funding agreements include any contract, grant, or cooperative agreement entered into between a federal agency and any small business for the performance of experimental, developmental, or research work funded in whole or in part by the federal government.

As indicated in the preceding section, DoD is by far the largest of the federal agencies participating in SBIR. The Office of Small and Disadvantaged Business Utilization in the Office of the Secretary of Defense coordinates these efforts, providing oversight and setting policy in coordination with the Director of Defense Research and Engineering (DDR&E).

DoD decentralizes most of the administration of the SBIR program to the three Services, four agencies, and one staff element with R&D budgets meeting the legislated requirements: Army, Navy, Air Force, the Defense Advanced Research Projects Agency (DARPA), the Ballistic Missile Defense Office (BMDO), the Defense Special Weapons Agency (DSWA), the U.S. Special Operations Command (USSOCOM), and DDR&E.

The Air Force SBIR program is larger than all of the SBIR programs at the other nine federal agencies, and the Army, Navy, DARPA, and BMDO programs each exceed the size of the programs at seven of the nine other federal agencies. Thus, how well DoD meets the goals of the SBIR program has a major impact on how well the overall program meets its goals.

The legislation requires agencies to issue a solicitation that sets the SBIR process in motion. The solicitation, a formal document issued by each agency, lists and describes the topics to be addressed and invites small businesses to submit proposals for consideration.

Twice a year, DoD issues a combined research solicitation for its eight component programs, indicating each program's R&D needs and interests and inviting R&D proposals from small companies. Companies apply first for a six-month Phase I award of up to $100,000 to test the scientific, technical, and commercial merit and feasibility of a particular concept. If Phase I proves successful, the company may be invited to apply for a two-year Phase II award of up to $750,000 to further develop the concept, usually to the prototype stage. Proposals are judged competitively on the basis of scientific, technical, and commercial merit. Following completion of Phase II, small companies are expected to obtain Phase III funding from the private-sector or non-SBIR military customers to develop the concept into a product for sale in military and/or private-sector markets.

The law requires the Small Business Administration (SBA) to issue policy directives for the general conduct of the SBIR programs within the federal government. These policy directives include elements such as simplified, standardized, and timely SBIR solicitations; a simplified, standardized funding process; and minimization of the regulatory burden for small businesses participating in the program. The most recent policy directive was issued in January 1993. Federal agencies are required to report key data to SBA, which in turn publishes annual reports on the progress of the program.

According to SBA's SBIR program policy directive, to be eligible for an SBIR award, a small business must be

- independently owned and operated,
- other than the dominant firms in the field in which it is proposing to carry out SBIR projects,
- organized and operated for profit,
- the employer of 500 or fewer employees (including employees of subsidiaries and affiliates),
- the primary source of employment for the project's principal investigator at the time of award and during the period when the research is conducted, and
- at least 51 percent owned by U.S. citizens or lawfully admitted permanent resident aliens.

Program Administration Within DoD

Commercialization success depends on many factors, some of which are influenced by the agency that makes the SBIR award. Does the topic describe a general need, or a very specific problem? The specificity of the topic may limit proposals and innovative approaches, which may reduce the private-sector appeal of proposals in response to a very specific DoD topic. On the other hand, such specificity may indicate a well-understood need that will result in DoD procurement of the solution to that need. Broad topics give more latitude for a firm to propose something with private-sector appeal; however, the agency may not select the proposal if it sees no clear payoff to DoD. How are topics selected? How are proposals evaluated—decentralized or centralized, rolling evaluation as received or batched evaluation, first solicitation of the Fiscal Year, second solicitation, or both? Are the procurement officials of the agency involved with topic selection? How rapidly are contracts let? Is bridge funding available between phases? Does the agency fund projects that exceed the Phase II funding limit? Is the agency using SBIR to supplement other research? How often does the agency provide follow-on Phase III R&D funding? What prior relationship has existed between the SBIR firm and the agency or command/laboratory within the agency? All of these factors, which may affect commercialization, vary within and among DoD agencies as described herein.

As indicated earlier, the program administration of SBIR is decentralized in DoD. The reason for this decentralization is that SBIR is part of R&D and, in DoD, R&D is decentralized because each of the agencies conducting R&D has a different mission, structure, and R&D focus. Two of the three Services have had 200 years and the third has had 50 years to evolve its structures in response to changing missions. Although each has the mission to recruit, train, organize, and equip forces for deployment under joint commanders, the differences in equipment needs between Army Divisions, Navy Carrier Groups, and Air Force Wings are dramatic. These differences lead to differences in the kinds of topics, and in the way that the Services have structured their own acquisition organizations and the research, development, and engineering organizations that support them. Certain needs common to all Services have been made the responsibility of a single Service, whose needs and capabilities were predominant. For example, among the Army's lead service, R&D responsibilities are small arms, food, clothing, and wheeled vehicles. Each Service has R&D organizations at various locations supported by contracting offices. In their areas of interest, Services conduct basic and advanced research, develop and demonstrate technology, and develop and engineer systems. Most of this effort is accomplished through universities and defense contractors. The Services also must provide support in maintaining and upgrading equipment that is already in the field.

DARPA focuses on high-risk, high-payoff critical defense technologies that may support any of the Services or other DoD needs. Most of its focus is on technology development and demonstration. It makes use of Service R&D organizations and contracting agencies to evaluate and support its efforts, which are largely contracted. Much of the DARPA organization is transient. The Services provide people to work at DARPA as program managers for two to three years (less than the life cycle of SBIR from topic generation to completion of Phase II). DARPA gauges success of an R&D project (including SBIR projects) on whether, at the end of the project, the technology is transitioned into one of the Services.

BMDO has the mission its name implies. Much of its R&D is supported out of Huntsville, Alabama, home of one of the Army's principal Research, Development, and Engineering Centers. The large expenditure in technology development over the past two decades and the national debate over the affordability and efficacy of fielding a missile defense system in the near future is indicative that BMDO is pushing the state of the art in many technologies. A larger, more structured and focused organization than DARPA, BMDO also uses the Services to help execute its R&D mission. Neither DARPA nor BMDO has responsibilities for system acquisition or support and upgrade of systems in the field.

DSWA focuses its R&D on the effects of nuclear, chemical, and biological weapons (the latter two for defense against such weapons). It does not actually develop or procure weapons, thus limiting the potential for SBIR Phase III funding for their topics. Private-sector sale of SBIR Phase II results tends to be limited to occasional spinoffs of the actual technology in the SBIR. The U.S. Special Operations Command has a small R&D program focused on near-term needs of Special Operating Forces provided by the Services. The Office of the Secretary of Defense DDR&E has a small SBIR program, which has attempted to establish topics with a high potential for dual use.

The SBIR process within a Service must operate within the organization and research, development, and acquisition (RDA) processes used by that Service. In decentralized systems such as those employed by the Navy, SBIR procedures vary among the systems commands. In general, SBIR is integrated into the R&D programs of each systems command. In the Navy, the Acquisition Program Executive Officers play a significant role in topic generation and selection of proposals, especially for Phase II. Acquisition Program Offices frequently fund Phase III or provide additional Phase II funding. The usefulness of the SBIR results to the Navy is an important part of the selection process. In many cases this may lead to selection of more mature technologies and less risk taking, trading a higher probability of success for a lower potential for payoff.

Air Force SBIR management is also very decentralized but nevertheless quite different from that of the Navy. The Air Force has program managers operating in two tiers. The top tier (at command level) reports to the Air Force SBIR program manager and the second tier (at lab level) reports to the first tier. Technical Officers within the labs write the topics. Approvals are made by the Lab Chief Scientist and generally supported by the Command Chief Scientist and the Air Force SBIR program managers although rewrites are sometimes required. Proposal approval is decentralized to the Directorate level. The Air Force awards Phase I for $100,000 for nine months rather than the nominal $75,000 for six months. The extra time and dollars help to bridge to Phase II. Technical officers are evaluated on their ability to transition technology; thus they have a bias toward helping SBIR to transition into the Air Force or commercially. Contractors claim that overall reductions in Air Force advanced development funding result in lab topics that cover work formerly done under the Air Force's R&D program. Such topics are alleged to be difficult to commercialize.

The Army, on the other hand, centralizes topic, Phase I, and Phase II proposal selection; this centralized process began with FY 1992 topics and Phase I proposals in late 1992. As in the Air Force, topics are allocated to the laboratories on the basis of the lab's R&D budget.[1] The Service calculates how much of the annual SBIR funding will be needed to fund new Phase II projects and to pay the second year of Phase II projects that were approved the prior year; then it determines how many Phase I contracts can be awarded with the remaining funds. Anticipating about 1.5 Phase I contracts per topic allows estimation of the number of topics that can be supported.[2] Labs are also allocated backup topics in the event that their primary topics do not survive the centralized selection process. The Army's 10 senior scientists/technologists, who receive input from evaluators and managers at the laboratories, head the centralized selection of topics and proposals. The Director of the Army Research Office heads the Source Selection Board for Phase I. The Army SBIR process is not as formally connected to its acquisition community as is the case in the other Services.

Unlike the Services and other agencies, BMDO has a small number of broad topics. The topics are largely the same each year, evolving gradually from year to year, and providing great flexibility to proposing firms. Topic development and proposal decisions are centralized. Volunteer evaluators from the Services, DSWA, and BMDO assist in the evaluation and serve as technical monitors. Commercialization plans are an important factor in evaluation, with co-investment considered a strong indicator that commercialization will occur. Nevertheless, BMDO has a reputation for funding only high-risk projects.

In DARPA, topic selection and proposal decisions are decentralized to the Technical Office directors. In DSWA, the five technical Directorates control the topics, but the proposal decisions are made by a board composed of the deputies from each directorate.

BMDO allows all Phase I winners to submit Phase II proposals. All of the Services and other agencies invite proposals from firms that are doing well in Phase I. In both DSWA and the Army, the centralized decision process for Phase II only meets once a year, which delays the award of Phase II.

DoD conducts two SBIR solicitations a year, the first closing in January and the second closing in July. Solicitations announce the topics and provide directions and formats for submission of proposals. The Air Force and BMDO participate only in the first solicitation. The Army participates only in the second. The Navy and the other agencies participate in both solicitations.

Prior Commercialization Studies

In 1991, the General Accounting Office (GAO) conducted a study across all federal agencies (including DoD) that fund SBIR, to evaluate the aggregate com-

[1]The Air Force also allocates the number of funded Phase I projects.

[2]The Navy allocates the money rather than the topics, allowing each command to determine how the money will be spent. Navy contractors sometimes experience a long time before Phase II approval or disapproval, implying that the awarding organization was waiting for the following year's funding.

mercial trends of products in the third phase of SBIR. The 1991 survey questionnaire was sent to all Phase II awardees from the first four years—1984 through 1987—in which the agencies made Phase II awards. GAO chose the earliest recipients because studies by experts on technology development concluded that five to nine years are needed for a company to progress from a concept to a commercial product. Their rationale for not including Phase II recipients from 1988 or later was that, in most cases, those project recipients had not had sufficient time to "make or break" themselves in Phase III. Responses to the GAO study indicated that 10 percent of the projects studied had not completed Phase II and even the earliest projects studied had had inadequate time to mature. Although upbeat about the overall early indications of commercialization, GAO expressed some concern over the rate of commercialization in DoD.

In 1996, the Deputy Director of Research and Engineering directed a study of commercialization of SBIR within DoD. The contractor[3] used the same methodology and survey instrument that GAO had used five years earlier, for example, surveying all Phase II awardees through at least four years prior to the 1992 study. Use of this methodology allowed direct comparison of results. In 1998, the SBA employed the same contractor to survey Phase II winners from the other (non-DoD) 10 federal agencies.[4] Once again the same methodology and survey instrument were used.

Several lessons from the earlier studies affected the survey conducted for this effort. In 1991, GAO allowed six months for responses and conducted three mailings of the survey as well as telephonic follow-up in selected cases. Despite this effort, 10 percent of the awardees could not be contacted because of bad addresses. In 1996, the second and third survey mailings were targeted after extensive phone contacts and use of the Internet to find firms who had not responded. Prior to any mailings, the 80 DoD offices that manage SBIR programs reviewed and updated SBIR firm addresses. After seven months of survey effort, one out of seven potential respondents still could not be located. Prior to the 1996 study, DoD was aware of a number of commercialization successes. Several had been uncovered by the GAO study and others were discovered by the Services in an unsystematic fashion. A number of these known successes did not respond to the survey, indicating that the absence of a response did not mean the absence of success.

The 1998 study required even more extensive effort to locate SBIR awardees. The SBA database of awardee addresses is not updated as frequently as that of DoD. Numerous mailings and phone calls and extensive Internet searches over 11 months could locate only three-fourths of the awardees.

The three earlier surveys demonstrated the mobility of award winners. Fre-

---

[3]BRTRC of Fairfax, Virginia. DoD has not published the final report.

[4]Projects surveyed included those from the Nuclear Regulatory Commission, which no longer participates in SBIR.

quently, firms had either grown or shrunken, ultimately requiring a move to a new facility. Often firms changed names because of new investment or products, or acquisition by a larger firm. Death and retirements contributed to economic reasons that firms went out of business. Principal investigators were often very mobile, reducing the value of their phone-contact information. Widespread changes to area codes and, to a lesser extent, zip codes, severed communications with firms that had not moved. The slow response by firms that did respond showed the value of follow-up calls and mailings.

Each of the prior surveys showed that most awards do not result in commercial sales. For the combined DoD and SBA studies, 61 percent reported no sales. At the other extreme, 4 percent of the projects were responsible for 75 percent of the commercial sales. The extreme divergence of these results raised questions about using a sampling scheme as a reliable measurement of sales. Each of the prior surveys measured the entire population being studied.

Fast Track

Subsequent to the GAO study, DoD initiated program changes designed to improve the rate of SBIR commercialization. Starting with the 1996 SBIR solicitations, DoD initiated a two year pilot policy—the SBIR Fast Track—under which SBIR projects that attract matching funds from third-party investors have a significantly higher probability of SBIR award, as well as expedited processing to reduce the delay in reaching the market. The purpose of the Fast Track policy is to concentrate SBIR funds on those R&D projects most likely to result in viable new products that DoD and others will buy.

At the conclusion of the initial pilot period, the Under Secretary of Defense for Acquisition and Technology extended the Fast Track pilot for two additional years, because of the promising early results. The Under Secretary also directed an independent analysis of Fast Track. The National Research Council was asked to conduct that analysis, part of which is documented herein.

SURVEY METHODOLOGY AND ADMINISTRATION

Sample Selection

The selection of projects or firms for interviews and government technical monitors who would receive e-mail surveys began with the selection of a broad sample of SBIR projects for a mailed survey. The projects selected for this sample included DoD Fast Track award winners and BMDO-sponsored projects that had received co-investment funding. Other DoD projects, which were neither Fast Track nor co-investment, were selected as controls.

As indicated earlier, implementation of the Fast Track program began with the 1996 DoD solicitations. Because Phase I and Phase II normally last 6 and 24

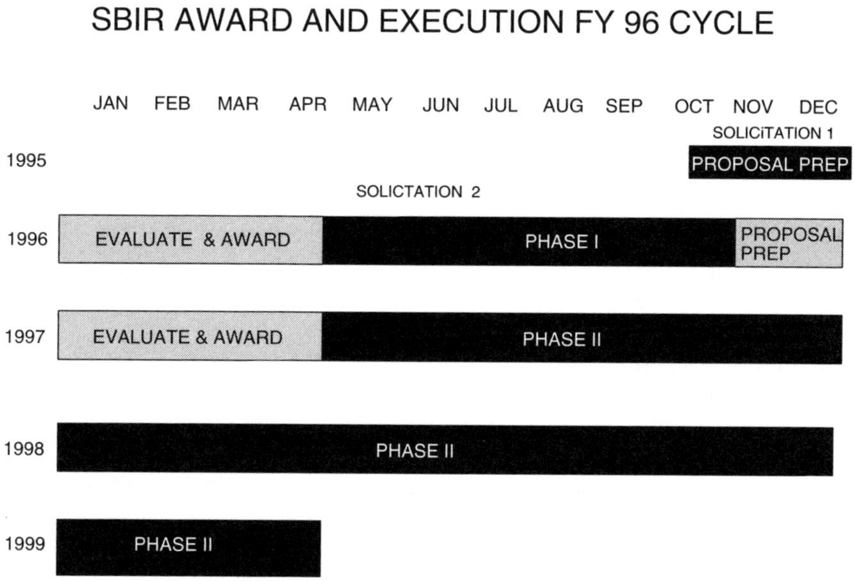

FIGURE 1 SBIR Cycle for Solicitation 1, FY96

months respectively, and because the targets for proposal evaluation and award of those two phases are 4 and 6 months respectively, completion of Phase II of an SBIR project can be expected to take at least 40 months from the closing date of the solicitation. Thus the earliest FY 1996 projects could be expected to complete Phase II 40 months after the January 1996 closing of the first solicitation (i.e., May 1999) and the July 1996 closing of the second solicitation (i.e., December 1999).

Clearly, little time was available prior to the February 1999 mailing of the survey for the commercialization (Phase III) of projects that had resulted from the 1996 solicitations. Projects awarded as a result of the 1997 or later solicitations would have barely entered Phase II if at all by January 1999; thus the only Fast Track projects selected for the survey were those resulting from the 1996 solicitations. All 48 Fast Track projects for 1996 were selected. The life cycle for a typical SBIR project from the FY 1996 solicitation is shown in Figure 1.

The prior studies of SBIR commercialization by GAO in 1991 and by DoD in 1996 showed that it often takes several years after completion of Phase II before significant sales are achieved.[5] There were no Fast Track firms that had

[5] In interviews of 150 firms representing the majority of the DoD SBIR projects that had achieved sales by 1996, 58 percent said that it took over two years after Phase II to their first sale. On average, these 150 firms said it was 3.2 years after Phase II before significant sales occurred.

had that much time to commercialize their Phase II.[6] However, one DoD agency—BMDO—had begun an SBIR policy in 1992 that had some of the characteristics of the Fast Track. Two principal advantages of Fast Track are the validation of the commercial worthiness of the project by an outside source and the ability of the firm to use the Fast Track leveraging of funds to attract investors.

In the case of Fast Track, the willingness of the third party to invest its funds in the project is a strong indicator that the project will be a commercial success. The high likelihood that Fast Track projects will be awarded Phase II serves as a means to attract investors who are thereby provided an opportunity to leverage their investment because the government is matching between one to one and four to one in dollars.

The BMDO co-investment program also gives priority to firms that include co-investment in Phase II. The standard to be met is easier than for Fast Track because in-kind investments are acceptable (versus cash-only for Fast Track) and the investment may come from the SBIR firm rather than a third party. Nevertheless, the willingness of the firm to invest substantially in its own SBIR carries a strong message that it believes the project will be a commercial success. Some of the BMDO awardees use the prospect of leverage as a means to attract third-party investment. BMDO began the co-investment policy with the 1992 solicitation. For 1992, slightly less than one-third of its SBIR Phase II awards involved co-investment. From 1993 to 1996, between 83 and 94 percent of the BMDO awardees included co-investment.

The 1996 DoD SBIR study showed that most SBIR projects require an additional investment at least equal to the amount of the combined Phase I and II to succeed commercially. Over 60 percent of the DoD projects surveyed in 1996 failed to commercialize. Few of those unsuccessful projects attracted additional investment (third-party or internal). Conversely, almost all of the successful projects had either internal or external additional investment. Commercialization success was better correlated to the size of additional investment than to the source (internal or external) of that investment. The study could not determine whether the additional investment caused the commercialization success or whether the commercialization potential of the project caused the additional investment to occur. Because both Fast Track and BMDO co-investment projects have additional investment, it can be hypothesized that they will be more likely to experience commercial success than the average DoD project.[7]

[6]As a general rule, Phase II awards occur one or two fiscal years after a solicitation. The GAO methodology for sample selection picked projects whose Phase II began four years prior to its 1991 survey; thus the solicitations for the projects in the GAO sample were released at least five years before the survey. Using the GAO methodology, projects resulting from both the 1996 and the 1995 DoD solicitations would have been considered too immature to be evaluated for commercialization results.

[7]Commercialization results of BMDO projects receiving Phase II awards from 1984 to 1992 were at least as good as the results of the rest of the DoD projects.

Although there are significant differences in the implementation of Fast Track and the BMDO co-investment policies, this study tested the assumption that BMDO co-investment projects could be used as a surrogate for Fast Track for the years 1992-1995 when Fast Track did not exist. Examination of projects from these earlier years provided a greater opportunity to find actual commercialization results. In 1996, ten BMDO Phase II projects used Fast Track and 22 others included co-investment. The survey sample included all 127 BMDO co-investment projects awarded from the 1992–1996 solicitations.

Determination of whether Fast Track or BMDO co-investment policies result in selection of projects that achieve greater commercialization success than the general population of DoD SBIR projects necessitated selection of a matched control group.

The Fast Track control group was structured to be similar to the 48 FY 1996 Fast Track awards, matching, to as great an extent as possible, the solicitation fiscal year, the number of prior Phase II awards received by the firm, the size of the firm, the location (state) of the firm, the awarding agency, the closing date of the solicitation, whether the firm was designated as woman- or minority owned, and the technology area of the project.

For BMDO co-investment projects in FY 1992, all matches were made with BMDO non-co-investment projects. Beginning with FY 1993, the vast majority of all BMDO awards were co-investment; thus matches for FY 1993 through FY 1996 had to be made with projects from other agencies. Generally, the match was made with the Service that was managing the Fast Track contract for BMDO[8] or with DARPA projects being managed by that Service. The latter—that is, DARPA—was the preferred match.

Inclusion of all Fast Track and BMDO co-investment Phase II awards from the 1996 or earlier DoD solicitations provided a nonrandom sample. Similarly, the one-for-one matching of a project in the control group based on the characteristics described earlier was not a random selection. The diversity of the population of SBIR awardees was such that it was impossible to match all characteristics project by project. The order in which the characteristics are listed above is the order of priority for matching.

In selecting the control-group projects, each solicitation year was drawn separately to ensure that the aggregate characteristics matched. Thus the 70 projects picked for 1996 in aggregate matched the 48 Fast Track plus 22 BMDO projects

---

[8]The projects of each Service (Army, Navy, and Air Force) are managed by that Service. Each Service has its own contracting agencies. Although the same rules apply, there are procedural differences between contracting agencies in the same Service, and especially between Services. BMDO does not have its own contracting agency; it often uses a technical monitor from one of the Services and always uses one of the Services' contracting agencies (normally the contracting agency that supports the technical monitor). DARPA similarly uses the Services for contracting and technical manager; thus its projects were often selected as a match for a BMDO project.

as closely as possible. A similar process was conducted for each solicitation year. In aggregate, the 175 control projects that were initially selected matched the Fast Track and BMDO co-investment projects fairly well in all the characteristics mentioned above except for agency (because of the lack of BMDO projects that were not co-investment).

Matching a Fast Track project with a non-Fast Track one that has similar characteristics was useful for comparing the performance of like firms, only one of which obtained matching funds before Phase II. Thus the initially constructed control group mirrored the projects under study. In several respects, however, Fast Track firms were different from the average firm that had been winning Phase II awards.

We postulated that the Fast Track policy might attract a different kind of firm to the program, replacing firms that would have won in the absence of Fast Track. If this were true, then the control group should be representative of the entire population of Phase II winners. Two related characteristics of the Fast Track winners and the initially selected control group that differed from those of awardees in the program as a whole were the number of prior Phase II awards received and the size of the firm: Thirty-four of the 48 Fast Track firms (71 percent) had had no previous Phase II awards. For the Phase II population, the number of first-time winners was about 40 percent. At the other extreme, the control group did not include the program's most frequent winner because the most prior awards by a Fast Track or co-investment firm was one-third the number of prior awards received by that most frequent winner.

First-time winners tend to be new, and thus small, firms. The high percentage of first-time winners among the Fast Track firms meant that the average size of the Fast Track firms tended to be smaller than the average size of the overall Phase II population of firms as a whole. After discussions with the SBIR Program Director in the DoD Office of Small and Disadvantaged Business Utilization, it was decided to increase the representativeness of the control group by adding projects that would result in characteristics similar to those of the Phase II population as a whole. In this fashion, the Fast Track performance could be compared to either that of similar firms (initial sample) or to that of the SBIR population as a whole (expanded sample).

Additional projects were added to each year of the control group in a modified random fashion. First, the average characteristics (number of prior Phase II awards, size of firm, state, minority or woman ownership, and agency) of all DoD Phase II projects for a specific year were determined. In an iterative fashion an estimate was made of how many projects would have to be added to the control group, and the expected distribution was calculated using the average characteristics. The expected distribution was compared to the initial control group to determine the variance between them. The goal was to reduce the variance to the minimum solely by adding projects to the initial control group. If the initial estimate of the number of projects to be added did not reduce the variance enough, a larger estimate was investigated in a similar fashion. The result was the number of projects that needed to be added to the sample and a compilation of their aggregate characteristics.

Next, all Phase II projects from that year that were not yet in the study sample or control group were listed in a random order by a random number generator. The analyst went through the list in that random order to select additional projects. Each project that met or nearly met the desired criteria was added to the sample until the desired number had been selected. This process was completed for each year in the sample, resulting in the addition of 29 projects to the original control group for a total control group size of 204.

The surveys (one per project) were mailed to the 298 firms conducting the 379 projects (control group plus study sample) on February 3, 1999. The characteristics of the firms in the sample are described in Appendix A. This sample was used to mail out the survey and as a basis for selection of firms to be interviewed. Project information, including addresses, principal investigators, and phone numbers, all of the characteristics used in matching, as well as other information in the database such as award amounts, dates, contract numbers, and scheduled durations were provided to the investigators to assist in selection of firms for interviews. Information was also provided to allow survey of the government technical points of contact.

Administration of the Survey

The questionnaire was derived from the one used in the 1996 and 1998 studies (which were substantially the same as the questionnaire for the 1991 GAO study). Fifteen of the 27 questions from the earlier DoD study formed the basis of this questionnaire. Questions that the earlier studies had answered and that were unlikely to yield new information were eliminated. The researchers added 14 new questions to attempt to find leading indicators of potential commercial success or to otherwise differentiate Fast Track firms and projects. Many of the resulting 29 questions had multiple parts.

The formatting, printing, mailing, and other administrative aspects of the survey including scanning results to a database was subcontracted to Questar in Eagan, Minnesota. The final questionnaire is contained in Appendix B.

An advance letter from the DoD SBIR/Small Business Technology Transfer (STTR) Program Director, Jonathan Baron, was sent to each selected firm one week prior to the survey. The letter described the purpose and importance of the study and requested cooperation in survey completion and participation in subsequent interviews. Selected firms also received a letter from the Board on Science, Technology, and Economic Policy of the National Research Council that gave further description of how projects were chosen and why the survey and interview input was needed.

As expected from the earlier studies, a number of advance letters and surveys were returned as undeliverable. As each was returned, the firm was looked up in Internet yellow pages. If a new address and phone number were found, the firm was contacted to verify that it was indeed the correct firm and, where possible, a point of contact (POC) was obtained. When the Internet did not list an address, an attempt was made to contact the principal investigator or other POC listed in the DoD database. Surveys were then re-addressed and mailed.

The very short time available for this survey necessitated sending a second mailing after only about 6 percent of the surveys had been received. About that time, researchers began reporting that many of the firms that they had tried to contact had new addresses or phone numbers. BRTRC then began calling every firm to verify that the survey had been received and to encourage that it be completed. Many contacts indicated nonreceipt, resulting in additional mailings targeted to the contact. Many phone numbers turned out to be dead ends, resulting in further Internet searches. Still other POCs had phone numbers that seemed correct, but they were never present and did not return calls. A targeted third mailing was sent out as well as additional individual mailings. During the final week of the survey, repeated calls were made to the Fast Track firms who had not responded. In some cases new surveys were sent by fax, and returns were accepted by fax.

By June 21, four and one-half months into the survey, 256 responses had been received. The original 182 bad addresses had been reduced to 16 undeliverables. Fifty-four percent of the responses showed new addressees. Four respondents indicated that no Phase II had been awarded. Using the same methodology as the GAO had used in 1992, undeliverables, non-Phase II, and out-of-business firms were eliminated prior to determining the response rate. Although 379 projects were surveyed, 20 were eliminated as described. This left 359 projects, of which 252 responded (256 – 4), representing a 70 percent response rate. Considering the length of the survey and its voluntary nature, this rate was relatively high and reflects both the interest of the participants in the SBIR program and the extensive follow-up efforts. This response rate provides a credible basis for evaluation. The six sample groupings and their address and response data are shown in Figure 2.

ANALYSIS OF SURVEY RESPONSES

This section displays the responses for four of the above sample groups: Fast Track, the Fast Track Population Match (referred to as 96 Control group), the BMDO Co-investment group, and the population-matched control group for BMDO (BMDO Control group). Complete survey responses are included as Appendix C.

SURVEY RESPONSE

GROUP	SURVEYS	BAD ADDRESSES	RESPONSES	UNDELIVERABLE
FAST TRACK	48	22	**45**	1
*FT MATCH**	*48*	*10*	*31*	*1*
FT MATCH (POP) (FY 96)	61	15	**42**	1
CO-INVESTMENT	127	47	**81 (77)****	6
*CO-INV MATCH**	*127*	*53*	*79*	*7*
CO-INV MATCH (POP)	143	55	**88**	8
TOTALS	379	139	**256 (252)****	16

RESPONSE RATE = 252/(379 - 20) = **70%**

** Included in Match (Pop)* ****4 Projects reported they did not receive a Phase II**

FIGURE 2 Survey Response Analysis

Phase II Completion

The short time frame from the solicitation for Fast Track to the completion of the survey had the expected result that most Fast Track projects had not yet com pleted Phase II. The 96 Control group fared even worse. As shown in Figure 3, only 16 percent of the Fast Track projects had completed Phase II prior to completion of the survey. For the 96 Control group, Phase II completion was only 5 percent.

The Phase II completion rate for the BMDO co-investment and its BMDO control group was slightly lower than expected. Part of this lower rate is due to a lower response rate from the older projects (more likely to be completed) than from more recent projects. Another factor contributing to the slow pace of these two groups was the delay between Phase I completion and the award of Phase II. For the BMDO Control group this delay averaged 8.9 months. For the BMDO Co-investment group, the delay averaged 10.3 months. Part of the delay is due to the government evaluation and award process and part is due to the time lag from completion of Phase I until the contractor submits its Phase II proposal. For these two groups the contractor proposal portion of the delay averaged three months.

Changes in the Phase II award process were instituted in 1996 in an attempt to reduce the average delay for evaluation and Phase II award to less than six

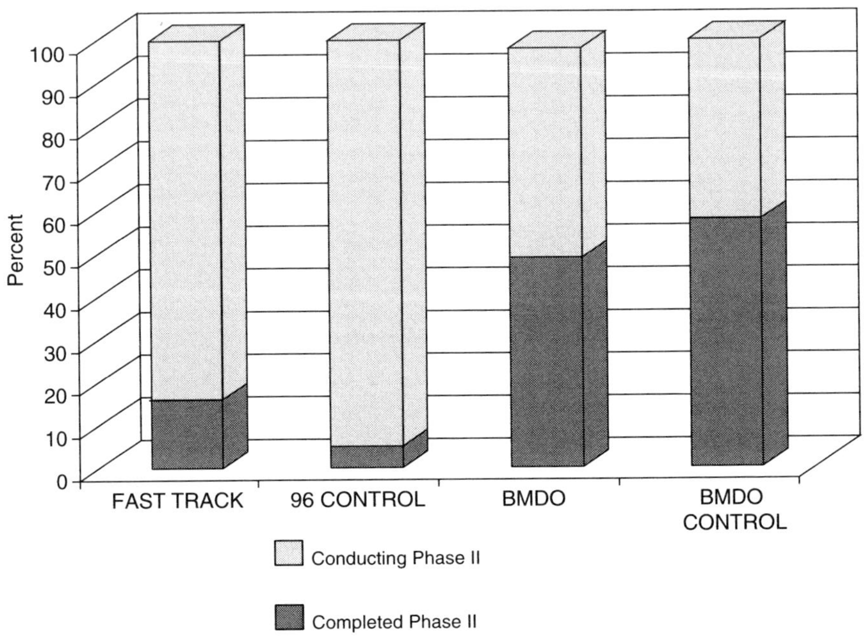

FIGURE 3 Status of Phase II

months. The total average delay for the 96 Control group was reported as 6 months. Proposal submission delay was less than one month. For Fast Track, Phase II proposals must be submitted during Phase I; thus there is no delay due to proposal submission. The reported delay for evaluation and award for Fast Track Phase II was 2.4 months. Government evaluation and award for FY 1996 Fast Track was accomplished in one-half of the time required for government evaluation and award of the 96 Control group, fulfilling the promised expedited treatment of Fast Track Phase II proposals.

Validation of Sample Selection Factors

Two of the factors used in sample selection were number of Prior Phase II awards and size of the firm. Both of these factors were examined in the survey. The DoD database tracked the number of DoD Phase II awards received by each firm. On the basis of this measure, 71 percent of the Fast Track firms had no prior Phase II awards. On average overall for DoD Phase II award winners, about 40 percent have had no prior DoD Phase II awards. In the survey, respondents were asked about prior Phase I and Phase II awards from any federal agency (including DoD). Results are displayed in Figure 4.

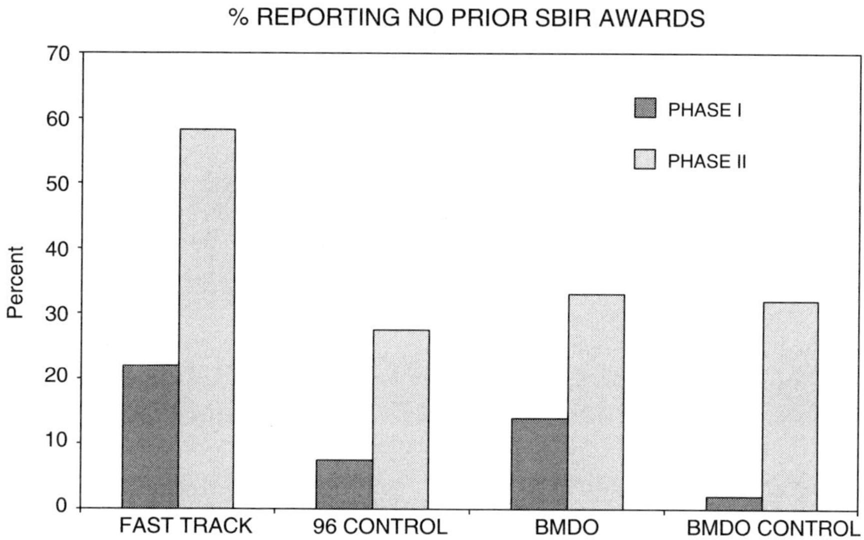

FIGURE 4 Prior SBIR Experience

Despite inclusion of awards from other agencies, the Fast Track group still reported that a high number of awardees (58 percent) had no prior Phase II award. The other sample groups showed more experience, with an average of about 30 percent reporting no prior Phase II award. Even the Fast Track firms were not inexperienced in SBIR. Seventy-six percent reported receiving a prior Phase I award. Forty percent of the Fast Track winners had three or more prior Phase I awards. One firm, which was awarded three Fast Track Phase II contracts, had had 14 prior Phase II awards. Over half of the average number of prior Phase II awards (1.8) per project for Fast Track are accounted for by this firm.[9] The 96 Control group reported the lowest percent of firms that had no prior Phase II, and the highest average number (14.7) of prior Phase II awards. Over half of that average can also be attributed to three projects, one awarded to each of three frequent winners.

The BMDO Co-investment group reported an average of 8.5 prior Phase II awards. Its control group averaged 10 prior Phase II awards. Each of these averages was also skewed by a few frequent winners.

A second selection factor was the size of the firm. The number of employees in the DoD database was used to measure firm size. Respondents identified firm

[9]Each of Digital System Resources' three projects reported 14 prior Phase II awards, thus the same 14 awards count three times in calculating the project average.

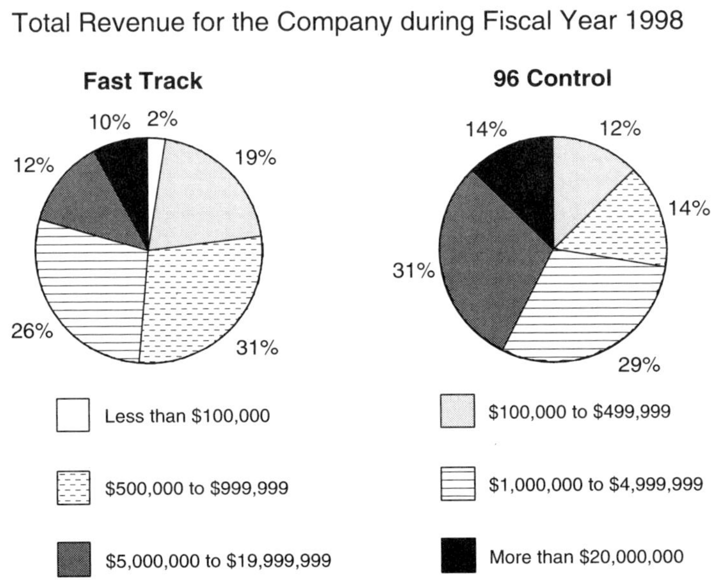

FIGURE 5 Annual Revenue Comparison

size using the metric of annual sales. Figure 5 shows the results for Fast Track and the 96 Control group.

Figure 5 shows that the annual revenue for Fast Track firms is considerably less than for the 96 Control group. Only 22 percent of the Fast Track firms reported over 5 million dollars in annual revenue compared to 45 percent for the 96 Control group. Fifty-two percent of the Fast Track firms reported less than 1 million dollars in annual revenue compared to 26 percent of the 96 Control group.[10]

The DoD database does not contain information on the age of the firm, but the survey elicited this information. The median founding date for the BMDO Co-investment respondents and for the BMDO Control group respondents was 1989. For the 96 Control group, 1985 was the median founding year. Fast Track firms were much younger; half were founded in 1994 or later.

[10]The 40 respondents of the 96 Control group consisted of 29 respondents originally matched to Fast Track projects and 11 from projects that had been added to make the control group match the population. The original 29 control group respondents had smaller annual revenues than the population, but larger than the Fast Track group. Of the 29 firms, 10 reported over $5,000,000 in annual revenues and 9 reported less than $1,000,000.

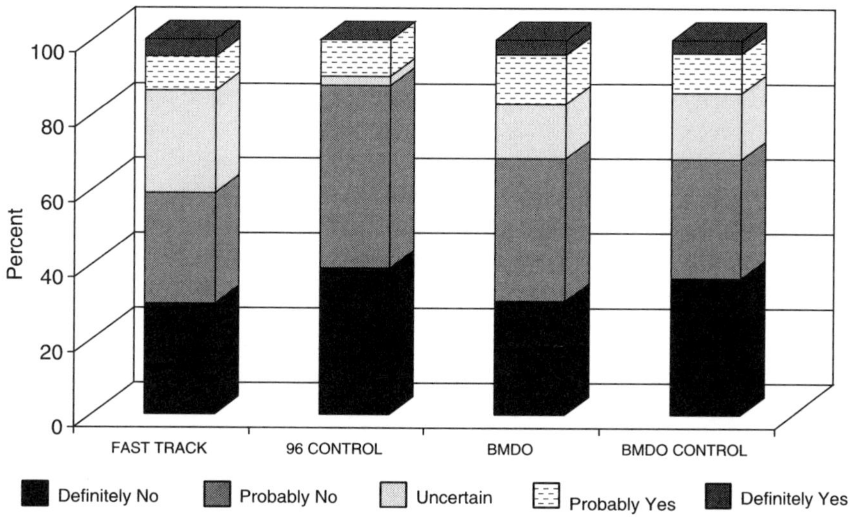

FIGURE 6 Would the Company Have Undertaken Project without SBIR?

Would They Have Done It Without SBIR?

When we attribute sales or other economic activity, we must consider whether these sales would have occurred in the absence of SBIR. Figure 6 displays responses to the question of whether the company would have undertaken the project in the absence of SBIR.

The Fast Track group is far less certain that they would not have undertaken the project without SBIR than any of the other three groups. Only 59 percent of the Fast Track firms state that they definitely or probably would not have undertaken the project without SBIR. Eighty-eight percent of the 96 Control group would not have undertaken the project. However, only 14 percent of the Fast Track respondents said that they definitely or probably would have undertaken the project in the absence of the SBIR. This percentage is surprisingly low given that each of the Fast Track awardees was able to raise third-party funding for their projects. The large number of uncertain answers from Fast Track firms may indicate that they were not sure the third party would have invested without Fast Track.

These answers indicate that the SBIR program is clearly funding projects that would not be undertaken without SBIR. Even projects that would have been undertaken are affected by SBIR. In earlier interviews, two awardees that had significant commercial success indicated that although they definitely or prob-

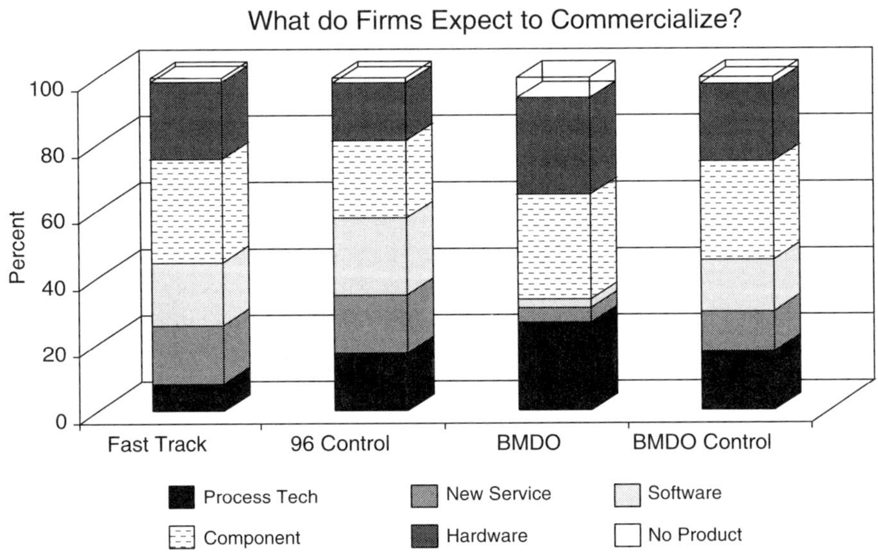

FIGURE 7 Expected Products of the SBIR

ably would have undertaken the project without SBIR, they would not have experienced the same level of success. One stated that the SBIR award gained at least two years for them and that a time delay would have substantially reduced their market share. Another attributed to SBIR the flexibility to try a higher-risk approach than would have been possible with private funding. That higher risk resulted in a lower-cost product and consequent higher sales.[11]

Expected and Achieved Commercialization

Respondents were asked to identify the product that they expected to commercialize from their SBIR project. Very few did not expect to achieve a product. This optimism stands in contrast to earlier studies that have shown that only about 40 percent of the SBIR projects achieve product sales. Figure 7 breaks out the types of products expected. Hardware and hardware components account for the largest number of expected products. Software is the second most anticipated product for all groups except for the Co-investment group, which expects little software, but a large amount of process technology.

[11]These respondents were Optiva Corporation of Washington State, which reported $240,000,000 in sales, and Genosys Biotechnologies of Texas, which reported $60,000,000 in sales during the SBA Commercialization Study.

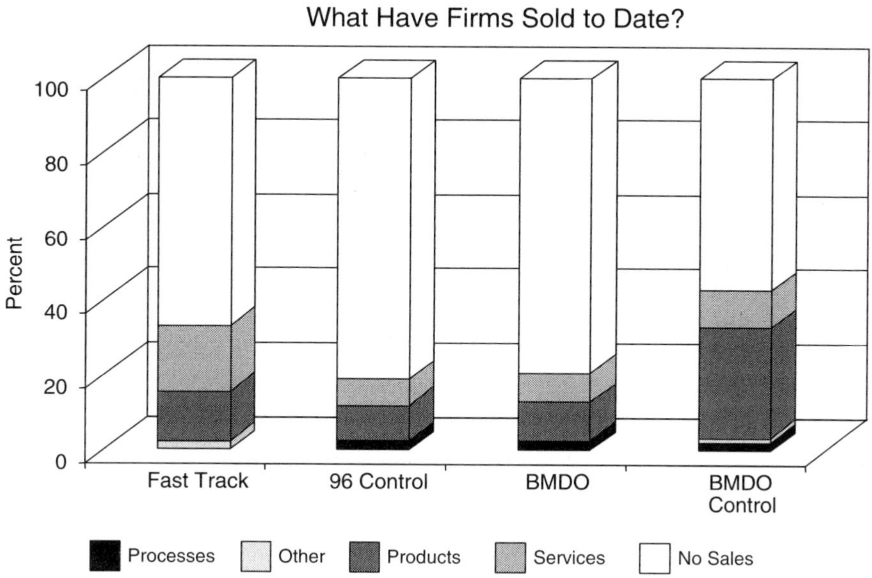

FIGURE 8 Sales through May 1999

Figure 8 displays the commercialization that has already been achieved for these projects. The much smaller achievement than projected is largely explained by the low number of these projects that have completed Phase II (Figure 2).

Products in Figure 8 include both hardware and software. The strong showing for services may imply that services can be commercialized faster than products or processes. It would appear in comparing Figures 7 and 8 that most, if not all, anticipated commercialization of services has already occurred. There is some distortion, however in each of these figures. Respondents were allowed more than one answer to each of these questions. For example, the 45 Fast Track respondents identified 74 products in 15 distinct combinations such as component, hardware product and software, software only, hardware and process, or software and service. The percentages shown are calculated against the total number of responses rather than the total number of respondents.

The performance levels in Figure 8 are consistent with the reported amount of completed marketing activities that are displayed in Figure 9. Fast Track projects reported the highest levels of completed marketing plans and the highest levels of marketing staff hired.

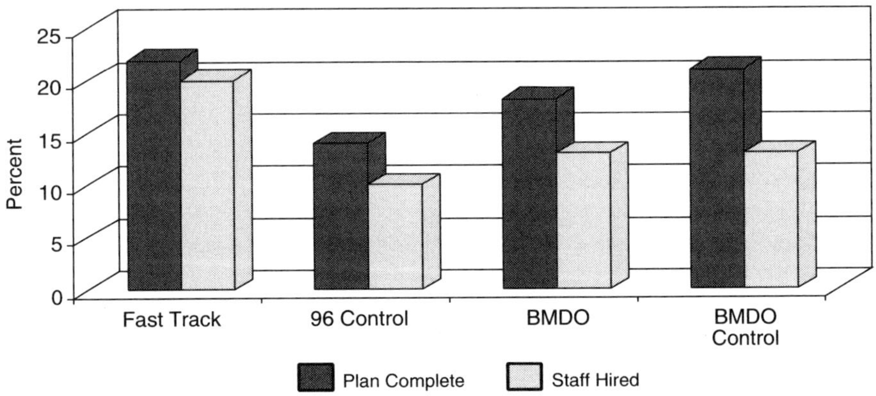

FIGURE 9 Percent of Firms that Completed Marketing Activities

Impact of Funding Gap

Fast Track was one of two reforms initiated in FY 1996 as a result of the 1995 DoD Process Action Team findings on the SBIR program that impacted the funding gap between Phase I and Phase II. The requirement for a Fast Track contractor to submit its Phase II proposal during Phase I and for the agency to expedite evaluation and award of Fast Track proposals should have reduced the Fast Track funding gap between the two phases. The second reform affecting the gap was establishment of a new standard of six months for the average time interval between receipt of an SBIR proposal and award in Phase II.

Funding gaps make it difficult for a small company to keep its research team together, and studies have shown that delay in time to market decreases the value of an innovation. The funding gap is addressed on the next two charts. Figure 10 shows the magnitude of the gap, and Figure 11 shows the number of firms experiencing a gap and how the firms dealt with the gap.

Figure 10 displays the lowest, highest, and average funding gaps (in months) reported for each group, for those projects that reported experiencing a funding gap. For Fast Track, 23 of the 45 respondents experienced no gap; thus the average gap for all Fast Track respondents was only 2.4 months. For the 96 Control group, ten projects reported no gap, resulting in an average gap for all 96 Control group respondents of 4.7 months. The percentage of projects that reported a gap in each group is shown in Figure 11.

Only four Fast Track projects (9 percent) reported that the funding gap caused them to stop work. Over half of the 96 Control group had to stop work because of the funding gap. The use of bridge funding was much more prevalent in Fast

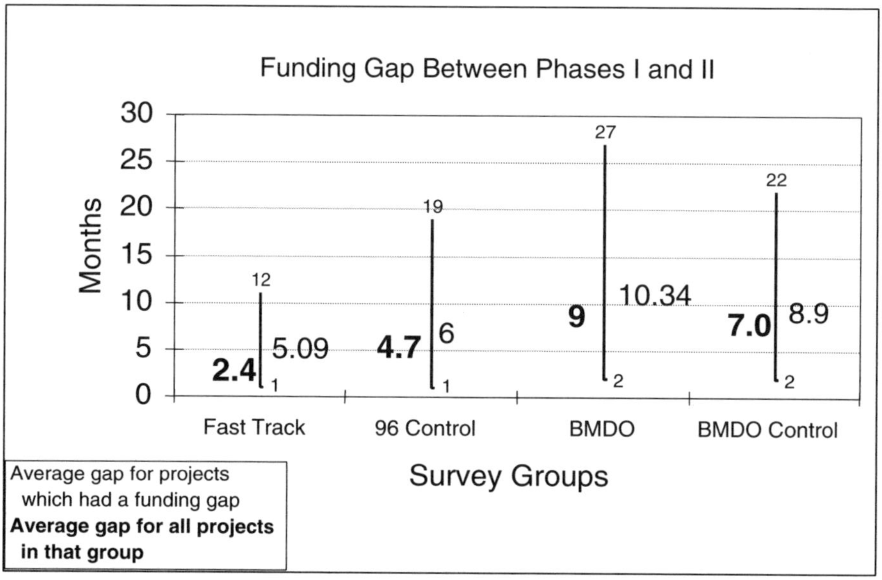

FIGURE 10 Duration of Funding Gap

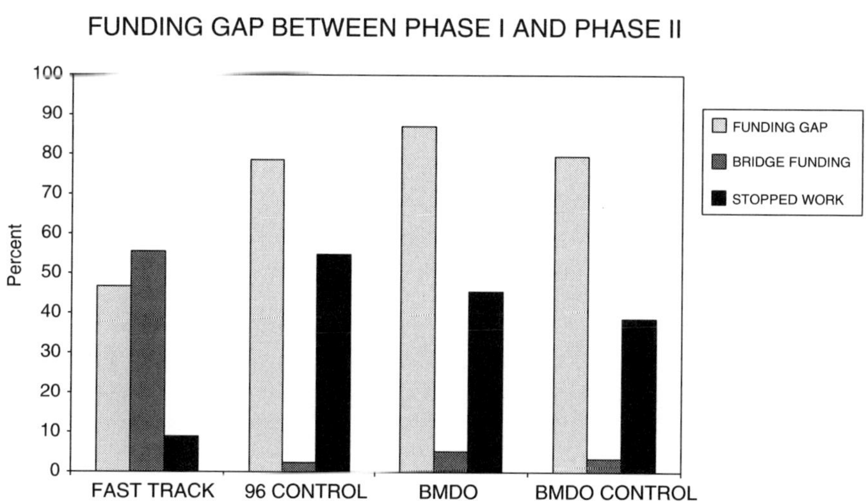

FIGURE 11 Effect of Funding Gap on Firm

Track than in any other group. According to program managers, the use of bridge funding is not limited to Fast Track. For the solicitation years 1992 to 1996, however, only 5 percent of the non-Fast Track projects reported that they received bridge funding, and over 43 percent of the non-Fast Track projects had to stop work awaiting the award of Phase II.

Additional Developmental Funding

SBIR does not provide federal funds to commercialize a product. Seldom can a firm go directly from Phase II to a commercial product. Further development, testing, standardization, producibility engineering, Beta testing of software, and similar activities usually are not performed until after Phase II. Activities related to market research, trade-show attendance, advertising, and sales are not allowable costs under SBIR. Generally, additional developmental funding must be at least as large as the SBIR awards for successful commercialization to occur. It is not unusual for an SBIR firm to use a prototype developed during Phase II to attract investors or customers who place an order for the further development and delivery of a product. Some necessary development activities may occur concurrently with early sales, with the revenue used to upgrade later versions of the product. However, most additional investment must happen early in the sales cycle. The reported additional non-SBIR development funding to date is shown in Figure 12.

Most private additional developmental funding is invested in anticipation of a return on investment. Thus the additional investment is a leading indicator of ultimate commercial sales.

Higher investment in Fast Track projects than in the projects of the 96 Control group is a foregone conclusion at this point. Fast Track projects had to bring third-party money to the table and have it invested early in Phase II. The control group projects did not. The earlier timing of investment in Fast Track would not necessarily mean higher ultimate commercialization. When compared to the co-investment projects, which also brought money to the table, and to the BMDO Control group, another conclusion emerges. The projects in these latter two groups, on average, were two years older than the Fast Track projects, yet their average additional developmental funding was half that of Fast Track projects. It would appear that in addition to occurring earlier, the ultimate investment in Fast track projects will be larger than in non-Fast Track ones.[12]

An examination of the sources of the additional funding reveals a significant difference between Fast Track and the other groups. Figure 13 looks at the percentage of projects that reported receipt of venture capital investment.

---

[12]The 1996 DoD Commercialization Study found that DoD projects awarded Phase II from 1984 to 1992 averaged $597,000 in additional non-SBIR developmental funding.

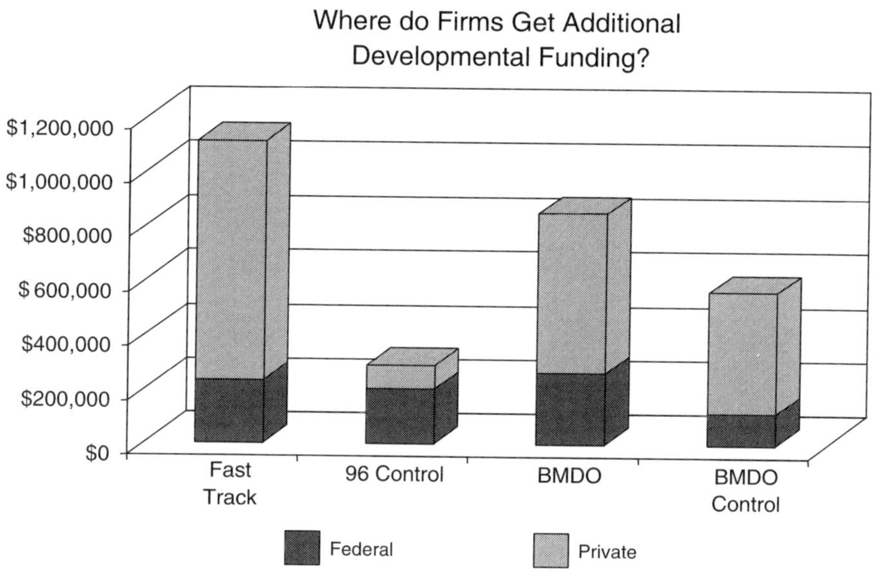

FIGURE 12 Source and Magnitude of Additional Development Funds

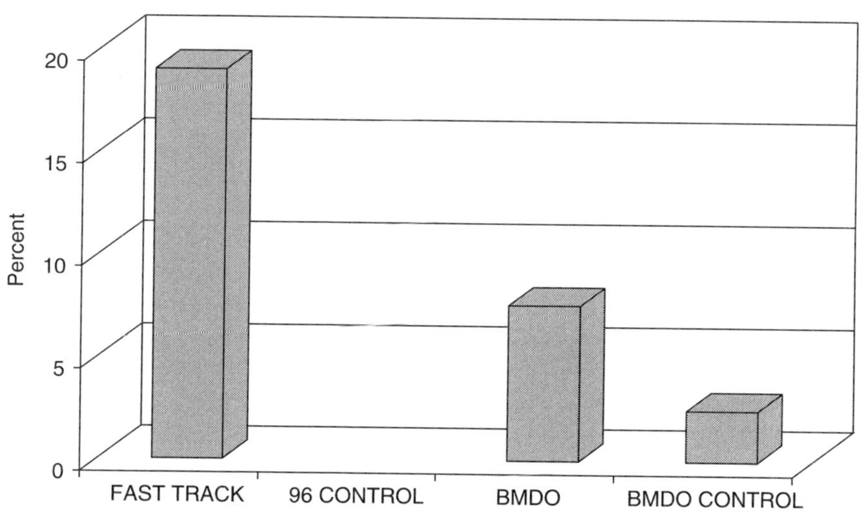

FIGURE 13 Percent of Projects with Venture Capital Funding

In the 1996 DoD SBIR Commercialization Study, only 3 percent of the projects reported receiving venture capital. The low reports of venture capital for the two control groups (0 and 2.5 percent) are consistent with the earlier study. For the BMDO Co-investment group, the level of venture capital (6 of 77) is significantly different from that in the earlier study at a 5 percent confidence level, but not at the 1 percent level. The level of Fast Track venture capital (8 of 45) is significantly different from that in the earlier study.[13]

Sales

Although it is too early for meaningful sales comparisons when so few projects have even completed Phase II, companies did report that some sales have already occurred. The sales and the customers for those sales are shown in Figure 14.

Three of the 11 Fast Track projects reporting sales account for two-thirds of the sales. All three projects are still in Phase II and are being performed by the same contractor, Digital System Resources. In each case the sales were entirely to DoD, and in each case, a Navy program office is providing the third-party funding.

The low level of sales for all groups is not unexpected because most of these projects are still in Phase II, and even those that completed Phase II have had little time to achieve sales. At this time, projections of sales are probably a better indicator than achieved sales. Figure 15 shows the projected sales through the end of the year 2001. Because most surveys were prepared in March and April 1999, this represents a sales projection for the next 11 quarters. In addition to the 11 Fast Track projects that have experienced sales, 8 expect their first sale in 1999, and 14 more expect sales in 2000. A total of 37 of the 45 expect to have had sales by 2001. For the 96 Control group, 8 have sales to date, with 6 more projecting sales in 1999 and 12 projecting sales in 2000. For this Control group, 30 of the 42 plan to have made sales by 2001.

On the basis of the 1996 and 1998 SBIR commercialization studies, it is not unreasonable to project an eventual average sale in excess of $1,000,000 per DoD SBIR project. To expect $1.2 million in sales in the first five years after the 1996 solicitation is probably overly optimistic for the 96 Control group. The more mature BMDO control group is likely to have higher sales in that time frame because to date it has reported more than double the investment and four times the sales of the 96 Control group. However, 3 million dollars in sales projected for the BMDO Control group appears to be even more optimistic. There is no reason to assume that these control groups would perform that much better than past SBIR winners. Such optimism, however, is normal among SBIR partici-

[13]Question 13 of the survey asked Fast Track recipients to identify the source of matching funds in the proposal. Thirteen of the Fast Track respondents marked the venture capital response. Five of these 13 used that response because there was no alternative listed for private investor in question 13.

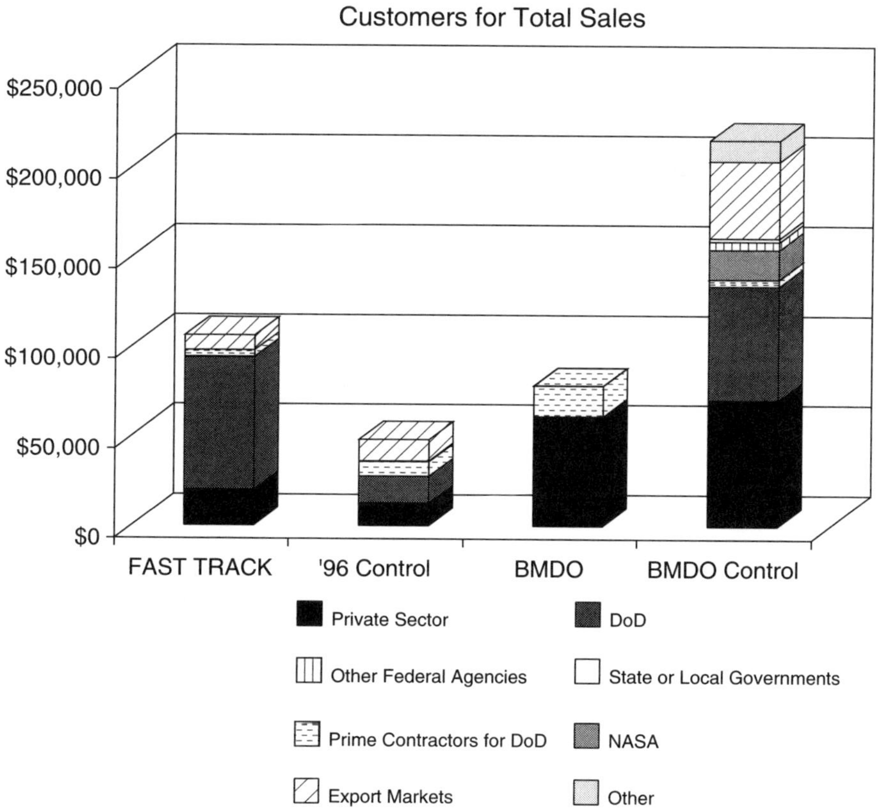

FIGURE 14 Sales as of May 1999

pants. Optimism in projected SBIR sales appeared in the 1991 GAO study. The level of sales projected to be achieved in two years in that study actually took five years to achieve.[14]

For Fast Track, the high levels of investment early in the development cycle and the marketing activities already completed add credibility to a projection of much higher than average sales; however, the $8\frac{1}{2}$ million dollars in projected average sales is unlikely to be achieved in the next three years. The relative relationship of the projected performance for the next three years of each of the groups is more likely to be valid than the absolute sales projected. The seven-to-one advantage projected for Fast Track over its control group in sales projected for the next three years should not be extrapolated to mean seven times as much in

[14]As measured in the 1996 study.

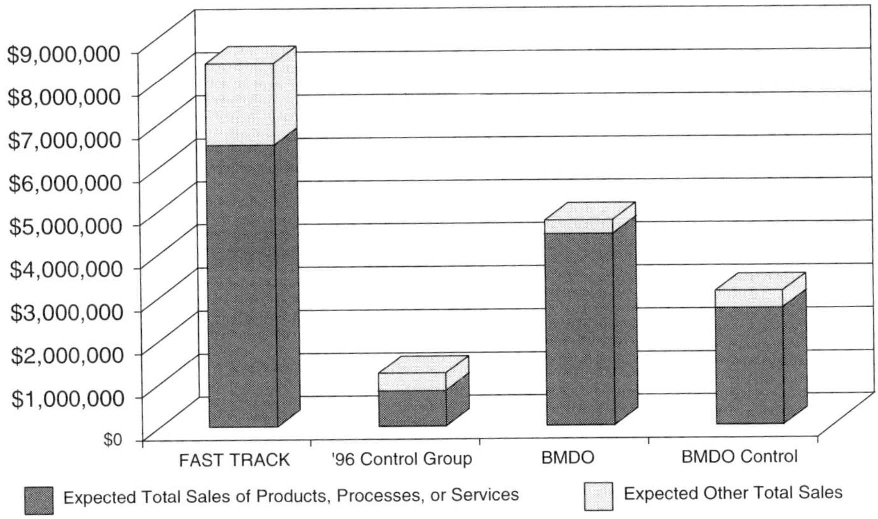

FIGURE 15 Expected Sales through 2001

eventual total sales. Fast Track projects are currently ahead of the 96 Control group in projects that have completed Phase II and projects that have had sales, and they are substantially ahead in additional developmental funding to date. Over time, some of the gap between the two groups should close.

Comparison of Performance and Group Characteristics

Several questions were asked of the respondents to identify differences among groups to see if the difference might correlate to performance of that group. The responses to Additional Developmental Funding (Figure 12), Total Sales (Figure 14), and Expected Sales (Figure 15) each show significantly better performance for Fast Track than for its 96 Control group. Because the greatest difference is in the Expected Sales, that parameter was chosen for this comparison.

Respondents provided information on their firm's founders, specifically on their business background and the number of other companies started by one or more of the founders. Figure 16 demonstrates that one or more founders was more likely to have a business background from firms in the Fast Track group than from firms in the other groups.

Examination of the number of other companies founded by one or more of the founders produces a very similar chart (Figure 17). On average, the founders

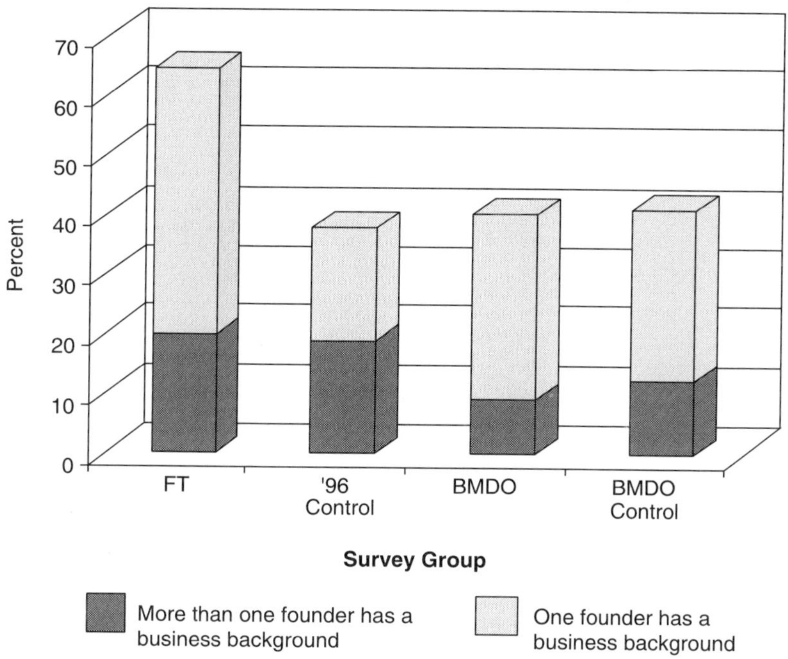

FIGURE 16 Percent of Founders with a Business Background

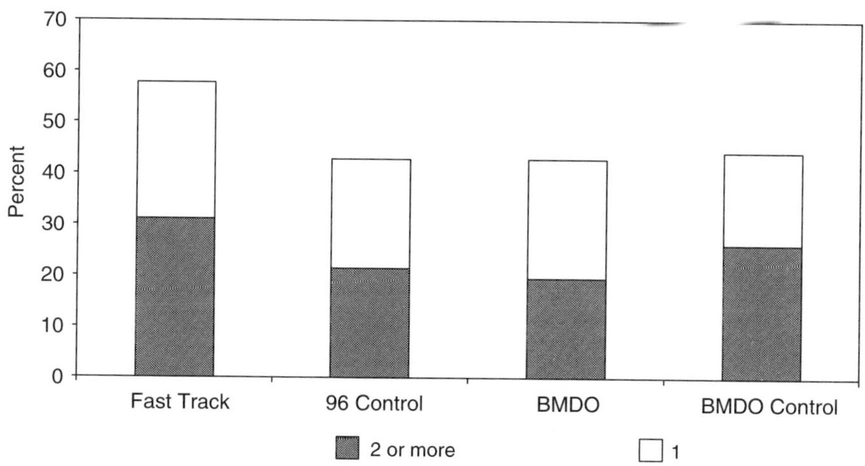

FIGURE 17 Percent of Firms Where a Founder has Started One of More Other Firms

of each Fast Track firm started 2.1 other companies. Founders of each BMDO co-investment firm started an average of 0.9 other companies while the average number of companies started by the founders of a firm in either of the two control groups was 1.3.

Comparison of the business background of the Fast Track firms to the reported expected sales yielded an interesting result. The 29 Fast Track firms where one or more founders have a business background expect average sales in the next three years of $11.4 million. The Fast Track firms that did not report having a founder with prior business background expect only $3.0 million in that time frame.

Fast Track firms whose founders had started one or more other firms report that they expect an even higher level of sales. Specifically, the 24 firms whose founders started other firms anticipate $13.5 million in sales in the next three years compared to the $1.6 million expected by firms whose founders had not started other companies.

These apparently strong correlations do not extend to the same extent to the other three groups. For the BMDO Co-investment group, average expected sales for the business-background firms is higher ($6.5 million) than for the firms whose founders have a nonbusiness-background ($2.7 million). However, for both control groups, there was essentially no difference between firms where the founders had a business background and those where they did not. The firms whose founders had started other firms in the 96 Control group predicted sales that were three times as great as other firms in that group, but for both BMDO and its control group, sales were essentially the same for the two types of firms.

For Fast Track, the business background and experience in starting other companies may have been quite useful in obtaining the third-party financing. Because such third-party financing was required to be in Fast Track, this could account for the high percentages on Figures 16 and 17. The business-background Fast Track firms attracted six times as much investment to date as the nonbusiness-background Fast Track firms. The Fast Track firms whose founders started other companies attracted two and one-half times as much investment as the Fast Track firms whose founders had not started other companies.

Patents, Copyrights, and Scientific Publications

Intellectual property is a primary product of the SBIR program to many firms. Patents and copyrights are a means to protect such property. The number of patents and copyrights applied for and issued is a measure of the intellectual property being generated. Since most scientific journals are refereed, the number of publications submitted and published measures to some degree the scientific merit of the SBIR. The numbers of patents, copyrights, and scientific publications to date, as measured by the survey, are shown in figures 18, 19, and 20, respectively.

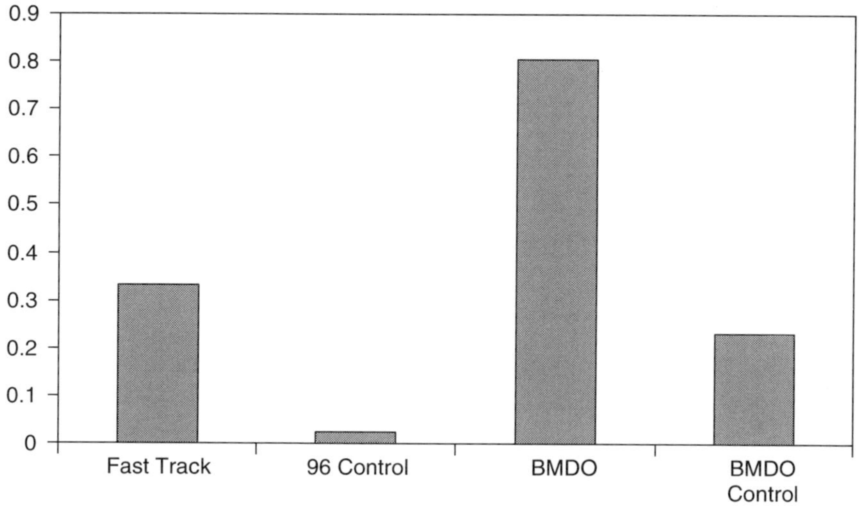

FIGURE 18 Average Number of Patents Received

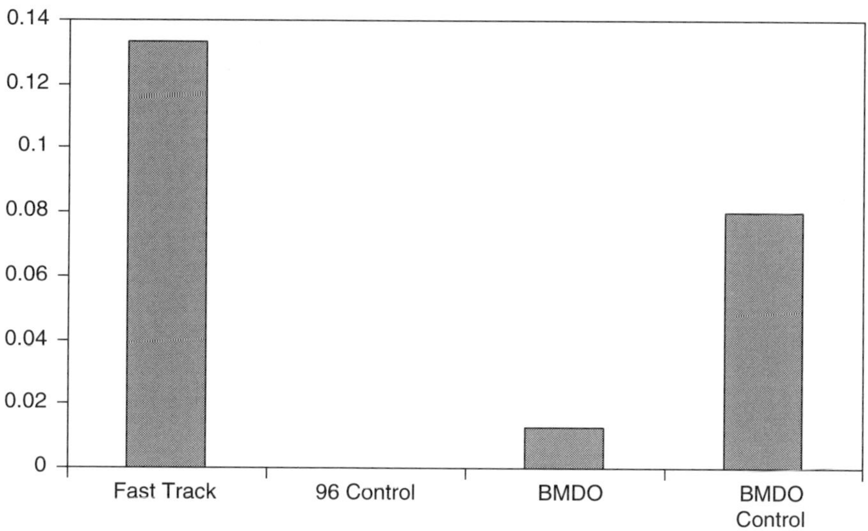

FIGURE 19 Average Number of Copyrights Received

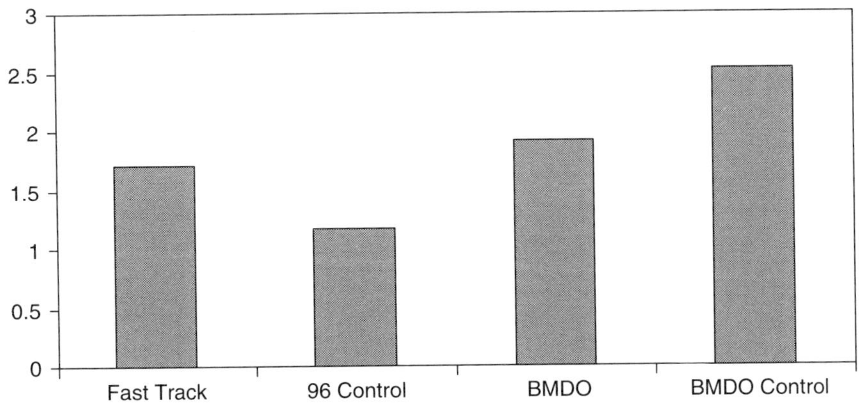

FIGURE 20 Average Number of Publications per Project

A concern expressed by some opponents of Fast Track is that, in order to attract third-party investors, there can be no true research in a Fast Track SBIR. The project must be much further down the development path. They contend that to obtain third-party financing innovation must be complete or nearly complete and the SBIR merely serves to validate it. If this is the case, one might expect a lower level of patent and copyright activity for Fast Track.

The first point to note in Figures 18 and 19 is the low average level of activity for all groups. Recall that only 16 percent of Fast Track projects have completed Phase II. Patenting is not a quick process. The BMDO and BMDO Control groups have had more time, but only half of those projects are out of Phase II. It is not unusual for a patent to take a year or more for approval after the completed application is submitted. (The application process is also time- and resource-consuming.) For Fast Track, 30 percent of the patent applications submitted so far have completed the approval process. For both BMDO groups, half of the applications have already been approved. The 96 Control group has submitted slightly over one-third as many patent applications as the Fast Track group, but none has yet been approved.

It would seem that Fast Track is outperforming its control group and BMDO Co-investment is outperforming BMDO Control in patents. The BMDO advantage over Fast Track is probably due to the extra time those projects have had. Conclusions about technical merit are premature. The limited time that projects have had for patent activities and the small numbers of patents issued so far makes any analysis at this time questionable. The number of copyrights issued also is too small for any meaningful comparison.

Activity in average scientific publications per project would seem to show Fast Track outperforming its control group (see Figure 20). Fifty-four percent of the Fast Track publications, however, come from a single project. When we eliminate this outlier and similarly eliminate the one 96 Control project that accounted for 30 percent of that group's publications, the averages are 0.8 and 0.8 publications per project, respectively. A similar elimination of the top 3 to 4 percent of the projections from the other two groups would halve their averages. Beyond concluding that those projects that have had longer time to mature have published more, it is difficult to draw more from these data.

Initial Public Stock Offerings (IPOs)

None of the Fast Track or 96 Control group firms has yet made an IPO. Four firms from the BMDO Co-investment group and one BMDO Control group firm have made IPOs. A logical explanation for the difference between Fast Track and BMDO groups is the age and revenue levels of the firms and the larger size of the BMDO groups. This explanation does not account for the 96 Control group, which is older than and has revenues equal to or higher than those of the BMDO groups. One Fast Track firm and one firm from each BMDO group plans an IPO this year.

Sale of Ownership

SBIR contractors are often reluctant to seek third-party financing, particularly from venture capitalists, for fear that they will lose control of their firms. Others welcome such cash infusions, preferring to have partial control over a large firm to full control of a small one. Some of this latter group intend to sell their interests as the firm gets large and roll the profits into starting a new firm. Those leery of venture capital are often heavily involved in advancing technology and less interested in production. They are afraid that outside investors would change the fundamental nature of the firm. The business expertise that venture capital insists on putting in place (if not already there) creates an environment likely to produce commercial success, but such success may not be the principal goal of the owners.

Respondents were asked to identify activities related to sale of partial ownership. Finalized agreements and ongoing negotiations are shown in Figure 21.

Figure 21 portrays the "cost" to firm owners of Fast Track firms. To obtain third-party funding, they may be limiting their personal share of the potential gain from their innovation by selling a share of their firm. However, they may also be increasing the ultimate gain from the innovation by infusing cash and business expertise at the critical point in development. Whether this is a cost or an opportunity is very much a personal evaluation.

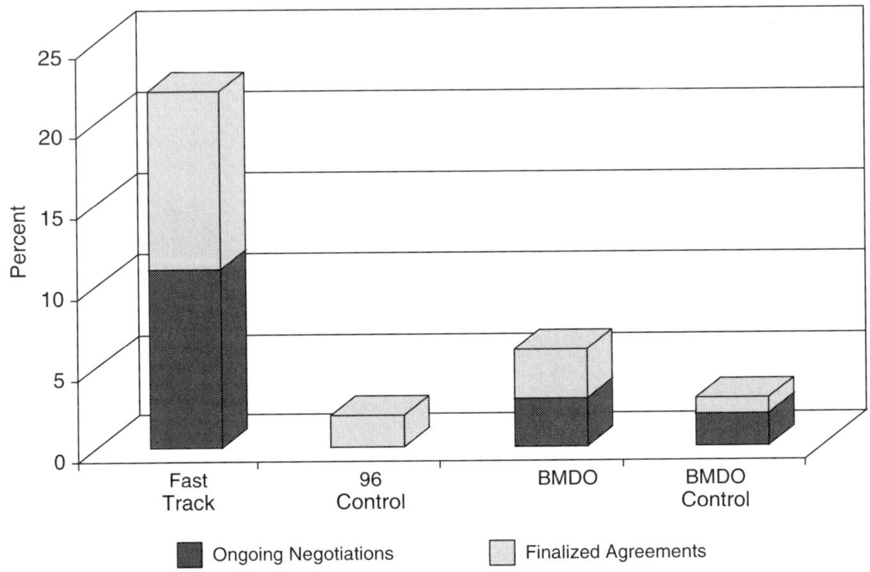

FIGURE 21 Sale of Partial Ownership

CONCLUSIONS

Past studies of SBIR commercialization has used three primary measurements of success: sales, additional developmental funding, and expected sales. By each of these measures, the Fast Track projects are clearly outperforming those in the control group.

Although Fast Track sales lag behind those of the BMDO Co-investment group (which on average started two years earlier), they exceed those of the BMDO Control group, which had a similar head start. Moreover, Fast Track projects have achieved double the additional developmental funding of either BMDO group, and Fast Track projects expect twice as much sales in the next three years.

Fast Track has been successful in nearly eliminating the funding gap between Phases I and II.

Firms that apply for Fast Track tend to be much younger than average SBIR firms. They have had far fewer Phase II awards than the overall population. Sixty percent have had no prior Phase II awards. The average annual revenue for Fast Track applicants is less. These characteristics may have little to do with success and may change if the cost-matching arrangement is changed.

The predominance of firms in Fast Track whose founders have business backgrounds and firms whose founders have started other firms may indicate that such firms have an easier time acquiring third-party funding.

Additional developmental funding is a leading indicator of commercial success. High expectations of sales are probably better than low expectations, but the bottom line is sales achieved, and it is several years too early to measure that bottom line. Given the inherent risk associated with research-driven business, today's high expectations for future sales may not be realized. A follow-up effort should be planned to verify the current conclusions.

Whether it is the validation of a third party to the commercial potential, the timing and magnitude of the additional funding, or merely the reduction in funding gap that contributes most to the apparent performance of projects selected for SBIR award under Fast Track, the program is working. Fast Track is selecting projects that should succeed in commercialization, and it is apparently contributing to their success.

APPENDIX A

Sample Characteristics

The survey sample of 379 projects consisted of a study group - 48 Fast Track Phase II (all of the Fast Track winners from 1996 DoD solicitations), 127 BMDO co-investment Phase II (all BMDO co-investment projects from the 1992–1996 solicitations) and 204 control group projects chosen from DoD 1992–1996 Phase II winners.

The Table below shows the firm size and the sponsoring agencies for the study sample projects and the control group compared to the expected distribution of 204 projects drawn at random from the Phase II award winners resulting from the 1992–1996 DoD solicitations.

TABLE A1 Firm Size and Awarding Agencies of Survey Sample

Firm Size # of Employees	# of Firms Expected from Population	# of Firms In Control Group	# of Firms In Study Group
1 to 5	45	48	57
6 to 20	63	61	72
21 to50	35	44	21
51 to150	38	27	18
150 to 500	23	24	7
	204	204	175
AGENCY			
AF	80	62	4
ARMY	38	24	20
BMDO	16	29	137*
DARPA	24	44	5
NAVY	41	36	6
DSWA	2	4	0
OSD	3	5	3
	204	204	175

* The 137 BMDO projects are 10 Fast Track projects awarded by BMDO and 127 BMDO co-investment projects. All projects listed for other agencies in the study group are Fast Track.

TABLE A2 Representation of Woman and Minority Owned Firms

Number of Firms Owned By	# of Projects Expected from Population	# of Projects In Control Group	# of Projects In Study Group
Women	18	22	17
Minority	31	34	33

The distribution of firms owned by women or minorities in the study sample and the control group was similar to what would be expected from the population.

TABLE A2 Number of Phase II Awarded Prior to the Selected Project

Number of Prior PH II Received By Firm	# of Projects Expected from Population	# of Projects In Control Group	# of Projects In Study Group
0	81	86	85
1	25	24	20
2	16	12	16
3	12	11	11
4	9	7	4
5	9	11	7
6	5	6	2
7	4	3	3
8	4	5	1
9	4	2	2
10	3	2	3
11	4	3	2
12	3	1	4
13	3	5	3
14	2	1	
15	2	2	3
16	2	2	4
17	2	6	2
18	1	1	1
19-29	3	4	1
30-74	7	7	1
>75	3	3	0
Total	204	204	175

Thirty-four of the forty–eight 1996 Fast Track Phase II awardees had no prior Phase II awards. They are included in the 85 shown at the top of the left column. For Fast Track, 71% of the winners had no prior awards. BMDO co-investment winners were far more likely to have prior awards. The percentage of BMDO co-investment with no prior Phase II was (85-34)/(175-48) equals 40% which is the average (40%) for all Phase II winners. The expected number with zero, one or two prior awards (122) is the same as in the control group and almost the same as in the study sample (121).

Award of SBIR are not spread uniformly across the country. They tend to cluster in states known for high technology firms, universities, and available venture capital. The table on the next page shows the geographical distribution of the sample compared to the expected distribution.

TABLE A3 Distribution of Samples by States

States	# of Projects Expected from Population	# of Projects In Control Group	# of Projects In Study Group
AL	5	6	6
AZ	3	4	4
CA	50	50	42
CO	7	10	10
CT	5	9	9
DE	1	1	4
FL	5	3	3
GA	2	2	2
IL	1	1	0
IN	1	0	0
KS	1	1	1
MA	31	31	25
MD	8	11	8
ME	1	0	0
MI	3	3	1
MN	4	2	1
MT	0	1	0
NC	2	2	2
NH	3	2	1
NJ	9	11	12
NM	5	3	2
NV	1	0	0
NY	9	12	11
OH	8	6	3
OK	1	1	1
OR	1	2	2
PA	7	6	3
RI	1	1	4
TN	2	2	3
TX	6	4	2
UT	2	1	1
VA	15	13	10
WA	3	3	1
WI	1	0	0
WV	0	0	1
	204	204	175

APPENDIX B

NAS Survey of Small Business Innovation Research

PROPOSAL TITLE

TOPIC NUMBER

PHASE II CONTRACT NUMBER

1. Is the above address complete and correct?
   a. Yes, Go to question 2
   b. No, Provide corrections below

2. What is the current status of this SBIR project? *(Select one.)*
   a. Project has not yet completed Phase II.
   b. Project completed Phase II.
   c. Project was not awarded a Phase II.

3. How do you expect to commercialize your SBIR award?
   a. No commercial product, process, or service is planned.
   b. Software
   c. Intermediate hardware product or component
   d. Final hardware product
   e. Process technology
   f. New or improved service capability

> Question 4 concerns whether the project is still active or if and when it was dropped.

4. When did your company drop the project?
   a. During or at the end of Phase II.
   b. Within one year after completing Phase II.
   c. More than one year after completing Phase II.
   d. Still active, not dropped.

If the technology developed during this project has led to no additional developmental funding and/or sales and neither of these is expected to occur, skip to Question 16.

ADDITIONAL DEVELOPMENTAL FUNDING AND SALES

PLEASE READ THE FOLLOWING DEFINITIONS

Additional Developmental Funds: Include funds from federal or private sector sources, or from your own company, used for further development of the technology developed during this Phase II project.

Sales: Includes all sales of a product, process, or service, to federal or private sector customers resulting from the technology developed during this Phase II project. *A sale also includes the sale of technology or rights, etc.*

5. To date, what has been the total additional developmental funding for the technology developed during this project? (*Do not include related SBIR funds received from DoD or other federal agencies. Enter dollars provided by each of the following sources. If none, enter 0 [zero].)*

Sources	Dollars
a. Non-SBIR federal funds.	
b. Your own company	
c. Other private company	
d. U.S. venture capital institution	
e. Foreign venture capital institution	
f. Private investor	
g. Personal funds	
h. State or local governments	
i. College or Universities	
j. Other sources (Specify)	

6. As a result of the technology developed during this project, which of the following describes your company's activities with other companies and investors? *(Select all that apply.)*

Activities	Ongoing Negotiations	Finalized Agreements
a. Licensing Agreement		
b. Sale of complete ownership		
c. Sale of partial ownership		
d. Sale of technology rights		
e. Joint venture agreement		
f. Marketing/distribution agreement		
g. Manufacturing agreement		
h. Other		

7. Which of the following, if any, describes the type and status of marketing activities by your company and/or your licensee for this project? *(Select one for each)*

Marketing activity	Planned	Under way	Completed	Not needed
a. Preparation of marketing plan				
b. Hiring of marketing staff				
c. Publicity/advertising				
d. Test marketing				
e. Other (Specify)				

> Questions 8 to 10 concern actual sales to date resulting from the technology developed during this project.

8. Has your company and/or licensee had any actual sales of products, processes, services or other sales from the technology developed during this project? (*Select all that apply.)*

   a. No sales to date ⟶ Skip to Question 10.
   b. Sales of product(s)
   c. Sales of process(es)
   d. Sales of services(s)
   e. Other sales (e.g., rights to technology, etc.)

9. For you company and/or your licensee, when did the first sale occur, and what is the approximate amount of total sales resulting from the technology developed during this project? If multiple SBIR contributed to the ultimate product, report only the share of total sales appropriate to this SBIR project. *(Enter dollars. If none, enter 0 [zero].)*

   Year when first sale occurred

   Total Sales Dollars of Product (s)
   Process(es) or Service(s) to date

   Other Total Sales Dollars (e.g.,
   Rights to technology, etc.) to date

10. To date, about what percent of total sales from the technology developed during this project have gone to the following customers? *(If none enter 0 [zero]. Round percentages. Answers should add to 100%)*

    Private sector
    Department of Defense (DoD)
    Prime contractors for DoD
    NASA
    Other federal agencies
    State or local governments
    Export markets
    Other (Specify)

11. Expected Sales. For your company and/or your licensee, what is the approximate amount of total sales expected between now and the end of 2001 resulting from the technology developed during this project? *(If none enter 0 [zero].)*

    If no sales to date, what year do you expect your first sale? 19__

    Total sales dollars of product(s), process(es) or services(s) expected between now and the end of 2001.

    Other Total Sales Dollars (e.g., rights to technology, etc.) expected between now and the end of 2001.

Questions 12 to 15 apply to activity prior to the Phase II SBIR award.

12. Did this award follow DoD Fast Track or a BMDO Phase II Co-investment procedures which identified Matching/cost shared/co-investment funds in the Phase II Proposal?
    a. Yes. This was a Fast Track proposal.
    b. Yes. BMDO co-investment was identified.
    c. Neither

If neither, skip to question 15.

13. Regarding sources of matching or co-investment funding that were proposed for Phase II, check all that apply.
    a. Venture Capital provided funding.
    b. Another company providing funding.
    c. Another company provided facilities, equipment and/or other in kind support.
    d. Own company provided funding.

14. How long in months did it take to obtain and finalize Agreement(s) for third party funding/in kind support?

    ☐☐ months.

15. Prior to your SBIR award, did your company receive funds for research or development of the technology in this project from any of the following sources?
    a. Prior SBIR
    b. Prior non-SBIR federal R&D
    c. Venture Capital
    d. Other private company
    e. Private investor
    f. Internal company investment
    g. State or local government
    h. College or University

OTHER COMPANY AND PROJECT RELATED INFORMATION

16. In you opinion, in the absence of this SBIR award, would your company have undertaken this project? *(Select one.)*
    a. Definitely yes   c. Uncertain   e. Definitely no
    b Probably yes   d. Probably no

17. What year was your company founded? 19 ☐☐

18. Information on company founders.
    a. Number of founders. ☐☐
    b. Number of other companies started by one or more of the founders. ☐☐
    c. Number of founders who have a business background. ☐☐

19. Most recent employment of founders prior to founding this company. *(Indicate all that apply.)*
    a. Other private company
    b. College or University
    c. Government

20. Total revenue for the company during your fiscal year (or calendar year) 1998. *(Select one)*
    a. Less than $100,000
    b. $100,00 to $499,999
    c. $500,000 to $999,999
    d. $1,000,000 to $4,999,999
    e. $5,000,000 to $19,999,999
    f. More than $20,000,000

21. How many SBIR awards your company received from federal agencies, including DoD, prior to this Phase II?
    a. Number of prior Phase I awards ☐☐
    b. Number of prior Phase II awards ☐☐

22. Please give the number of patents, copyrights, and/or scientific publications for the technology developed during this project. *(Enter numbers. If none, enter 0 [zero].)*

Number Applied for/ Submitted		Number Received/ Published
☐☐☐	Patent(s)	☐☐☐
☐☐☐	Copyrights	☐☐☐
☐☐☐	Scientific Publications	☐☐☐

23. Which, if any, of the following has your company experienced as a result of the technology developed during this project? *(Select all that apply.)*
    a. Made an initial public stock offering in 19
    b. Planned an initial public stock offering for 1999.
    c. Established one or more spin-off companies
       How many companies?
    d. None of the above.

24. If you experienced funding gap between Phase I and Phase II for this award, select all answers that apply
    a. Duration of gap in months.
    b. Stopped work on this project during funding gap.
    c. Continued work at reduced pace during funding gap.
    d. Continued work at pace equal to or greater than Phase I pace during funding gap.
    e. Received bridge funding between Phase I and II.

25. Administrative information needed for assessment. Please fill in for all dates that have occurred.

Date Phase I ended

Date Phase II proposal submitted

Date Phase II started

Date Phase II ended

26. Employee information. *(Enter number of employees. You may enter a number less than 1; i.e., decimal numbers.)*
    Number of employees (if known) when Phase I was submitted

    Current number of employees

    Current number of employees hired as a result of the technology developed during this Phase II project.

27. Did the technology developed during this SBIR project provide knowledge that aided your subsequent research

    a. No, this project was separate from our other research. The knowledge gained has not been helpful elsewhere.
    b. Yes, the project was one step in the development that has continued (or will continue) after the project.
    c. Yes, the knowledge developed has helped other research projects.
    d. Other. The above options do not fit our situation.

28. Please provide the following information about the person who completed this questionaire.
    Name
    Title
    Telephone Number
    E-mail Address
    Company URL, if available

29. Would you like an electronic version of the final report?
    a. Yes b. No

APPENDIX C

NAS Survey Analysis

Responses Received:	
Fast Track	45
96 Control	42
BMDO	77
BMDO Control	88
Total projects	252

(excludes four projects where no phase II awarded)

Q1	Fast Track		'96 Control		BMDO		BMDO Control	
2	15	33%	8	19.05%	27	35.06%	30	34.09%
1	28	62%	30	71.43%	49	63.64%	54	61.36%
0	2	4%	4	9.52%	1	1.30%	4	4.55%

Q2	Fast Track		'96 Control		BMDO		BMDO Control	
2	7	16%	2	4.76%	38	49.35%	51	57.95%
1	38	84%	40	95.24%	38	49.35%	37	42.05%
0	0	0%	0	0.00%	1	1.30%	0	0.00%

Q3	Fast Track	'96 Control	BMDO	BMDO Control
3-1	1	1	7	3
3-2	14	16	3	22
3-3	23	16	36	43
3-4	17	12	33	34
3-5	6	12	30	26
3-6	13	12	5	17
blank/NA	1	1	0	0

Q4	Fast Track		'96 Control		BMDO		BMDO Control	
4	42	93%	38	90.48%	64	83.12%	72	81.82%
3	0	0%	0	0.00%	3	3.90%	5	5.68%
2	0	0%	1	2.38%	2	2.60%	1	1.14%
1	0	0%	0	0.00%	3	3.90%	3	3.41%
blank/NA	3	7%	3	7.14%	5	6.49%	7	7.95%

Q5 Avg	Fast Track	'96 Control	BMDO	BMDO Control
A	$233,067	$201,651	$261,567	$116,126
B	$170,644	$28,476	$60,795	$124,023
C	$91,700	$44,238	$259,443	$11,386
D	$345,556	$0	$130,195	$242,046
E	$88,889	$0	$0	$0

Appendix C—continued

F	$156,433	$0	$123,636	$56,818
G	$6,556	$3,310	$4,221	$7,500
H	$20,778	$8,333	$8,208	$5,523
I	$0	$0	$896	$0
J	$0	$2,381	$2,758	$693
BJ	$880,556	$86,738	$590,152	$447,989
AJ	$1,113,623	$288,389	$851,719	$564,115

Q6-A	Fast Track		'96 Control		BMDO		BMDO Control	
2	1	2%	0	0.00%	5	6.49%	7	7.95%
1	12	27%	11	26.19%	21	27.27%	18	20.45%
0	32	71%	33	78.57%	51	66.23%	63	71.59%

Q6-B	Fast Track		'96 Control		BMDO		BMDO Control	
2	0	0%	0	0.00%	0	0.00%	0	0.00%
1	0	0%	0	0.00%	2	2.60%	1	1.14%
0	45	100%	42	100.00%	75	97.40%	87	98.86%

Q6-C	Fast Track		'96 Control		BMDO		BMDO Control	
2	5	11%	0	0.00%	2	2.60%	2	2.27%
1	5	11%	1	2.38%	2	2.60%	1	1.14%
0	35	78%	41	97.62%	73	94.81%	84	95.45%

Q6-D	Fast Track		'96 Control		BMDO		BMDO Control	
2	1	2%	1	2.38%	2	2.60%	1	1.14%
1	5	11%	3	7.14%	7	9.09%	4	4.55%
0	39	87%	38	90.48%	68	88.31%	83	94.32%

Q6-E	Fast Track		'96 Control		BMDO		BMDO Control	
2	1	2%	0	0.00%	3	3.90%	1	1.14%
1	6	13%	5	11.90%	11	14.29%	12	13.64%
0	38	84%	37	88.01%	63	81.82%	75	85.23%

Q6-F	Fast Track		'96 Control		BMDO		BMDO Control	
2	1	2%	1	2.38%	2	2.60%	6	6.82%
1	9	20%	8	19.05%	6	7.79%	21	23.86%
0	35	78%	33	78.57%	69	89.61%	61	69.32%

Q6-G	Fast Track		'96 Control		BMDO		BMDO Control	
2	1	2%	0	0.00%	5	6.49%	1	1.14%
1	9	20%	3	7.14%	10	12.99%	10	11.36%
0	35	78%	39	92.86%	62	80.52%	77	87.50%

Appendix C—continued

Q6-H	Fast Track		'96 Control		BMDO		BMDO Control	
2	0	0%	1	2.38%	1	1.30%	3	3.41%
1	2	4%	1	2.38%	2	2.60%	7	7.95%
0	43	96%	40	95.24%	74	96.10%	78	88.64%

Q7-A	Fast Track		'96 Control		BMDO		BMDO Control	
4	0	0%	2	5%	7	9%	3	3.41%
3	10	22%	6	14%	14	18%	19	21.59%
2	21	47%	17	41%	19	25%	31	35.23%
1	7	16%	3	7%	11	14%	11	12.50%
0	7	16%	14	33%	26	34%	24	27.27%

Q7-B	Fast Track		'96 Control		BMDO		BMDO Control	
4	7	16%	10	24%	20	26%	18	20.45%
3	9	20%	4	10%	10	13%	12	13.64%
2	10	22%	4	10%	3	4%	9	10.23%
1	7	16%	5	12%	5	7%	7	7.95%
0	12	27%	19	45%	39	51%	42	47.73%

Q7-C	Fast Track		'96 Control		BMDO		BMDO Control	
4	2	4%	3	7%	12	16%	8	9.09%
3	3	7%	1	2%	3	4%	8	9.09%
2	16	36%	14	33%	22	29%	21	23.86%
1	14	31%	5	12%	9	12%	15	17.05%
0	10	22%	19	45%	29	38%	36	40.91%

Q7-D	Fast Track		'96 Control		BMDO		BMDO Control	
4	5	11%	9	21%	13	17%	13	14.77%
3	7	16%	1	2%	2	3%	8	9.09%
2	11	24%	8	19%	21	27%	18	20.45%
1	15	33%	4	10%	9	12%	13	14.77%
0	12	27%	20	48%	32	42%	36	40.91%

Q7-E	Fast Track		'96 Control		BMDO		BMDO Control	
4	0	0%	3	7%	6	8%	5	5.68%
3	2	4%	0	0%	0	0%	0	0.00%
2	0	0%	0	0%	2	3%	5	5.68%
1	1	2%	1	2%	0	0%	1	1.14%
0	40	89%	38	91%	69	90%	77	87.50%

Q8	Fast Track	'96 Control	BMDO	BMDO Control
8-1	28	23	56	38
8-2	6	4	8	26

Appendix C—continued

8-3	0	1	2	2
8-4	8	3	6	9
8-5	1	0	0	1
blank/NA	6	11	8	17

Q9-A	Fast Track	'96 Control	BMDO	BMDO Control
99	0	2	0	5
98	8	4	3	9
97	3	1	7	9
96	0	0	1	7
95	0	0	2	1

Q9-BC Avg	Fast Track	'96 Control	BMDO	BMDO Control
B	$106,000	$47,945	$61,961	$209,722
C	$89	$0	$16,571	$375
BC	$106,089	$47,945	$78,533	$210,097

Q10*9BC Avg	Fast Track	'96 Control	BMDO	BMDO Control
A	$19,900	$12,446	$60,918	$68,157
B	$74,043	$15,130	$130	$60,514
C	$3,885	$8,107	$16,835	$3,636
D	$0	$0	$130	$16,563
E	$0	$357	$130	$5,331
F	$0	$0	$0	$991
G	$8,261	$11,905	$390	$41,672
H	$0	$0	$0	$11,473
SUM	$106,089	$47,945	$78,533	$208,337

Q11-A	Fast Track	'96 Control	BMDO	BMDO Control
2003	1	0	1	2
2002	0	2	8	1
2001	4	4	7	5
2000	14	12	13	17
1999	8	6	17	14
Sales no yr	13	7	14	28

Q11-BC Avg	Fast Track	'96 Control	BMDO	BMDO Control
B	$6,546,111	$817,500	$4,445,584	$2,657,500
C	$1,895,778	$411,905	$316,883	$384,034
BC	$8,441,889	$1,229,405	$4,762,468	$3,041,534

Appendix C—continued

Q12	Fast Track	'96 Control	BMDO	BMDO Control
12-1	40	0	0	4
12-2	2	1	36	0
12-3	2	33	34	71
blank/NA	3	8	7	14

Q13	Fast Track	'96 Control	BMDO	BMDO Control
13-1	13	0	5	1
13-2	19	1	17	0
13-3	3	1	15	2
13-4	7	0	22	7
blank/NA	10	40	37	78

Q14	Fast Track	'96 Control	BMDO	BMDO Control
23	0	0	0	0
22	0	0	1	0
21	0	0	0	0
20	0	0	0	0
19	0	0	1	0
18	0	0	3	0
17	0	0	0	0
16	0	0	1	0
15	0	0	0	0
14	0	1	0	0
13	0	0	1	0
12	3	0	6	2
11	0	0	0	0
10	1	0	1	0
9	3	0	1	1
8	2	0	2	0
7	0	0	0	0
6	10	2	7	0
5	1	0	0	0
4	8	0	3	0
3	8	0	2	0
2	4	0	0	0
1	0	1	2	0
0	5	38	45	85

Q15	Fast Track	'96 Control	BMDO	BMDO Control
15-1	6	12	28	23
15-2	5	8	5	13
15-3	2	0	2	1
15-4	3	2	3	1
15-5	14	1	6	3
15-6	10	7	15	25

Appendix C—continued

15-7	2	2	5	6
15-8	0	1	3	3
blank/NA	18	20	35	40

Q16	Fast Track		'96 Control		BMDO		BMDO Control	
5	13	29%	16	38%	23	30%	31	35%
4	13	29%	20	48%	29	38%	29	33%
3	12	27%	1	2%	11	14%	15	17%
2	4	9%	4	10%	10	13%	9	10%
1	2	4%	0	0%	3	4%	3	3%
0	1	2%	1	2%	1	1%	1	1%

Q17	Fast Track	'96 Control	BMDO	BMDO Control
98	0	1	2	0
97	1	0	0	0
96	9	2	1	1
95	7	3	6	5
94	5	1	4	3
93	3	0	6	7
92	2	6	11	11
91	4	2	6	3
90	3	0	2	7
89	0	0	5	7
88	1	3	3	9
87	0	0	2	2
86	0	1	6	5
85	1	2	3	1
84	0	2	0	2
83	0	3	8	3
82	5	2	1	1
81	1	0	3	0
80	0	1	1	1
pre-80	2	12	7	20
N/A	1	0	0	0

Q18-1	Fast Track	'96 Control	BMDO	BMDO Control
10	1	0	0	0
9	0	1	0	0
8	0	0	0	0
7	0	0	0	0
6	0	0	0	0
5	0	1	2	2
4	3	2	1	6
3	16	5	9	13
2	9	18	25	30
1	13	15	39	36
N/A	3	0	1	1

Appendix C—continued

Q18-2	Fast Track		'96 Control		BMDO		BMDO Control
> 5	1		3		1		5
5	1		0		1		1
4	1		1		4		0
3	2		2		1		6
2	9		3		8		11
1	12		9		18		16
0	19		24		44		49
45	45	42	42	77	77	88	86

Q18-3	Fast Track	'96 Control	BMDO	BMDO Control
> 5	0	0	0	0
5	1	0	0	1
4	0	0	1	0
3	1	2	2	3
2	7	6	4	7
1	20	8	24	25
0	16	26	46	52

Q19	Fast Track	'96 Control	BMDO	BMDO Control
19-1	39	32	55	69
19-2	15	8	26	23
19-3	2	5	5	4
blank/NA	3	2	5	3

Q20	Fast Track		'96 Control		BMDO		BMDO Control	
6	4	9%	6	14%	8	10%	10	11%
5	5	11%	13	31%	16	21%	18	21%
4	11	24%	12	29%	28	36%	36	41%
3	13	29%	6	14%	10	13%	11	13%
2	8	18%	5	12%	13	17%	9	10%
1	1	2%	0	0%	2	3%	4	5%
blank/NA	3	7%	0	0%	0	0%	0	0%

Q21-1	Fast Track	'96 Control	BMDO	BMDO Control
>100	0	3	4	6
75-100	0	1	2	1
50- 75	0	4	2	2
40-49	0	0	5	4
30-39	2	2	1	1
20-29	3	4	3	6
19	0	0	0	0
18	0	0	0	2
17	0	0	1	1
16	0	1	0	1

Appendix C—continued

15	0	0	1	1
14	0	1	1	0
13	1	2	0	1
12	0	0	3	1
11	0	0	2	0
10	1	0	0	4
9	0	0	1	3
8	0	2	1	2
7	1	1	5	3
6	0	1	4	1
5	4	1	7	6
4	1	2	2	6
3	5	5	12	11
2	3	5	6	8
1	13	4	5	15
0	11	3	9	2

Q21-2	Fast Track	'96 Control	BMDO	BMDO Control
>100	0	1	0	2
75-100	0	2	1	0
50- 75	0	0	3	3
40-49	0	0	0	1
30-39	0	0	1	1
20-29	0	5	6	2
19	0	1	0	0
18	0	2	0	1
17	0	0	0	1
16	0	0	3	0
15	0	0	1	3
14	3	0	1	2
13	1	1	0	0
12	0	0	0	0
11	0	0	0	0
10	0	2	2	3
9	0	0	1	2
8	0	1	1	2
7	0	1	2	0
6	1	2	0	3
5	1	0	3	2
4	2	1	3	2
3	1	4	5	8
2	3	3	9	9
1	7	4	10	13
0	26	12	25	28

Appendix C—continued

Q22-A1	Fast Track	'96 Control	BMDO	BMDO Control
>10	1	0	2	0
10	0	0	1	0
9	0	0	0	0
8	1	0	0	0
7	0	0	0	0
6	0	0	3	0
5	0	0	1	0
4	3	0	1	0
3	2	1	3	3
2	1	1	11	5
1	8	7	12	18
0	29	33	43	60

Q22-B1	Fast Track	'96 Control	BMDO	BMDO Control
>10	0	0	0	0
10	0	0	2	0
9	0	0	0	0
8	0	0	0	0
7	1	0	0	0
6	0	0	2	0
5	0	0	0	0
4	0	0	0	0
3	0	0	2	2
2	3	0	6	2
1	2	1	12	10
0	39	41	53	74

Q22-A2	Fast Track	'96 Control	BMDO	BMDO Control
>10	0	0	0	0
10	0	0	0	0
9	0	0	0	0
8	0	0	0	0
7	0	0	0	0
6	0	0	0	0
5	0	0	0	1
4	0	0	0	0
3	1	0	0	0
2	1	0	0	0
1	2	0	1	2
0	41	42	76	84

Q22-B2	Fast Track	'96 Control	BMDO	BMDO Control
>10	0	0	0	0
10	0	0	0	0
9	0	0	0	0

Appendix C—continued

8	0	0	0	0
7	0	0	0	0
6	0	0	0	0
5	0	0	0	0
4	0	0	0	0
3	1	0	0	0
2	1	0	0	0
1	1	0	1	2
0	42	42	76	86

Q22-A3	Fast Track	'96 Control	BMDO	BMDO Control
>10	1	0	0	3
10	0	1	0	1
9	0	0	0	0
8	0	0	0	0
7	0	0	2	0
6	0	1	1	1
5	1	0	2	5
4	1	0	2	4
3	4	2	5	7
2	4	3	3	14
1	5	5	11	6
0	29	30	49	47

Q22-B3	Fast Track	'96 Control	BMDO	BMDO Control
>10	1	1	2	3
10	0	0	2	2
9	0	0	0	0
8	0	2	0	0
7	0	0	2	1
6	0	0	1	1
5	1	0	3	4
4	1	0	1	7
3	5	3	6	9
2	6	3	9	19
1	4	5	12	6
0	28	28	39	36

Q23	Fast Track	'96 Control	BMDO	BMDO Control
23-1	0	0	4	1
23-2	1	0	1	1
23-3	4	3	4	4
23-4	40	34	66	77
N/A	0	2	1	5

Appendix C—continued

Q23-YR	Fast Track	'96 Control	BMDO	BMDO Control
98	0	0	4	1
97	0	0	0	0
96	0	0	0	0
95	0	0	0	0
94	0	0	0	0

Q23-CO	Fast Track	'96 Control	BMDO	BMDO Control
5	0	0	0	0
4	0	0	0	0
3	0	0	0	0
2	1	0	1	0
1	3	3	3	4

Q24	Fast Track	'96 Control	BMDO	BMDO Control
24-1	21	33	67	70
24-2	4	23	35	34
24-3	12	6	26	33
24-4	4	2	3	4
24-5	25	1	6	4
N/A	5	7	4	7

Q24-Months	Fast Track	'96 Control	BMDO	BMDO Control
>15	0	1	11	7
15	0	0	1	3
14	0	0	3	0
13	0	1	3	0
12	1	1	8	9
11	1	2	2	2
10	0	0	4	3
9	2	3	3	8
8	1	2	10	5
7	0	2	1	1
6	3	5	12	17
5	1	3	1	5
4	4	3	5	4
3	4	2	2	4
2	3	5	1	1
1	1	2	0	0

Q26 Avg	Fast Track	'96 Control	BMDO	BMDO Control
A	32.42	61.90	21.82	34.67
B	53.51	122.17	58.39	65.07
C	6.88	4.67	2.15	3.51

Appendix C—continued

Q27	Fast Track	'96 Control	BMDO	BMDO Control
27-1	1	2	4	2
27-2	31	28	50	54
27-3	31	31	51	68
27-4	0	1	1	3
blank/NA	2	0	2	3

An Assessment of the Small Business Innovation Research Program in New England: Fast Track Compared with Non-Fast Track Projects*

John T. Scott
Dartmouth College

EXECUTIVE SUMMARY

This paper provides case studies for 14 research and development projects funded in 13 New England companies by the Department of Defense Small Business Innovation Research (SBIR) program. The performance of the six Fast Track projects, each conducted by a different company, is compared with the performance of eight non-Fast Track projects. The primary conclusions from the study of the New England cases are

- The collection of 14 New England SBIR projects studied here exhibited, at the outset of Phase I, high risk—both technical and market risk, high capital costs, and often expectation of a long-time before commercialization of the resulting technology.
- In the absence of the SBIR funding, the research projects would not have been undertaken in the same way or at the same pace. Outside investors, at the outset of Phase I, would have required too high a rate of return to make it possible for the project to proceed with only private financing.
- On the whole, the projects, both Fast Track and non-Fast Track, met both the funding agency's mission and the company's strategy. All fit the general scenario for socially valuable research projects that would have been underfunded in the absence of the SBIR program. In particular, the projects appear to be ones for which the private rates of return in the absence of SBIR funding would have fallen short of the private hurdle

*This paper was prepared for presentation at the National Academy of Sciences Symposium on the Assessment of the SBIR Fast Track Program, May 5, 1999.

rate required by outside financiers to whom the small businesses would have had to turn for financial support. Yet the social rates of returns to the projects are large and exceed the hurdle rates. The funding from the SBIR program changes the ordering of rates of return anticipated at the outset of Phase I. With the SBIR program providing funds, the expected private return relative to just the private portion of the total project costs is sufficient to move the private rate of return above the hurdle rate, and then the socially valuable research investment is undertaken.

- Taken as a group, the Fast Track projects show higher prospective lower-bound social rates of return—a measure that is based upon expected profits to the innovator and other producers benefiting from the innovation.
- The average duration of additional development beyond Phase II and before commercialization is somewhat less for the Fast Track projects, suggesting that at least on average they are somewhat closer to commercialization at the end of Phase II than the non-Fast Track projects.
- The respondents were unanimous in their appreciation of the SBIR program and in their belief that the program generally works well. They did have several recommendations to improve the working of the program, and those recommendations are listed in this paper. Among other things, the respondents cautioned that the Fast Track program is often simply not useful for companies pursuing socially valuable high-risk research, because at the end of Phase I, such projects often do not yet have the characteristics of projects that allow outside private investors to be attracted.
- In summary, the SBIR program has funded innovative projects with high social rates of return that would not have been undertaken in the absence of the program. Further, the non-Fast Track as well as the Fast Track projects appear to be quite valuable, although the non-Fast Track projects typically do not exhibit private commercial potential as quickly as the Fast Track ones.

INTRODUCTION

As part of a National Academy of Sciences study of the Department of Defense (DoD) Small Business Innovation Research (SBIR) Program, six SBIR Fast Track projects from six companies in New England are studied here along with SBIR non-Fast Track projects from different New England companies matched by similarity of location, size, and project duration. A total of seven projects from six non-Fast Track companies are studied—one project for each of five companies and two for the sixth. Additionally, the study includes a non-Fast Track project of a thirteenth company, Foster-Miller, which is much larger than the other companies in the sample and has been successful with an unusually large number of SBIR awards. In all, the study covers 14 projects at the 13 firms shown in Table 1. All of the SBIR projects studied were awarded both Phase I and Phase II funding. The goal of the study is to describe the SBIR projects and compare the Fast Track projects with the non-Fast Track projects, determining the effect that Fast Track has had on SBIR performance and firm behavior.

The 14 projects are high-risk research projects performed by small businesses, or with Foster-Miller in the sample, what the technology literature calls SMEs—small and medium-sized enterprises. The study finds that these risky SBIR-funded projects have high prospective, expected social rates of return. The social rates of return are calculated as lower bounds based solely on anticipated innovative investment profits for companies rather than on the sum of those profits (producer surplus) and consumer surplus (economists' measure of the value above and beyond what they actually pay that consumers receive from a product or service). Thus, the study's finding that the Fast Track projects as a group have

TABLE 1 The Firms

Company Name	Date Founded	Initial Size[a]
Brock-Rogers Surgical	1995	3
Cape Cod Research	1982	18
Foster-Miller, Inc.	1956	260
Hyperion Catalysis International	1982	20
Lithium Energy Associates, Inc.	1989	3
Materials Technologies Corp.	1986	5
Mide Technology Corp.	1989	3
Optigain, Inc.	1991	8
QSource, Inc.	1982	3
SEA CORP (Systems Engineering Associates Corp.)	1981	93
Spectra Science Corp.	refounded 1997 (originally 1989)	7
Synkinetics, Inc.	1994	8
Yardney Technical Products, Inc.	1940	155

[a]Employees at the time of application for Phase I.

higher social rates of return supports the perception that their prospects for generating profits for innovating firms are especially good. However, some non-Fast Track projects have higher lower-bound expected social rates of return than some Fast Track projects, despite the fact that consumer surplus is not measured. Fast Track and non-Fast Track projects alike have lower-bound social rates of return exceeding the private rates of return in the absence of SBIR funding. Each of the studied projects is the type of research project in which the market would fail to invest in a socially valuable innovation in the absence of SBIR or similar public funding. For the 14 New England SBIR projects studied, the average value of the lower bound for the prospective (i.e., at the start of Phase I) expected social rate of return is estimated to be 60 percent. The estimate would be much higher if consumer surplus could be measured.

HISTORY OF THE FIRMS AND THE PROJECTS

Table 2 provides some background information about the 13 companies as a group. The sampled firms have similar histories in the ways reviewed in the table, except that the Fast Track respondents are less likely to have had a previous SBIR award. Not surprisingly, the companies are not Advanced Technology Program (ATP) award winners; the ATP projects require substantial contributions of private funds from the outset of the projects. As seen in Table 2, the sampled companies are typically small businesses facing severe capital constraints for internal financing of research. Somewhat more than half of the respondents indicated locational advantages. A variety of other competitive advantages were cited; representative examples include "large patent base," "patented core tech-

TABLE 2 History of the Firms. (The number of respondents indicating each category)

Characteristic	Fast Track	Non-Fast Track
Locational advantages		
Near universities	2	3
Near corporate research centers or research parks	1	2
Previous SBIR awards prior to Phase I of current award*[a]		
Yes	3	6
No	3	1
Previous or current ATP Aawards		
Yes	0	0
No	6	7

[a]One company discussed the details of two awards. For purposes of this table, the two awards are considered as one award; their periods of performance are essentially concurrent, and both are non-Fast Track projects.

nologies," "small and lean," "twenty five years experience in the underlying technology," "trade secrets and know-how."

As discussed in detail later, the research projects on the whole are characterized by both high technical risk and high market risk. At the outset of Phase I, there is considerable uncertainty about whether the research will resolve outstanding technical problems. Furthermore, the acquisition plans of DoD are not typically firm at the outset of the research, and although the potential for spillovers to the nonmilitary commercial sector is present, many uncertainties remain about the form of the nonmilitary applications and about the market success of those applications.

Table 3 lists the companies along with the titles of the SBIR projects studied in this paper and their Fast Track status. The following paragraphs are brief overviews of the technologies being created by the sampled SBIR projects, along with discussion about the relationship of the project to the mission of the funding agency and to the strategy of the company.

TABLE 3 The Projects

Company	Project Title	Fast Track Status
Brock-Rogers Surgical	Development of a Force-Reflecting Laparoscopic Telemanipulator	Fast Track
Cape Cod Research, Inc.	Multilayer Capacitors Based on Engineered Conducting Polymers	Non-Fast Track
Foster-Miller, Inc.	Tunable Sting Net	Non-Fast Track
Hyperion Catalysis International	Ultracapacitors Based on Nanofiber Electrodes	Fast Track
Lithium Energy Associates, Inc.	Lithium Copper Chloride Inorganic Electrolyte Battery for More Electric Aircraft Systems	Non-Fast Track
Materials Technology Corp.	Life Prediction of Aging Aircraft Wiring Systems	Non-Fast Track
Mide Technology Corp.	Development of Distributed Area Averaging Sensor	Non-Fast Track
Optigain, Inc.	Single Longitudinal Mode Distributed Feedback Fiber Optics Laser	Non-Fast Track
QSource, Inc.	Multiple Rectangular Discharge CO_2 Laser	Fast Track
SEA CORP (Systems Engineering Associates Corp.)	Rapid Prototype Portable Combat and Launch System	Non-Fast Track
	Second project also discussed: Modular Gas Generator Launch Canister	Non-Fast Track
Spectra Science Corp.	Quantum Dots: Next Generation of Electronic Phosphors	Fast Track
Synkinetics, Inc.	High Precision Gimbal System	Fast Track
Yardney Technical Products, Inc.	Low Cost, Lightweight, Rechargeable Lithium-ion Batteries	Fast Track

Brock-Rogers Surgical. Development of a Force-Reflecting Laparoscopic Telemanipulator. Fast Track. The technology merges electronics, mechanics, computer networking and software to create a telerobot to be used for surgery. The technology allows the surgeon to feel as if he or she were one inch tall and inside the patient. DoD is interested in such computer-augmented remote connections to allow medical personnel to operate on the front lines from remote locations. Beyond the military applications, such technology will change the face of surgery. A deep infrastructure technology is being created—a sophisticated electronic, mechanical, software-networked machine. In that sense, the technology is an enabling one with wide applications outside of medicine. The robot no longer needs to "see"—recognition and reception problems are handled by the human controlling the process.

Cape Cod Research, Inc. Multilayer Capacitors Based on Engineered Conducting Polymers. Non-Fast Track. The technology uses electrically conductive polymers to store energy to power electric cars. The project involves the development of novel and useful materials, and it provides the funding agency with improved energy storage for a variety of applications.

Foster-Miller, Inc. Tunable Sting Net. Non-Fast Track. The technology is the latest in a line of "NETS"—nonlethal entanglement technology systems—developed as SBIR projects by Foster-Miller in response to DoD's interest in funding research about capture mechanics. The family of nets developed by Foster-Miller are compact, light-weight, far-ranging, fast, and can be fired from conventional weapons. The "Sting Net" delivers a remotely controlled electric charge for use with especially aggressive targets and is anticipated to have military applications only. Less physically active versions range from nets that simply entangle to nets using pepper irritant powder to subdue more dangerous targets. The less harsh nets will have use in nonmilitary police operations The Sting Net project fits with Foster-Miller's highly successful corporate strategy of inventing and licensing patented technologies, and spinning off subsidiary companies to manufacture and market the innovations. Numerous SBIR projects have contributed to that strategy, although the company gets only about 20 to 25 percent of its revenues from the SBIR awards.

Hyperion Catalysis International. Ultracapacitors Based on Nanofiber Electrodes. Fast Track. Electrochemical capacitors, sometimes called ultracapacitors or supercapacitors, are being developed for potential applications in hybrid electric vehicles and other automotive electronic and military systems. To be cost- and weight-effective compared to batteries, these "supercaps" must have adequate energy and power with a long life cycle and must meet cost targets. Hyperion has a proprietary line of nanofibers that have desirable properties and a cost advantage over competing materials. During Phase I, the Hyperion nanofibers showed great promise regarding their power and now in Phase II the nanofibers are being used to design, fabricate, and test electrochemical capacitors. Hyperion would make the nanofiber electrodes and sell them to the manufacturers of supercaps. Beyond the potential for a large commercial market for supercaps and the fact that the military has specialized needs for them that explain the DoD funding of the research, there are other potential applications including uses in boom boxes, electric motor starters, defibrillation medical devices, and in cell phones in combination with batteries where power from a small supercap can allow the use of a smaller battery and a better product than results using a large battery alone.

Lithium Energy Associates, Inc. Lithium Copper Chloride Inorganic Electrolyte Battery for More Electric Aircraft Systems. Non-Fast Track. The batteries developed by Lithium Energy Associates are rechargeable and have high energy density and extraordinary low-temperature performance. They have military applications in small, light-weight, remote-controlled reconnaissance aircraft equipped with TV cameras and in solar planes that fly to high altitudes, charge during the day, and then keep flying at night. The batteries have other military applications as well; for example, after using conventional power to get equipment to a battlefield, the engines could be turned off and the batteries could reposition vehicles quietly and without infrared detection. The batteries will have applications for a variety of military electronics applications such as radios. The low-temperature performance of the batteries also makes them the potential power source for applications in space, such as powering robot stations on the moon or Mars, and research in progress will push the low-temperature capabilities of the batteries into the range making them suitable for lunar or Mars missions. Customers, apart from DoD and NASA, should include original equipment manufacturers of military electronics or civilian police equipment.

Materials Technology Corporation. Life Prediction of Aging Aircraft Wiring Systems. Non-Fast Track. The technology allows safe, accurate, and efficient diagnostic tests of the wiring in airplanes to ensure that the wiring is defect free. With the current technology the inspector opens a panel door and examines bundles of wires with the naked eye. If the 12- to 18-inch section of wire that can be seen looks okay, then the entire wire is judged to be safe. In some cases, the inspector may use a mirror to try to look at the back side of the wires, but because of visibility and space limitations it is rare that the back side is inspected well. The wires themselves are rarely a problem; instead, the insulation on the wires is what degrades; becoming brittle with age, it begins to crack. The plasticizer vaporizes and, over time, the insulation degrades, becomes brittle, and begins to fall apart, exposing bare wire; if two wires are exposed, a short circuit is possible.

The new technology developed by Materials Technology Corporation is the first approach to inspecting for damaged insulation of wiring that allows viewing of all sides of the wiring and does not risk damaging the wires as typically occurs if the wires are bent or disturbed in trying to examine their back side. The technology uses embedded optical sensors in a device that can be put around the bundle of wires and used to get a 360-degree view of the wires. The information gathered by the handheld device is signaled to a computer that pinpoints and displays precisely where on the 360-degree surface a crack is located. New optical imaging technology is used. With the press of a button the image can be recorded and the data transported for use at other sites. It is expected that the system will allow the entire wiring history of the aircraft to be stored on a zip drive that will be carried in the aircraft. Planes will not have to return to a home base to be inspected and repaired. Historical data, supplemented with a visible image, will allow the inspector to see what the wiring looked like at the last inspection and calculate the progression of changes. In addition to examining aircraft wires and cable, the technology can be used to examine the connections and to detect corrosion more generally in aircraft and other objects.

There are many applications beyond those for military and commercial aircraft. The optical scanning procedure is expected to be relevant for dealing with vision problems caused by macular degeneration. And, of course, what is good for aircraft inspection is also good for inspecting bridges and other infrastructure.

Mide Technology Corporation. Development of Distributed Area Averaging Sensor. Non-Fast Track. The technology eliminates harmful vibrations in structures by use of active materials that respond to stimuli; for example, if voltage is applied, the active material expands or contracts. The vibrations of structures have several natural frequencies, and the technology developed by Mide Technology Corporation uses shaped sensors to filter out noise, focusing on a desired frequency to eliminate the associated vibrations. The area averaging sensor simplifies a higher-dimension multi-input/multi-output information problem to a lower-dimension control system that characterizes more simply the necessary information about the natural frequencies causing vibrations, despite a complex set of underlying information. The frequencies that really transmit the noise through the structure of interest can be isolated using a control system with active fiber composite actuators; the smart material is used to simplify the control problem and, ultimately, to allow the elimination of the vibrations from the structure.

The immediate application of the technology is to protect launch satellites from damage from structural vibrations. Alternatively, one could protect the launch satellite with blankets—thin ones to protect against high-frequency noise, and thick ones to protect against lower-frequency noise. The Mide technology is the active way of dealing with the problem. Commercial potential extends beyond the protection from vibration of components in space launch vehicles. The commercial potential comes from using area average sensors with flexible circuitry, and Mide has four commercial products using that technology. The products range from generic technology such as sensors on a flexible circuitry for signal conditioning, a high-voltage amplifier to drive active fiber composites, and sensors connected in various ways on a small matrix board, to a specific application that uses sensors on a the shaft of a golf club to detect club head speed and provide feedback. The generic applications range from military uses such as protecting the launch of a spacecraft or quieting torpedoes in a submarine to nonmilitary commercial uses such as vibration control for the blades of a gas turbine or in air-conditioning ducts. Anything that vibrates and has a dynamics problem with the vibration and noise can potentially benefit from the technology.

Optigain, Inc. Single Longitudinal Mode Distributed-Feedback Fiber Optics Laser. Non-Fast Track. Optigain's technology provides a fiber version of a signal source that is similar to a semiconductor laser. Optigain's fiber laser is a distributed laser induced in a fiber rather than a semiconductor. Transmission systems need a high-quality signal source, and Optigain's technology provides a narrow, high-quality low-noise laser that can potentially capture some of the market for semiconductor lasers used in communications markets. The company's strategy is to develop various fiber-based devices, and the product here is a fiber-based component that can be put into other systems. Several lasers, each a different wavelength, have been developed, and communications markets where Optigain's fiber-based laser will be preferred over the semiconductor lasers are being sought. The superior performance of the fiber-based laser is in the linewidth of the laser and its spectral purity, which should lead to applications in sensor markets as well.

Regarding the relationship between the project and the mission of the funding agency, in this case the agency was quite open about different topics, with awards going to further technology for high-speed communications networks quite generally. The goal of the funding was to enable new technologies for such networks, and the concern was with the overall strength of the solutions rather than a specific set of narrow requirements.

QSource, Inc. Multiple Rectangular Discharge CO_2 Laser. Fast Track. QSource's CO_2 laser technology generates high power and efficiency and has specialized military uses. There are also nonmilitary commercial applications with large market potential. QSource's laser features higher power, smaller size, and an advantage in cost. CO_2 lasers are used in laser radar to bounce a pulse off an object, its high sensitivity allowing detailed information to be obtained about a tank or an aircraft many miles away. The laser system along with a DC battery source can be built in the size of a small suitcase. The CO_2 laser has very high efficiency, transmitting substantial distances with very little power loss; it is compact, uses a simple gas, and is very efficient.

The technology is dual use. For example, the basic transmitter unit in the laser radar has applications for heating, cutting, and trimming, for example, in conjunction with one of the lasers used in eye surgery that was developed initially in another DoD SBIR project trying to track objects at great distance. The laser is inherently sterile, and so, it is ideal for cutting tissue. It can be used for cutting teeth, working on teeth, and as a mechanical drill. It is more expensive than a drill, but it eliminates the risk of transmitting hepatitis or other viruses. A laser dental system has a detachable head in the optical system that delivers the laser and is easy to clean. The surgery is painless; there is no need for anesthesia. The dental market alone over the next 10 years is projected to sell 100,000 dental laser systems once the procedure for hard tissue is approved. A CO_2 laser dental system will sell for about $20,000. The energetic CO_2 lasers that QSource technology improves upon should have a market of $2 billion in the dental market alone. The medical therapeutic uses include the dental applications, skin resurfacing, and microsurgery in the ear. Further markets have been identified for sealed CO_2 lasers in materials processing and various research applications.

There are a large number of CO_2 lasers available and, over the past decade, they have become more functional. The cost of producing them in terms of dollars per watt is not great, but more than half of that cost is in the basic power source needed to energize the laser (i.e., powering the basic laser itself, not the entire system). The big advance provided by QSource technology is to reduce the cost of the power supply. Some of the older technology can achieve the same level of efficiency as the new QSource rectangular discharge laser, but those technologies result in products that are very big and not very sturdy.

SEA CORP (Systems Engineering Associates Corp.). Rapid Prototype Portable Combat and Launch System. Non-Fast Track. The technology is a software based-fire control system that allows a submarine to fire various types of torpedoes. Modern submarine systems are not compatible with all types of torpedoes. SEA CORP has created a system in a suitcase that can be plugged in and will allow the submarine to use different types of torpedoes.

Modular Gas Generator Launch Canister. Non-Fast Track. This second technology developed by SEA CORP is a launcher for torpedoes that uses automotive air-bag technology rather than a conventional gas system. It is modular, environmentally friendly, and uses a commercial off-the-shelf item to meet a specialized military purpose. Other commercial applications of both technologies are being considered, and the technologies will allow SEA CORP to diversify its activities into profitable new lines of business that are very different from its historical focus.

Spectra Science Corporation. Quantum Dots: Next Generation of Electronic Phosphors. Fast Track. The technology centers on better phosphor that results in a brighter image on large-screen projections. The technology combines the three core technologies of Spectra Science. First, the company has laser paint technology using disordered lasers. Conventional lasers use mirrors as the gain source, but laser paints use scatters such as titanium particles. So, a composite system is used to create the gain source; the laser excites the material and the feedback is from materials rather than from mirrors. The laser paint technology is used for identification or authentication, for example, via a label on a fabric or in a document.

The second core technology came from Phase I of this SBIR award. In that research, Spectra developed the ability to make smaller phosphor particles with surfaces for the composite systems that could exhibit gain and could be used in a laser paint. The difficulty to be surmounted was that the surfaces of the particles have a large number of defect sites, which trap light, preventing its emission.

The third core technology is the focus of the SBIR project's Phase II. It is a combination of the first two technologies. Phase I resulted in development of quantum dot phosphors for display applications—better phosphors that could be driven harder with the result of a brighter image. Phase II then shifted gears and focused on developing what had been discovered. Spectra Science has merged its work on display technologies and materials for laser paints to develop a lasing projection system. With previous technologies, large-screen projections can be viewed only in the dark. With Spectra Science's new technology, the phosphors are excited and emit higher energies than previously, overcoming this limitation. DoD's SBIR award here meets their mission in terms of improved images for large-screen projection systems. That goal remained the same even when Phase II was refocused, and the project clearly satisfies Spectra Science's strategic mission of seeking potential applications for its core technologies.

Synkinetics, Inc. High Precision Gimbal System. Fast Track. The Synkinetics technology is an innovative system of gears that provide cost-effective, sturdy, precision devices for positioning and pointing armaments. Such devices are used, for example, in missile control systems. Synkinetics's new speed conversion technology improves current high-precision pointing and positioning transmission equipment at a reasonable cost. The technology features flat-plate cam gears in an in-line mechanism that combines the rolling aspects of bearings with the transmission aspects of gears to obtain a versatile, robust, compact, reduced-weight, high-precision, efficient, and cost-effective drive mechanism. The technology is generic and has countless applications. The transmissions will have uses for pointing and precision positioning of various payloads for industry and the military. Applications of reliable, low-cost, and low-maintenance precision positioners are expected in the medical, electronics, marine, mobile satellite communications, and aerospace industries.

Yardney Technical Products, Inc. Low Cost, Lightweight, Rechargeable Lithium-Ion Batteries. Fast Track. Yardney has developed the battery using the prismatic cell technology identified in Phase I of the project and plans to deliver a prototype to its sponsoring agency. The battery has a 25 percent improvement in capacity compared to the battery that the military now uses and would represent a major jump in performance for the DoD uses for the particular style of battery. The market for the lithium-ion battery has grown rapidly from nothing in 1990 to current sales of $1.2 billion. Currently, the market is growing at 30 percent a year, and there is much opportunity for new technologies. The technology will be useful to other governments; with approval, sales to armies of U.S. allies are expected. Nonmilitary commercial applications are expected as well.

COMPANIES' EXPECTATIONS FOR SBIR PROJECTS AND REASONS THAT SBIR SUPPORT WAS NEEDED

As Table 4 shows, the SBIR awards made possible research that otherwise would not have been undertaken or would have been done on a smaller

TABLE 4 The SBIR Award and the Company's Strategy (The number of respondents indicates the total for the category)

Impact of SBIR Award	Fast Track	Non-Fast Track
Expected an increase in company size (sales or employment) or more diversified product line	6	7
Would the company have undertaken the research without the SBIR award?		
No	5	6
On a lesser scale	1	1
Yes	0	0

scale at a slower pace. The awards are expected to expand the businesses of the SBIR winners.

In answer to the question of whether the company would have undertaken the research without the SBIR award, representative comments included the following ones paraphrased from the interviews.

> No. To support our Phase I project, we tried to find support from other companies and venture capitalists. The venture capitalists want too high a rate of return and want returns too quickly. Joint ventures don't work either. You need their money, so they want lots of rights. You must sell your soul to them. These partner companies are providing capital basically and sometimes distribution networks.

> No. Working on a particular DoD program enabled us to do further work on our technology and gain insight into commercialization. The SBIR project is an incubation period of sorts to new start-up companies with new technology, an innovative way of approaching a problem. After the SBIR project, the development work that remains is a reorientation of the technology, looking at how to manufacture and commercialize for nonmilitary applications, to come up with a low-cost way to mass produce for less sophisticated requirements. But at the outset, the SBIR award is the lifeblood of new entrepreneurial ventures when new technology is to be advanced. We came up with something worthwhile for DoD, but we also advanced our own technology to another level without going crazy looking for outside investors. The lessons learned in the SBIR project provide the database that allows us to extrapolate intelligently and succeed in nonmilitary, commercial applications of our technology.

> We would have devoted some resources to the project, but it is questionable whether we would have gotten this far. We would have sought assistance from other companies and from universities. We would have proceeded on a smaller scale and sought a partner down the road.

> No. There is no guarantee that such high-risk research will pan out. And the SBIR program understands that, and it therefore requires not necessarily a commercial product, but instead a good effort. It understands that in many cases the value will be to prove the technological approach taken is *not* the right path. So the [funding agency within DoD] will not go down that path again. In fact, in the case of our award, [the funding agency] gave two awards. So, it spends $1.5 million on two projects running parallel, and the probability of at least one success is increased. We'll have a cookoff . . . with the other company to see which box is the better one to go with.

> Probably not. We have no means of acquiring capital except through loans or from investors. But being honest with them, we could not raise the necessary funds—at least not at the outset of the SBIR projects. DoD does not select its highest priority acquisition projects to develop through the SBIR program. Instead, it uses projects that are interesting and have great potential value and the possibility for acquisition. But they are lower priority, high-risk projects. It is difficult to attract outside investment for such projects. These are projects for which we could not show an outside investor definite acquisition plans. If we could, DoD would not use the SBIR program to fund the projects. The SBIR projects are ones for which the acquisition plans are fuzzy.

> No. We would not have done the project without the SBIR award. We devote about 6 percent of our income to research. This would not have been a project to get those funds. Without SBIR help, our research would have been more near-term and less challenging.

Table 5 shows the reasons why the SBIR funding was needed. The projects entailed substantial technical and market risks, and the projects required substantial amounts of capital from the perspective of the small businesses doing the research. Clearly, many respondents are concerned about the possibilities for opportunistic behavior by sources of external financing in the early stages of the research that the SBIR program funds. On the other hand, once the small business can finance its early-stage research and has moved beyond the initial research and development (R&D) stage, the company is, on the whole, comfortable with the degree to which it can protect its property rights. Whether from patent protection or from carefully negotiated licensing agreements, despite the fact that the firms typically do not anticipate capturing all of the profits that their research will create, they do expect to capture a sufficient amount of those profits to make their investments worthwhile. Table 5 focuses on the reasons that the companies would not have been able to carry out the research without public funding. Although of course the SBIR program is not expected to change the technical risk or opportunistic behavior, as explained later, it does increase the private expected rate of return above the private hurdle rate. With public funding, despite the risks, the firm will undertake the research.

In addition to their comments accompanying Table 4, respondents offered further insights about their needs for SBIR funding when discussing the list of

TABLE 5 Reasons SBIR Funding Was Needed: Why the Company Would not Have Done the Research, or Would Have Delayed the Research, or Would Have Done It on a Lesser Scale, Without SBIR Funding

Reason[a]	Number of Respondents Indicating the Reason	Rank-Weighted Score[b]
Fast Track Companies		
1. High technical risk	6	43
2. High capital costs	5	35
3. Long time to market	4	27
4. Spillovers to multiple markets	1	1.5
5. Uses technologies in different industries	2	9
6. Property rights	2	8
7. Compatibility and interoperability	1	4.5
8. Opportunistic behavior	2	11
Non-Fast Track Companies		
1. High technical risk	3	23
2. High capital costs	5	35.5
3. Long time to market	4	28.5
4. Spillovers to multiple markets	0	0
5. Uses technologies in different industries	1	7
6. Property rights	2	9
7. Compatibility and interoperability	1	3
8. Opportunistic behavior	4	22

[a]Detailed descriptions for the reasons: (1) High technical risk associated with the underlying research, (2) high capital costs to undertake the underlying research, (3) long time to complete the research and commercialize the resulting technology, (4) underlying research spills over to multiple markets and is not appropriable, (5) market success of the technology depends on technologies in different industries, (6) property rights cannot be assigned to the underlying research, (7) resulting technology must be compatible and interoperable with other technologies, (8) high risk of opportunistic behavior when sharing information about the technology. Reasons are based on Tassey (1997) and Link and Scott (1998).

[b]If a respondent ranks a reason first, that counts for 8 points, second implies 7 points, and so on down to eighth which would count 1 point. However, if a respondent ranks, say, only three reasons, then only those three reasons would receive any rank-weighted score—they would receive 8, 7, and 6 points, respectively, while all other reasons would receive 0 points. Thus, if 10 respondents ranked the first reason most important, and 2 ranked it second, and 1 respondent ranked it fourth, the rank-weighted score for the first reason would be $(8 \times 10) + (7 \times 2) + (5 \times 1) = 99$. Ties split evenly the points assigned for the number of tied reasons. For example, if two reasons tied for first for a respondent, then each would receive scores of 7.5. Finally, note that the rank-weighted scores for the sample of six Fast Track companies should be multiplied by 7/6 to make them comparable with the rank-weighted scores from the sample of seven non-Fast Track companies.

reasons in Table 5. For example, the difficulties faced by small businesses when raising funds from large corporations or venture capitalists are reflected in the following comments that were made when discussing opportunistic behavior.

> In one of our earlier SBIR projects, after we had used Phase I for risk reduction we became convinced that the technology would work, but then only after we had a patent were we willing to approach the large companies for a partnership. A small business needs to have a patent in hand in our area of technology. The big companies, in our area, will say: "We do not sign nondisclosure agreements with small companies."

> The eighth reason, opportunistic behavior is also important. It is what kills Fast Tracks. The outside partner wants to claim rights to the technology. That is what killed our Fast Track. It did not fly because our partner wanted more complete rights. The outside partner would provide one-third of the money but wanted over one-half of the rights.

The second of the two comments is different from the first. The first comment reflects the concern that the outsider will steal the small business's intellectual property and use it for its own purposes. The second comment reflects the fact that because the SBIR projects are high-risk projects, outside investors demand very high expected rates of return. Many comments like the second comment were made when the interviews turned explicitly to discussion of outside finance, and those comments are reported later.

THE COMPANIES' PLANS FOR FUTURE SBIR PROPOSALS

Because of the barriers to complete private funding of small business innovation research that are emphasized in Table 5, all of the small businesses plan to apply for SBIR awards in the future to support additional high-risk research. Table 6 shows that fact and also notes the range of responses to the question of whether previous awards were important for winning the current award. There are two issues here. One is a substantive issue of whether the technology pursued in the present award has evolved from technology developed in previous awards. For some firms, past awards were not directly relevant to the present one, but for others the current award was for further development of technology developed with earlier SBIR awards. A second issue is a procedural one: Would the fact that SBIR awards were won previously have affected the chances, *ceteris paribus,* for winning the current award? There was no general perception among many of the respondents that their chances were affected one way or the other by having won previous awards, apart from the substantive benefits when the technologies were linked and evolving sequentially through time as new SBIR projects were begun. However, a couple of respondents expressed views that previous awards sometimes can reduce a company's chances for winning subsequent awards.

TABLE 6 Plans for SBIR Awards (The number of respondents indicates the number in the category)

	Fast Track	Non-Fast Track
Previous awards were important for winning current award		
Yes	0	2 (reduced chances) 3 (increased chances)
No	1	1
Planning applications for SBIR awards in future		
Yes	6	7
No	0	0

Responses about the impact of earlier awards include the following comments illustrating views about positive and negative effects.

> Yes. Previous success is a very, very negative factor. Managers of the SBIR program at the highest levels are frustrated by what they perceive as the lack of "success" stories from their program. They have difficulty accepting that their definition of success (commercialization of products from an SBIR program) is an extremely unlikely outcome given the structure of the SBIR program. Thus, there is a built-in bias to award Phase IIs to small companies who already have in place commercial successes not supported by the SBIR program and for which they can take credit. These companies are very rare because they normally do not participate in the SBIR program.

> With the Fast Track Program, there is a two-tier standard favoring firms new to the SBIR program, with 25 percent cost-sharing for a new company and 100 percent cost-sharing for companies like us.

> Previous awards helped us; we learned what the funding agencies' needs were. We learned that to have a successful proposal, we need to understand what is wanted by the agency. Previous SBIR awards helped us learn how to have a successful proposal.

Responses to the question about future applications for SBIR awards (Do you anticipate applying for SBIR awards in the future? Why?) include the following and reflect the reasons that innovative small businesses are enthusiastic about the SBIR program..

> Yes. Although the developments for the military from the SBIR awards will not be directly applicable for nonmilitary, commercial products, indirectly the nonmilitary technology is being advanced. While coming up with something worthwhile for DoD, we also advance our own nonmilitary commercial technology to a higher level without going crazy looking for outside investors. The lessons learned in these SBIR projects provide the data base that allows us to extrapolate intelligently. The SBIR program makes it possible for us to learn and develop our technology. The program is fantastic for young vibrant entrepreneurs.

Yes. It's the only way to keep the lights on, given the high risk and high capital costs for the research we are doing as we try to get into a different technology. Our existing line of business generates very little revenue and we cannot fund R&D ourselves. For the type of research we are doing, neither venture capitalists nor large companies will work as sources of outside funding. Both the venture capitalists and the large companies want too high a rate of return—too many rights to the future returns relative to the investment they would make in our company.

Yes, because we have created an efficient infrastructure to generate prototypes in response to requests for Phase I and Phase II proposals.

Yes, selectively. We've got a pilot line now, so we are beyond the SBIR-type project in our current work. Maybe we will find a wrinkle appropriate for another SBIR project in our current technology.

Yes. The SBIR program is the way to get funds for truly innovative high-risk small business projects that cannot effectively be financed by outside private funds, given the opportunistic behavior by companies or lack of understanding of the technology by venture capitalists.

Right now, we're in the midst of Phase II, so no immediate plans. But, yes, because the SBIR program gives us the ability to develop a technology we would not have been able to develop on our own ,given the technical risk and capital costs and the long time from initial research until commercialization.

Yes. We like the challenge and broad scope of the topics.

Yes. Our DoD customers have identified several areas where our technology can be developed further and applied to their needs.

Had it not been for this Phase I, neither we nor the [sponsoring agency] would have been at this stage. We believe we can make similar breakthroughs with future SBIR awards.

Yes. The SBIR awards help us research new technologies, given technical risk and the risk of opportunistic behavior by large companies if we go to them with our ideas before they are developed.

Yes. SBIR awards let us accept the risk of good projects.

Yes. The SBIR program has been very successful for us. It has allowed us to develop a product line for eventual commercialization and growth of our company. The new product line will be more profitable than our existing product lines. We understand the SBIR program is a start-up program, not intended to be used over and over. But it allows us to do high-risk research with commercial potential and to expand our business into new product lines.

THE COMPANIES' COMMERCIALIZATION PLANS

Table 7 shows the ranges of responses about commercialization plans across the projects. Also shown is the range of responses regarding the use of patents

TABLE 7 Commercialization Plans (The number of projects for the indicated category)

Plan	Fast Track (6 projects for 6 firms)	Non-Fast Track (8 projects for 7 firms)
Expected time until commercialization		
≤ 1 year	4	4
> 1 year and ≤ 2 years	1	2
> 2 years and ≤ 3 years	1	0
> 3 years and ≤ 4 years	0	1
> 4 years and ≤ 5 years	0	1
> 5 years	0	0
Anticipating strategic alliances for production		
No	0	4
Yes	5	4
Uncertain	1	0
Patents		
Yes or expected soon	5	5
No	1	3
Scientific papers		
Yes	4	1
No	2	7

and scientific papers that help to disseminate technology as well as protect rights to intellectual property. The patents, of course, can help to create and protect intellectual property rights, while the papers disseminate information and may even bolster the effectiveness of patent rights by making ancillary nonpatented materials common knowledge that cannot be the basis for competing patents. Finally, discussion revealed that the respondents see their SBIR-funded research in a different light from their other technologies with commercial potential. On the whole, the projects are different, entailing more technical and market risk, and they are not generally the sort of research projects that the companies would have pursued without SBIR support. Not surprisingly, then, the respondents report that commercialization plans are different than what would have been the case without SBIR support. On the other hand, given that the SBIR project has proven the commercial potential for what was an extraordinarily risky project at the start of Phase I, the firms often report that with the commercial potential now established, the commercialization plans look very much as they would for any project that had reached the stage of making prototypes and gearing up for production. Nonetheless, it is also true that in many cases the respondents are in the position of needing a "Phase III" to provide the bridge from highly promising technology with great commercial potential to successful development of the manufacturing technology and the final product for the market.

Of course, as seen in Table 4, without the SBIR program, the research projects typically would not have been as close to commercial results, because the projects would not have been undertaken or would have proceeded in a different way. However, here the respondents commented further not only about the delay that might have resulted, but also about the remaining difficulties that they faced as they looked for financial support for additional periods of development before commercializing their technology.

Respondents' comments here about the impact of the SBIR award on commercialization plans included the following.

> I was an academic. I never, never would have done this commercialization research without the SBIR program. The SBIR program spawned a development that would not have happened in the same time frame, and the development will result in commercialization.

> We never would have gotten this far; we would not have taken the $100K look [in Phase I]. We would have put this project aside to work on something else. The SBIR award allowed us to take a six-month look at a promising idea. We got some good results, and we can now justify a million dollar investment ourselves.

> Without the SBIR program, we would have, at a slower pace, tried to bring the technology along so far, to a point, and then we would have tried to generate interest to bring in a partner. But without the SBIR award, we would not have been so far along.

> The SBIR program let us develop our own technology while we created something worthwhile for DoD. We will be able to use the understanding and data we developed in significant ways to further our commercial, nonmilitary technology.

> We would not have done the project without SBIR, but if we had done the project, our commercial plans would be the same.

> The SBIR program put us in a position to develop this new product.

> Assuming that we could have gotten to this point without SBIR, our commercialization plans would have been the same. But we could not have gotten to this point without SBIR. The technology was too unproven. The SBIR program was willing to take the risk when alternative sources of investment funding were not available.

COMPANIES' PERCEPTIONS OF THE RELATIONSHIP BETWEEN THE SBIR AWARD AND PRIVATE THIRD-PARTY FINANCING

Table 8 shows that the Fast Track program *does* address what most respondents see as a difficult period for SBIR projects—namely, the gap between Phase I and Phase II funding. Both Fast Track and non-Fast Track respondents emphasize the difficulties created by the gap: Employees must be paid and the project

TABLE 8 Financing and External Partners (The number of respondents indicating the category.)

	Fast Track	Non-Fast Track
Expressed difficulties bridging a gap in time between Phase I and Phase II		
Yes	0	4
No opinion expressed	6	3
Did the SBIR award facilitate the attraction of outside investors?		
Yes	4	1
No	1	6
No opinion expressed	1	0

kept afloat and progressing from the Phase I stage; yet many small businesses, for the reasons discussed above, find it very difficult to acquire financing while waiting for a Phase II award. The effectiveness of the Fast Track program in this regard is clear from Table 8, because the Fast Track winners did not report "gap" difficulties on their Fast Track projects. Table 8 also shows that winning an SBIR award (for our sample of projects that all won Phase II awards) facilitated the attraction of outside financing to further commercialize the technology developed under the SBIR project, in the perceptions of most Fast Track respondents. With one exception, the non-Fast Track respondents did not find that the SBIR award helped them to secure outside funds. This reflects the difference in prospects for commercialization early in the SBIR-funded research of Fast Track projects as compared with non-Fast Track projects.

Comments of the respondents about the financing issues and their relationship to the SBIR award included the following.

> One Fast Track respondent reported: During Phase I, there was full funding by the government, but beyond that we have had one-to-one matching. Although we did not use the SBIR award as a marketing tool to attract outside funding, when we went to an investor, the award was part of the whole package. The matching funds to go with the outside investor's funds were there, so yes, that helped.

> Another Fast Track respondent said: Yes, the SBIR award was used as leverage, as a marketing tool. That is when it is most helpful. Having the SBIR Fast Track award and the financing that it helped attract served as a bridge between Phase I and Phase II; the gap between Phase I and Phase II was eliminated.

> A non-Fast Track respondent observed: There was an eight-month delay [between Phase I and Phase II]. This allowed our competitors . . . to take our work and get a head start on improvements. This left us in a difficult position when Phase II was eventually awarded.

The same non-Fast Track respondent said: Personal friends seem the best source of outside financing for us. The SBIR award is not a useful marketing tool because most investors do not wish to have copies of proposals, Phase I final reports and Phase II progress reports wandering over the desks of unknown reviewers. Some agencies use outside reviewers who take the best ideas and have their graduate students pursue them. This practice is not supposed to happen, but is the rule rather than the exception.

Another non-Fast Track respondent said: After Phase I ends, you have trained people and your staff is waiting around. You are holding your breath until Phase II begins and the needed funds are available.

Another non-Fast Track respondent commented about the gap between Phase I and Phase II in this way: Considering the public's $750,000 investment, the rights demanded are acceptable; the processing takes a little longer than it should; by the time reports are submitted for Phase I, four months go by. That gap is not a problem for a large corporation, but for us a gap that lasts well over a quarter, and probably six months in the end, is a problem. But overall the process works well.

Another non-Fast Track respondent stated: We used internal funds only. We did not consider using Fast Track. That would have required an outside investor. But the project is a high-risk project and DoD's acquisition plans are not yet clear. It is highly unlikely that we could get the outside investors required for Fast Track. At the end of Phase II, it would be possible to get outside investors, but not at any time prior to that and not even now. Fast Track is a good idea in theory, but in execution there is a problem. At the end of Phase I, a company typically has just a concept. Later when there is a prototype, then you can do something with outside investors.

Another non-Fast Track respondent said: In a small business setting, we are wrapped up in technological issues and production issues and decisions about how to market our product. It would not be productive to get into the specialized activity of fundraising, given the circumstances. There must be somebody willing to put money into the project. Plenty of people will give lip service to the idea and take your time. But the probability of actually getting the money is less than 5 percent. To spend 80 percent of the time for the 5 percent chance of financial support is not a good use of our time.

COMPANIES' VIEWS ABOUT ADMINISTRATION OF THE SBIR PROGRAM

The respondents report that they are highly satisfied with the SBIR program; they are overwhelmingly positive in their overall impressions of the program, as shown in Table 9. The responding companies believe that the SBIR program made it possible for them to do significant research that they otherwise would have been unable to do. They clearly believe that the research has furthered not only their company's strategy but the mission of the sponsoring DoD agency as well.

TABLE 9 SBIR Program Administration: Overall Impressions (The number of respondents indicating the category.)

Impression	Fast Track	Non-Fast Track
Favorable	6	7
Unfavorable	0	0

Nonetheless, the companies offered many suggestions for improving the SBIR program. The suggestions for improvement and the general concerns that were expressed include the following from the Fast Track companies.

> I have no recommendations to improve the program itself, but one recommendation could be made based on the success of my company. I did the technical work, but I brought in others to do the accounting and business administration. The SBIR program could promulgate information about how to proceed: Small business principals should recognize their strengths and weaknesses, and they should bring someone in to do the administrative work rather than having the scientist have to do it all. The SBIR program could encourage small businesses to bring in outside expertise to ensure competence in business administration to go along with the competence in the scientific work.

> I want to promote the SBIR concept. It provides a wonderful opportunity for us to develop our own technology and at the same time do something worthwhile for DoD. SBIRs are the lifeblood of new entrepreneurial ventures when new technology is to be advanced. We are all appreciative and thankful and grateful that there is this highway that allows us the opportunity to develop our technologies. However, one must remember that there is a dichotomy between requirements of the armed forces and requirements of nonmilitary commercialization. Typically, there is very little direct overlap. The military development will often have very costly requirements for high precision, and the results will not often be directly applicable for nonmilitary commercial use, although indirectly the nonmilitary technology is being advanced. One sees a correlation from the program itself into a commercial project, but the commercial project typically will not need to be as sophisticated, as accurate, as costly. It won't need the special materials. One cannot take the thing developed for military use and say it is a commercial product. And, if there are no direct commercialization results, one cannot say the SBIR project failed to pay off. The merits of the SBIR program should not be defined and based on commercialization. There is a gray area here. There may ultimately be commercial products that might not be obvious. The procedure and testing and designing for DoD is a bulwark for the work that follows in the nonmilitary commercial market, work that follows in nonobvious ways. When and where the experience pays off commercially is not always clear.

> The SBIR award did not really help us find our outside financing. We have a sister company that uses the same venture capitalist as the one we have brought

into the project, and the coincidence of the venture capitalist interests in the sister company and the research we are pursing allowed us to attract the third-party investment. So we were lucky getting the Fast Track designation and priority for a Phase II award. The company with equally good research prospects but no luck finding a financial partner will have lower priority. Why should we have priority just because we were lucky finding a partner for third-party investment. The advantage of a Fast Track is that it almost guarantees that you get a Phase II award. Getting a leg up on Phase II is very attractive to us. But there is a problem here if worthwhile projects get low priority because they are either unlucky in seeking outside funding or because they entail research that is not at a stage allowing the small business to attract third-party funds.

I like the SBIR program. It's easy to use. You get early warning on the web. It works. The problems are in the implementation, not the program itself. Some RFPs are so detailed, they are clearly written for one firm. Some RFPs no one can understand. Some are simply silly, asking for something that is not doable. But, overall, there is no real problem with the SBIR program. It works well.

Fast track is a great innovation. One of the lessons, I think, can be discovered by looking at the solicitations. Projects solicited will include at times requests for work on a very specialized technology that already is in existence and for which an upgrade is solicited. I'm not sure why an upgrade should be needed in many of these cases, but in any case why does such a thing appear under the SBIR program? It looks as if the DoD is using SBIR to get little companies to handle what used to be done with routine R&D and procurement at big companies. Now those companies are out of the defense-related business and their former employees are in little companies. It looks like the DoD program solicitation is designed to get the little companies to do what used to be done with DoD procurement. The little companies are perhaps easier to drive a good bargain with, but such projects are not an appropriate use of the SBIR pool of dollars. Fast Track is a great innovation because it takes money out of that pot and puts it into truly innovative small business projects with a high chance of commercialization. The money should be going to finding people with great ideas; lots will crash and burn, but the technology goes into our U.S. technology data bank, and that's where our good jobs come from. Fast Track makes good scientific industrial policy based on innovation and technology. It can find the truly innovative projects. The success here comes from experts reviewing proposals for DoD and making judgments better than cigar-chomping venture capitalists who know nothing about technology. Fast Track takes money out of the general pot of dollars and gets it away from procurement and to truly innovative people. Fast Track helps the SBIR program work as it should.

It is not clear how to structure the Fast Track partnerships so that they fly. A third party is dumping in hundreds of thousands of dollars, and the partnership includes that third party, the government, and us. It becomes very difficult to negotiate the deal. One suggestion might be for the SBIR program to incorporate a Phase III focused on manufacturing technology. When Phase II is successfully completed, there is an interesting product. But it is then up to us to get

> the money for equipment. A Phase III for developing manufacturing technology, for ramping up for production, might be quite helpful given the difficulties negotiating the third-party investments.
>
> The prospects for commercialization could be improved if the SBIR program provided funding for a Mentor/Consultant as a part of Phase II. The SBIR firm would identify in the Phase II proposal a large corporation or marketing consulting firm that would work with the SBIR firm during Phase II and provide expertise about commercializing the technology. The small firm knows the technology, but the larger firm would act as a mentor during Phase II and would be able to help the small firm understand how to market the technology. The big company with the marketing channels and capabilities needed would look at the small company's innovative device and advise it on how to proceed. It would watch the small company and see what was going on and make recommendations and guide the small company, so that at the end of Phase II the small company is not left wondering what to do next. The funding for the Mentor/Consultant need not be an overly large amount. A cross section of the mentoring company would be needed. Someone from marketing, someone from engineering, someone from administration, finance, and management. Three or four people, maybe 100 hours each, to oversee and mentor the small company so that at the end of Phase II they have a direction and a good feel for the market potential and what to do. The SBIR program is now open ended; it is not realizing the fruits of what the program's projects are developing. Providing the opportunity of mentoring from and consulting with a large corporation could improve the prospects for commercialization of SBIR results.

The critical comments from the non-Fast Track respondents were as follows but, again, these are all comments made in the context of an overwhelmingly favorable impression of the SBIR program. The respondents were simply offering these thoughts as ideas that might be used to make a fine program even better.

> We participated in earlier versions of the SBIR program. Fast Track is not useful for us. If we had technology in house during the first few weeks of the Phase I (which we never do), we would not go through the SBIR system at all.
>
> The SBIR program is administered at the top by hard-working and well-meaning people who are really trying to improve our national technology base. As a suggestion, they might rethink how to best go about this task. We need improvements in certain key technology areas and there is widespread agreement as to which areas. However, we do not need uncoordinated Army, Navy, Air Force, DoE, NSF, and NIH SBIR programs, each trying to achieve the same broad goals. These key technology need areas should be assigned a lead agency that should fund all proposals in this area.
>
> Of the 25 pages in the application, only about 5 are needed for technical evaluation. The other 20 could be filed separately, electronically, and be used only in the event the application is being considered for award. This would greatly reduce the complexity of the application process.

The original concept of Phase I followed by Phase II, and then Phase II leading to commercialization is probably flawed. I would recommend revising that concept. The SBIR program could help to complete the process of commercialization. From a successful Phase II project, the SBIR project could go to a stage where the Phase II success is developed further. The SBIR program could support such a bridge to commercialization. The SBIR program now is aimed at establishing technical feasibility, not commercial feasibility. A stage subsequent to Phase II, with government and the company sharing the costs of continuing development work, would be a good policy. Note that it is such sharing of cost for development work that the ATP projects entail. If big companies feel the government needs to help them with such projects, then small businesses need such support too. The Fast Track program is flawed because the end of Phase I is too soon to be ready to establish commercial potential. Fast Track will drive things more toward implementation rather than toward research. Such projects in themselves are fine as long as there is a limit on the amount of the SBIR program that goes to support that sort of effort. I would recommend setting a limited percentage of the SBIR funds that could go to Fast Track projects.

One of the biggest problems we have faced is that our program managers are not able to travel to us because of a lack of funding. Along with the funding, include 30K to 50K for the project monitor to do his job with the specific program for which he is the project manager. It would be easier to interact with the DoD manager if the manager could travel to our location. Also, we can spend the funds for Phase II in two years, but to physically accomplish all of our goals takes time. It would help if there were the latitude to make the Phase II projects three or four years in length rather than just two years.

We thought about Fast Track, but it was not right for us. It was too soon in our research to go to outside investors. We're too inexperienced for Fast Track. To use Fast Track, a company must be in a position to negotiate. Then there is a substantial cost for lawyers.

I recommend that the SBIR program ensure that the technical monitor is involved in the project. When the technical monitor is involved, things go much better than when the technical monitor is not involved.

There is a conceptual problem with Fast Track. The typical SBIR project will not be to the point by the end of Phase I to allow a commitment from an outside investor. Rarely would a venture capitalist think of funding a project unless the research is already done. Such research gets done during Phase II.

Overall the program works very well; things move fairly smoothly.

We did not consider using Fast Track. That would have required an outside investor. But the project is a high-risk project and DoD's acquisition plans are not yet clear. It is highly unlikely that we could get the outside investors required for Fast Track. At the end of Phase II, it would be possible to get outside investors, but not at any time prior to that and not even now. Fast Track is a good idea in theory, but in execution there is a problem. At the end of Phase I, a company typically has just a concept. Later, when there is a prototype, then you can do something with outside investors.

ESTIMATION OF LOWER-BOUND SOCIAL RATES OF RETURN FOR THE SBIR PROJECTS

The data gathered during the interview with the 13 respondents allowed estimation of the private and social rates of return for the 14 projects discussed. The procedure is explained in much greater detail by Link and Scott (1998, 1999). Here, let me simply emphasize that these are *prospective* expected rates of return, even though the estimates of the investment costs are gathered subsequent to the beginning of Phase I (and arguably reflect what would have been the rational expectations for the costs when the project began). That is, the expected rates of return are, by the logic of the approach used to calculate them, estimations of the expectation of the rates of return *at the time that Phase I began.* At that time the SBIR projects were extraordinarily risky; they had upside potential, but also extraordinary downside risk. That is not only because the projects had great technical risks, and not only because of the market risk—even DoD procurement plans are not clear at the beginning of a Phase I SBIR project, but also because of the issues that make it difficult for small businesses to finance innovative investment. Impacted informátion, moral hazard, potential for opportunistic behavior on both sides of the financial transaction—all combine to result in a market failure. Indeed, we expect an incomplete market here.

The expected rate of return required by the potential outside investor exceeds the rate that the small business is willing to promise for the project and, as a result, the small businesses in our samples would not have proceeded with their innovative investments without the support of the SBIR program. Regarding the difficulties of raising outside financial capital, the respondents made the following observations.

> We would not agree to sell our souls to the venture capitalists or a large company

> . . . the project is a high-risk project and DoD's acquisition plans are not yet clear. It is highly unlikely that we could get the outside investors required for Fast Track.

> We would not agree to an arrangement where we would lose control of our company and our intellectual property.

> The outside investors wanted half of the rights to profits in return for providing one-third of the financing.

> For the type of high-risk research funded by the SBIR program, a small company cannot go to the large companies with an interest in the projects because in our area of technology the large companies will not sign nondisclosure agreements with small companies. Only after the technical risk has been reduced and a prototype and a patent are in hand would the small company have the ability to negotiate a partnership.

> We came up with something worthwhile for DoD, but we also advanced our

> own technology to another level without going crazy looking for outside investors.
>
> There must be somebody willing to put money into the project. Plenty of people will give lip service to the idea and take your time. But the probability of actually getting the money is less than 5 percent. To spend 80 percent of the time for the 5 percent chance of financial support is not a good use of our time.
>
> . . ., to support our Phase I project, we tried to find support from other companies and venture capitalists. The venture capitalists want too high a rate of return and want returns too quickly. Joint ventures don't work either. You need their money, so they want lots of rights. You must sell your soul to them. These partner companies are providing capital basically, and sometimes distribution networks.
>
> We cannot use large companies or venture capitalists to fund our research. We protect our intellectual property with trade secrets rather than patents. We must stay out of the grips of the venture capitalists in order to protect our intellectual property.

Thus, at the outset of the SBIR project, the required rate of return for outside financing is not met. Had the expected rate of return exceeded the rate of return required to secure outside financing, the deal for outside financing could have been struck. However, uniformly, the respondents explain that, at the outset of the SBIR, such funding could not be obtained. The SBIR award allows the SBIR project to proceed and ensures that socially valuable research is not lost because of imperfect financial markets, incomplete appropriability, and substantial downside risk. The required rate of return for the outside investors is simply not expected at the *outset* of the project. *Now*, as Phase II draws to an end for the sampled projects, upward of a million dollars or much more has been spent to resolve uncertainties—technical and market uncertainties and also uncertainties about the small business doing the research. Now, after Phase I and Phase II, the logic of our construction of the expected cash flows below would not necessarily hold. *We have estimated prospective rates of return at the outset of Phase I, and these show the market failure and show the reason for the SBIR awards. Without the SBIR funding, socially valuable research would not be undertaken because the required rate of return for outside private investors could not be expected to be achieved, and the small business would not have been able to finance the research itself.*

The calculation of the lower bound for the social rates of return uses the information summarized in Table 10; the information was developed from the interviews that were conducted with the SBIR award winners. Some of the information is also available in the DoD files; however, the information was verified with the respondents and updated to reflect any changes from the DoD files. Variables for duration, total costs, and SBIR funding were combined into one figure for both Phase I and Phase II of the project. Typically, there is an extra

TABLE 10 Definition of Variables for Determining the Prospective Expected Social Rate of Return

Variable	Definition
d	Duration of the SBIR project in years
C	Total cost of the SBIR project
A	SBIR funding
r	Private hurdle rate
z	Duration of the extra period of development beyond Phase II in years
F	Additional cost for the extra period of development
T	Life of the commercialized technology in years
L	Lower bound for average expected annual private return to investing firms
U	Upper bound for average expected annual private return to investing firms
v	Proportion of value appropriated

period of development after Phase II is completed and during which further work with prototypes and initial production lines is done. The length of that additional development period and the extra costs that the company would incur were obtained in the interviews.

Companies cannot expect to appropriate all of the value created by their innovations. First, the innovations will generate consumer surplus that no firm will appropriate, but that society will value. Our estimates of the social rate of return are conservative because we do not attempt to estimate the value of consumer surplus generated by the SBIR projects. Second, some of the profits generated by the innovations will be captured by firms other than the innovators. Larger companies, for example, will observe the innovation and some will successfully imitate it and produce the commercial product in competition with the small business innovator. Respondents were asked to estimate the proportion of the returns generated by their anticipated innovation that they expected to capture. Then, in an extended conversation, other possible applications of the technology developed during the SBIR project were explored. The respondent was then asked to estimate the multiplier to get from the profit stream generated by the immediate applications of the SBIR project's technology to the stream of profits generated in the broader applications' markets that could reasonably be anticipated. Finally, the responding company estimated the proportion of the returns in those broader markets that it anticipated capturing. From the discussion, we were then able to estimate the proportion of value appropriated by the innovating SBIR award winner.

The lower bound L for the average annual private return is found by solving Eq. (1) for L, because that will be the value for L such that the private firm just barely earns the hurdle, or required, rate of return on the portion of the total investment that the private firm must finance. The firm would not invest in the SBIR project unless it expected at least L for the average annual private return.

$$-\int_0^d \left(\frac{C-A}{d}\right) e^{-rt}dt - \int_d^{d+z}\left(\frac{F}{z}\right)e^{-rt}dt + \int_{d+z}^{d+z+T} Le^{-rt}dt = 0$$

$$\Rightarrow -\left(\frac{C-A}{d}\right)\left(\frac{-1}{r}\right)e^{-rt}\Big|_0^d - \left(\frac{F}{z}\right)\left(\frac{-1}{r}\right)e^{-rt}\Big|_d^{d+z} + (L)\left(\frac{-1}{r}\right)e^{-rt}\Big|_{d+z}^{d+z+T} = 0 \tag{1}$$

$$\Rightarrow \left(\frac{C-A}{dr}\right)(e^{-rd}-1) + \left(\frac{F}{zr}\right)(e^{-r(d+z)} - e^{-rd}) - \left(\frac{L}{r}\right)(e^{-r(d+z+T)} - e^{-r(d+z)}) = 0$$

To find the upper bound U for the average annual private return, solve Eq. (2) for U, because any expected annual return greater than U would imply that the rate of return expected by the private firm would be more than its hurdle rate in the absence of SBIR funding, and therefore SBIR funding would not be required for the project.

$$-\int_0^d \left(\frac{C}{d}\right) e^{-rt}dt - \int_d^{d+z}\left(\frac{F}{z}\right)e^{-rt}dt + \int_{d+z}^{d+z+T} Ue^{-rt}dt = 0$$

$$\Rightarrow -\left(\frac{C}{d}\right)\left(\frac{-1}{r}\right)e^{-rt}\Big|_0^d - \left(\frac{F}{z}\right)\left(\frac{-1}{r}\right)e^{-rt}\Big|_d^{d+z} + (U)\left(\frac{-1}{r}\right)e^{-rt}\Big|_{d+z}^{d+z+T} = 0 \tag{2}$$

$$\Rightarrow \left(\frac{C}{dr}\right)(e^{-rd}-1) + \left(\frac{F}{zr}\right)(e^{-r(d+z)} - e^{-rd}) - \left(\frac{U}{r}\right)(e^{-r(d+z+T)} - e^{-r(d+z)}) = 0$$

Our estimate of the average expected annual private return to the firm is $(L + U)/2$. The average expected annual private return to the firm equals v times the average expected annual return that will be captured by all producers using the technology (producer surplus). Knowing the average expected annual private return is $(L + U)/2$ and knowing the portion of producer surplus that is appropriable, v, then we find that the total producer surplus equals $(L + U)/2v$ and hence this value is a lower bound for the average expected annual social return. It is a lower bound because consumer surplus has not been measured.

The private expected rate of return without SBIR funding would be the solution to i in Eq. (3):

$$-\int_0^d \left(\frac{C}{d}\right) e^{-it}dt - \int_d^{d+z}\left(\frac{F}{z}\right)e^{-it}dt + \int_{d+z}^{d+z+T} \left(\frac{L+U}{2}\right)e^{-it}dt = 0$$

$$\Rightarrow \left(\frac{C}{di}\right)(e^{-id}-1) + \left(\frac{F}{zi}\right)(e^{-i(d+z)} - e^{-id}) - \left(\frac{L+U}{2i}\right)(e^{-i(d+z+T)} - e^{-i(d+z)}) = 0 \tag{3}$$

The lower bound on the social rate of return is found by solving Eq. (4) for i:

$$-\int_0^d \left(\frac{C}{d}\right) e^{-it}dt - \int_d^{d+z}\left(\frac{F}{z}\right)e^{-it}dt + \int_{d+z}^{d+z+T} \left(\frac{L+U}{2v}\right)e^{-it}dt = 0$$

$$\Rightarrow \left(\frac{C}{di}\right)(e^{-id}-1) + \left(\frac{F}{zi}\right)(e^{-i(d+z)} - e^{-id}) - \left(\frac{L+U}{2iv}\right)(e^{-i(d+z+T)} - e^{-i(d+z)}) = 0 \tag{4}$$

The private expected rate of return with SBIR funding would be the solution to i in Eq. (5):

$$-\int_0^d \left(\frac{C-A}{d}\right) e^{-it} dt - \int_d^{d+z} \left(\frac{F}{z}\right) e^{-it} dt + \int_{d+z}^{d+z+T} \left(\frac{L+U}{2}\right) e^{-it} dt = 0$$

$$\Rightarrow \left(\frac{C-A}{di}\right)(e^{-id} - 1) + \left(\frac{F}{zi}\right)(e^{-i(d+z)} - e^{-id}) - \left(\frac{L+U}{2i}\right)(e^{-i(d+z+T)} - e^{-i(d+z)}) = 0 \quad (5)$$

Table 11 provides the various prospective expected rates of return for the New England projects as a group and for the Fast Track and the non-Fast Track projects. Tables 12, 13, and 14 provide the summary statistics for the data.

It seems clear that the Fast Track cases are much different from the non-Fast Track cases. Although they begin with a Phase I where a small business needs outside support, they exhibit sufficient commercial potential to attract outside funding quickly, and as a result these are likely to be projects that, relative to non-Fast Track projects, have higher *lower bounds* for social rates of return (recall that the social rates of return measure only producer, not consumer, surplus). Furthermore, because there will be more of the project investment cost paid by private funds, the private rates of return given SBIR support will be lower for the Fast Track projects.

CONCLUSIONS

In all, the collection of 14 New England SBIR projects studied here exhibited high risk at the outset of Phase I—both technical and market risk, high capital costs, and often a long expected time before commercialization of the resulting technology. Comments suggest fairly substantial appropriability problems for some projects, even within the narrower applications of the technology. Appropriability problems typically are greater when broader potential applications are considered. Uniformly, in the absence of the SBIR funding, the research projects would not have been undertaken in the same way or at the same pace.

TABLE 11 Prospective Expected Rates of Return (ROR) for New England SBIR Projects

Region	Number of Cases	Private ROR Without SBIR (prvnosbr)	Social ROR, lower bound (soclwrbd)	Private ROR with SBIR (prvsbr)
New England	14	0.31	0.60	0.58
Fast Track	6	0.33	0.68	0.53
Non-Fast Track	8	0.30	0.55	0.61

TABLE 12 Data for New England Observations (Fast Track and Non-Fast Track)

Variable	Obs	Mean	Std. Dev.	Min	Max
d	14	777,857	0.3833893	2.17	3.5
C	14	1,285,053	692,912.4	717,873	3,450,000
A	14	785,715.6	144,072.6	507,873	1,099,966
T	14	16.32143	8.111263	5	30[a]
z	14	1.535714	1.456442	-0.375[b]	5
F	14	1,063,929	2,597,688	0	1.00e+07
r[c]	7	0.3821429	0.1222312	0.2	0.5
v	14	0.3053929	0.1992511	0.025	0.6
L	14	1,009,056	1,185,032	79,185	4,486,450
U	14	2,370,090	2,174,385	413,400	8,666,330
prvnosbr	14	0.31	0.0689481	0.19	0.43
soclwrbd	14	0.605	0.1754445	0.28	0.82
prvsbr	14	0.5757143	0.2172455	0.21	1.03

[a]One company responded that *T* would be several decades, and another reported that *T* would be forever. In both cases, *T* was conservatively entered as the value 30 years. However, because the relevant discount rates are so high, the difference between 30 years and "forever" is not significant. In the integrals, the term with *T* entered negatively as an exponent would become zero, but with a large value of *T*, the term is very small in any case.

[b]This observation has a negative value because commercial returns started before the end of Phase II.

[c]Half of the respondents were uncomfortable estimating the private hurdle rate that outside financiers would apply to their projects at their outset. For those, the average value of *r* was used in the calculations.

TABLE 13 Data for the New England Fast Track Observations[a]

Variable	Obs	Mean	Std. Dev.	Min	Max
d	6	2.578333	0.3279888	2.17	3.17
C	6	1,659,609	926,636.1	850,000	3,450,000
A	6	855,436.5	170,450.7	598,700	1,099,966
T	6	18.75	9.585145	7.5	30
z	6	1.208333	0.7486098	0.25	2.5
F	6	500,000	411,096.1	100,000	1,000,000
r	3	0.4	0.0901388	0.325	0.5
v	6	0.2379167	0.1296767	0.1575	0.5
L	6	971,784.8	673,530.7	533,085	2,237,640
U	6	1,962,022	654,443.7	1,125,230	3,036,360
prvnosbr	6	0.3266667	0.0388158	0.3	0.4
soclwrbd	6	0.6783333	0.1553598	0.44	0.82
prvsbr	6	0.53	0.1749286	0.35	0.86

[a]See notes to Table 12.

TABLE 14 Data for the New England Non-Fast Track Observations[a]

Variable	Obs	Mean	Std. Dev.	Min	Max
d	8	2.9275	0.3693527	2.5	3.5
C	8	1,004,136	260,581.3	717,873	1,419,895
A	8	733,424.9	102,491.9	507,873	820,000
T	8	14.5	6.907553	5	25
z	8	1.78125	1.838028	-0.375	5
F	8	1,486,875	3,454,596	0	1.00e+07
r	4	0.36875	0.1546165	0.2	0.5
v	8	0.356	0.2342849	0.025	0.6
L	8	1,037,010	1,510,587	79,185	4,486,450
U	8	2,676,142	2,867,886	413,400	8,666,330
prvnosbr	8	0.2975	0.0856488	0.19	0.43
soclwrbd	8	0.55	0.1784857	0.28	0.78
prvsbr	8	0.61	0.2503141	0.21	1.03

[a]See the notes to Table 12.

Not surprisingly, then, respondents reported that outside investors, at the outset of Phase I, would have required too high a rate of return to make it possible for the project to proceed with private financing. For example, one respondent reported that the outside investor wanted one-half of the rights to the profits for contributing one-third of the investment cost. Another reported that the outside financiers wanted so much of the company that he would have lost control of the company and ultimately of its intellectual property. Many other comments along those lines are provided in detail earlier in this paper.

The projects on the whole met both the funding agency's mission and the company's strategy. All fit the general scenario for socially valuable research projects that would have been underfunded in the absence of the SBIR program. In particular, the projects appear to be ones for which the private rates of return in the absence of SBIR funding would have fallen short of the private hurdle rate required by outside financiers to whom the small businesses would have had to turn for financial support. Yet the social rates of returns to the projects are large and exceed the hurdle rates. The funding from the SBIR program changes the ordering of rates of return anticipated at the outset of Phase I. With the SBIR program providing funds, the expected private return relative to just the private portion of the total project costs is sufficient to move the private rate of return above the hurdle rate, and then the socially valuable research investment is undertaken.

In the foregoing ways, the Fast Track and non-Fast Track projects are essentially similar. Nonetheless, taken as a group the Fast Track projects show higher prospective expected lower-bound social rates of return—just as we would expect, because the measure includes only expected profits to the innovator and other producers, rather than including consumer surplus as well. Thus, the Fast Track projects have higher expected private profits, and we expect that to be the case because these are the projects that attracted outside investors at an early stage in the research. Furthermore, the average duration of additional development beyond Phase II is somewhat less for the Fast Track projects, suggesting that at least on average they are somewhat closer to commercialization at the end of Phase II than the non-Fast Track projects.

The respondents and the rate-of-return calculations make clear that although the Fast Track program selects projects that are different from SBIR projects more generally, projects that do not qualify for the Fast Track designation are typically no less deserving of SBIR support, but rather are high-risk projects with potentially great social value that would go unfunded in the absence of the SBIR program. The respondents suggest that, typically, the Fast Track program is simply not useful for companies pursuing socially valuable high-risk research because at the end of Phase I, most such projects do not yet have the characteristics of projects that attract outside private investors.

Finally, two things must be emphasized in conclusion. First, the high social rates of return estimated and reported for the SBIR projects are very conservative, lower-bound estimates because they do not include consumer surplus in the benefit stream. Consider, for example, the non-Fast Track innovation of Materials Technologies that will allow safe, accurate, and efficient diagnostic tests of the wiring in airplanes. The profits that will be generated by the technology are obviously a tiny proper subset of the social benefits that the technology will generate, but the estimation method used measures only the returns in the form of profits to the innovator and to other producers of the technology. Second, some readers will be skeptical about the SBIR award recipients' earnest belief that without SBIR funding the projects would not have been undertaken or at least would not have been undertaken to the same extent or at the same speed. With the SBIR program in place, certainly the pursuit of SBIR funding would perhaps be a path of least resistance. However, if the research would have occurred without the public funding, the estimated upper bound and hence the average of the upper and lower bounds for expected private returns would be too low, and the actual lower bounds for the social rates of return would be even higher than we have estimated. Further, the gap between the social and private rates of return would remain, although that would not in itself justify public funding of the projects.

To summarize in a concise manner, Tables 15 offers a comparison of costs and benefits of Fast Track and non-Fast Track projects over the same time frame.

Other differences between the Fast Track projects and non-Fast Track Projects in the New England comparison groups include the following:

- A smaller proportion of Fast Track companies have had previous SBIR awards (3 of 6 vs. 6 of 7).

TABLE 15 Fast Track and Non-Fast Track Projects: New England Comparison Groups

Variable	Averages for Timeline of Costs and Benefits	
	Fast Track	Non-Fast Track
Total SBIR project cost	$1.7 million	$1.0 million
SBIR funding	$0.9 million	$0.7 million
Additional period of development	1.2 years	1.8 years
Costs for additional development	$0.5 million	$1.5 million
Lower bound rate of return to society (including benefits to SBIR firm and its investors and also to other firms)	68%	55%

- A smaller proportion of Fast Track companies expressed difficulties bridging a gap in time between Phase I and Phase II (0 of 6 vs. 4 of 7).
- A larger proportion of Fast Track companies said that the SBIR award facilitated the attraction of outside investors (4 of 6 vs. 1 of 7).
- Fast Track projects show commercial potential earlier and, by the end of Phase I, outside third-party investors are found.
- Fast Track projects have a higher lower bound for the social rate of return (based on the benefits for the collection of firms using the technology created by the SBIR project).

Similarities between Fast Track and non-Fast Track projects in the New England comparison groups include the following:

- Barriers to investment (such as high technical risk and high capital costs) imply the need for partial public funding to carry out the SBIR projects.
- None of the companies has received ATP awards.
- All of the companies expect long-run strategic benefits from the SBIR award in the form of increased company size (sales or employment) or a more diversified product line.
- The SBIR projects are socially valuable: The social rate of return is greater than the rate of return needed for a worthwhile project.

Respondents in the New England comparison groups expressed concerns about and recommendations for improving the SBIR program. From the Fast Track project respondents came the following:

- Small businesses should be encouraged to acquire expertise to ensure proper business administration to go along with the competence in scientific work.

- Nonmilitary commercialization should not be the defining basis for the merits of the SBIR program because many valuable projects develop information with narrow applications within DoD.
- The Fast Track program may cause worthwhile projects to have a low priority for Phase II awards simply because they entail research that does not by the end of Phase I reach the stage that attracts outside funding.
- Some SBIR projects appear to be the sort of routine R&D and procurement that used to be done at large companies. Fast Track is a great innovation because it puts money into truly innovative small business projects with a high chance of commercialization.
- A Phase III for developing manufacturing technology, for ramping up production, might be quite helpful given the difficulties in negotiating the third-party investments.
- Funding should be provided for a Mentor/Consultant as a part of Phase II, with the SBIR firm identifying in the Phase II proposal a large corporation or marketing consulting firm that would work with the SBIR firm during Phase II and provide expertise about commercializing the technology.

From non-Fast Track project respondents came these observations:

- Fast Track is not useful when the SBIR funding is needed to support high-risk research that does not result in a commercially viable technology before Phase II. Without having such an early result, attraction of outside funding is not possible in time for a Fast Track award.
- Key technology areas should be assigned to a lead agency, which would fund all proposals in that area. There is agreement that improvements are needed in certain key technology areas. However, better coordination of the efforts of various agencies administering SBIR awards, each trying to achieve the same broad goals, is needed.
- Of the 25 pages in the application, only about 5 are needed for technical evaluation. The other 20 could be filed separately, electronically, and would be used only in the event the application is being considered for an award.
- The SBIR program could help to complete the process of commercialization. Continuing support for a successful Phase II project, the SBIR program could support a bridge to commercialization. The SBIR program now is aimed at establishing technical feasibility, not commercial feasibility.
- Phase II funding for the DoD project monitor to travel to our location and interact with us should be provided. This would ensure that the technical monitor is actively involved in the project.
- Phase II awardeed should be allowed to spend funds over three or four years instead of just two years.

TABLE 16 Fast Track and BMDO-Matching Projects Compared with Others in the New England Sample[a]

Type of Project[b]	Averages		
	d (years)[c]	*z* (years)[d]	soclwrbd (%)[e]
F	2.43	1.00	69
B	2.72	2.04	67
Both F & B	2.72	1.42	67
Neither	3.05	1.62	48

[a]Sample consists of 14 New England projects; 3 F, 3 B, 3 F & B, 5 neither F nor B.

[b]F denotes Fast Track and B denotes BMDO-matching.

[c]*d* is the duration of the SBIR project (Phase I + Phase II) without including the gap between the two phases and hence *d* is the duration of performance.

[d]*z* is the length of the additional period of development beyond the end of Phase II and until commercialization.

[e]soclwrbd is the lower bound rate of return to society, including benefits to the SBIR firm and its investors as well as to other firms.

Table 16 provides additional insight by distinguishing the projects of the Ballistic Missile Defense Office (BMDO), where matching funds are required although, unlike Fast Track, the matching funds can come from the SBIR company itself. Fast-Track and BMDO-Matching SBIR projects are of shorter duration than other projects, even ignoring the gap between Phase I and Phase II. The additional period of development beyond the end of Phase II and until commercialization is less for Fast Track projects than for BMDO-matching projects. The lower-bound rate of return to society (including benefits to the SBIR firm and its investors and also to other firms) is greater for Fast Track and BMDO-matching projects. In sum, Fast Track Projects take less time to reach commercialization; both Fast Track and BMDO-matching projects have more commercial potential in the sense that they are expected to generate greater returns to the SBIR firm and its investors and also to other firms. Further investigation, available on request from the author, showed that the qualitative differences among the projects remain the same when controls for technology categories are added in a regression model.

The conclusion is that the SBIR program has funded innovative projects with high social rates of return that would not have been undertaken in the absence of the program. Further, the non-Fast Track as well as the Fast Track projects appear to be quite valuable, although the non-Fast Track projects typically do not exhibit private commercial potential as quickly as the Fast Track projects.

ACKNOWLEDGMENTS

I would like to thank the following individuals for generously giving of their time for the interviews underlying the study. David Brock of Brock-Rogers Surgical; Myles Walsh of Cape Cod Research; Adi Guzdar of Foster-Miller, Inc.; Bob Hoch of Hyperion Catalysis International; Frederick Dampier of Lithium Energy Associates, Inc.; Yogesh Mehrotra of Materials Technologies Corporation; Marthinus C. van Schoor of Mide Technology Corporation; Steven P. Bastien of Optigain, Inc.; Peter Chenausky of QSource, Inc.; David Miller, Larry Willner, and Gregory Lane of SEA CORP (Systems Engineering Associates Corp.); Timothy J. Driscoll of Spectra Science Corp.; Frank Folino of Synkinetics, Inc.; and Grant M. Ehrlich of Yardney Technical Products, Inc. I am grateful to David B. Audretsch, William L. Baldwin, Albert N. Link, and Nancy A. Scott for helpful comments and suggestions.

REFERENCES

Link, Albert N., and John T. Scott. 1998. *Overcoming Market Failure: A Case Study of the ATP Focused Program on Technologies for the Integration of Manufacturing Applications (TIMA).* Draft final report submitted to the Advanced Technology Program. Gaithersburg, MD: National Institute of Technology. October.

Link, Albert N., and John T. Scott. 1999. "Estimates of the Social Returns to SBIR-supported Projects," this volume.

Tassey, Gregory. 1997. *The Economics of R&D Policy* (Westport, Conn., and London: Quorum Books.

Patterns of Firm Participation in the Small Business Innovation Research Program in Southwestern and Mountain States

Reid Cramer
University of Texas at Austin

EXECUTIVE SUMMARY

The conclusions of this paper are derived from interviews with firms that have received Small Business Innovation Research (SBIR) contracts in the southwestern and mountain regions. The firms selected in this set of case studies are diverse in terms of their technological expertise, but they are more fundamentally differentiated from each other by organizational characteristics. Primary among these characteristics are the firm's size, relationship to capital, and business development strategy. Depending upon these characteristics, firms are classified by the author in one of three categories: contractor, technologist, or scientific.

The SBIR program fills a different role for each type of firm. Furthermore, features of the Fast Track program make it more or less appealing for a firm depending upon its organizational characteristics. Where a firm is located in its own developmental cycle appears to be a major factor in how it will structure its participation in the SBIR program. An SBIR grant can be invaluable to a young start-up company, but less essential after that same firm has attracted significant outside resources.

Firms in all three categories are enthusiastic about the SBIR program and identify a range of positive impacts that the program has had on their work. The program provides resources to allow firms to conduct expensive research and development activities, and to expand their firms through the acquisition of capital equipment, facility maintenance, and the addition of staff. The SBIR program is valuable in that it provides an accessible revenue stream. It provides research support to move technology into core commercial directions. SBIR benefits are not always expressed in commercial sales, but are related to expanding basic research, responding to government needs, and developing applications for technology.

The Fast Track program offers a valuable tool to firms looking to attract investors because it allows the firms to offer prospective investors a means to leverage their investment. The Fast Track program does give an advantage to firms with a higher potential for commercialization and an interest in producing for the market. Firms that are interested in commercialization prior to their application for an SBIR award appear most likely to succeed as Fast Track firms. This may be because these firms are more likely to have their sights set on pursuing technological innovations that will attract market attention.

INTRODUCTION

The Department of Defense (DoD) has requested that the National Research Council conduct an evaluation of the Small Business Innovation Research Program (SBIR). The conclusions of this paper are derived from interviews with 12 firms that have received SBIR contracts in the southwestern and mountain region, with particular attention given to the experience of firms participating in the SBIR program according to their Fast Track status. Although the firms selected in this set of case studies are diverse in terms of their technological expertise, common experiences are apparent. This paper summarizes some of the distinct patterns among firms participating in the SBIR program.

A list of selected firms and their SBIR Phase II projects appears in Appendix A. Approximately 30 Phase II SBIR projects from the southwestern and mountain region were included in a representative data set developed by BRTRC for the National Research Council's evaluation. Individual firms could be represented on this list more than once if they had received multiple contracts. Four projects were identified as participating in the Fast Track program, and all were selected for case studies. Additionally, geography was taken into account as a selection criterion, and firms that were located in the Austin-San Antonio, Boulder-Denver, and Albuquerque metropolitan areas were contacted to learn about their SBIR experience.

Interviews were designed to elicit information about how each firm was affected by the SBIR program, with a focus on a particular Phase II award. In this manner, information was gathered at the firm and the project levels. The case studies are organized following the template agreed upon by the Board on Science, Technology, and Economic Policy of the National Research Council and its team of SBIR researchers. This template is organized into sections, corresponding to a description of the firm, a summary of the SBIR project, an analysis of the impact of the SBIR award, and a critique of SBIR program administration. Cross-cutting questions that were designed to focus the analysis on the broader experience of the SBIR program were also considered. The survey template appears in Appendix B.

DESCRIPTION OF THE FIRMS

The firms described in the SBIR case studies have diverse areas of expertise, but they are more fundamentally differentiated from each other by organizational characteristics. Primary among these characteristics are the firm's size, relationship to capital, and business development strategy. With the combination of these characteristics, firms typically can be classified in one of three categories of firm type.

The first category comprises firms that are committed to the business of research and development (R&D). Companies in this category can be called *contractor firms*. For this group, the SBIR program represents a source of contract revenue. The performance of research is a major business activity, and the SBIR program supports this work. The SBIR solicitation topics are interpreted as the articulated research needs of various government agencies. When a firm has the capabilities to respond to these topics through existing staff and resources, it generally will apply for an SBIR award. These firms generally are well-established and support larger staff than many SBIR participant firms. Typically, they have over 50 employees, and perhaps as many as 300 to 400, on their payrolls. These firms are multiple recipients of SBIR awards and the SBIR program represents a significant portion of the firm's annual revenue. Other sources of revenue are pursued, such as marketing commercial products, signing licensing agreements, or forging partnerships with larger firms, building upon intellectual property rights and research activities.

A second category is exemplified by firms focusing on the development of a specific product or promising enabling technology. These companies can be labeled as *technologist firms*. They may have applied for an SBIR award early in their history when the product or technology was at an early stage in the innovation cycle. For these firms, the SBIR program represented a means to develop technology and demonstrate its viability to the market. If the product is attractive to investors—either venture capitalist or angel investors—these firms can be transformed quite dramatically. The infusion of capital allows them to expand their staff, purchase capital equipment, and attract senior management. These firms are oriented toward the market and their growth decreases the relative importance of the SBIR program as an ongoing revenue source. The SBIR program represents an invaluable building block for these companies even as work on the SBIR contract loses importance relative to the goal of marketing a product.

A third category is represented by firms interested in pursuing basic research outside of a university setting, and thus may be labeled as *scientific firms*. These firms are generally small and were founded by scientists to explore whether a particular research area can generate ideas or products that might attract interest. The SBIR program enables these firms to work on specific projects whose outcomes are unclear. They may have the chance to develop a commercial product, but they are not currently attracting interest from outside investors. For firms in

TABLE 1 Typology of SBIR Firms by Organizational Characteristics

Type of Firm	Primary Activity	Firm Size	Relationship to Capital	Business Strategy
Scientific	Basic research	Small (0-15)	Distant	Pursue research
Technologist	Applied research	Small but growing (0-30)	Seeking investors	Commercialize
Contractor	R&D contracts	Larger (30-350)	Partnerships and contracts	Develop expertise

TABLE 2 Characteristics of Case Study Firms

Firm	State	Founded	Employees	Phase IIs	SBIR Project	Firm Type
AvPro	OK	1990	10	2	Fast Track	Technologist
Bolder Technologies	CO	1991	100	1	SBIR	Technologist
Chorum Technologies	TX	1996	60	2	Fast Track	Technologist
Coherent Technologies	CO	1984	90	27	SBIR	Contractor
Lipitek International	TX	1988	15	1	SBIR	Scientific
Mission Research	NM	1970	350	45	SBIR	Contractor
Picolight	CO	1995	26	5	Fast Track/ BMDO	Technologist
Radiant Research	TX	1994	12	8	BMDO Co	Scientific
SPEC	TX	1986	60	25+	BMDO Co	Contractor
TRAC	CO	1991	4	1	Fast Track	Scientific
TPL, Inc.	NM	1990	90	12	BMDO Co	Contractor

this category, the SBIR contract represents a significant portion of revenue, and the size of a Phase II award provides enough resources to acquire equipment and people to perform the work. The firms in this category provide a vehicle for scientists to pursue a focused research agenda.

These three categories represent types of firms distinguished by their organizational characteristics. Firms develop these characteristics across time in response to evolving conditions. As these characteristics change over time, a particular firm may need to be reclassified in this typology or may exhibit characteristics of more than one firm type. Still, it is instructive to use this classification scheme to distinguish the firms in this analysis according to their own organizational characteristics.

The SBIR program fills a different role for each type of firm. Furthermore, features of the Fast Track program make it more or less appealing for a firm depending upon its organizational characteristics. Additionally, where a firm is located in its own developmental cycle appears to be a major factor in how it will structure its participation in the SBIR program. Thus, a firm's relationship to the program will not be static. An SBIR grant can be invaluable to a young start-up company, but less essential after that same firm has attracted significant outside resources.

TABLE 3 Classification of Case Study Firms by Type

Scientific Firms	Technologist Firms	Contractor Firms
Lipitek International	AvPro	Coherent Technologies
Radiant Research	Bolder Technologies	Mission Research Corp.
TRAC	Chorum Technologies	SPEC
	Picolight	TPL, Inc.

IMPACT OF THE SBIR PROGRAM

Firms in all three categories are enthusiastic about the SBIR program and identify a range of positive impacts that the program has had on their work. The program provides resources to allow firms to conduct expensive R&D activities, and promote growth of their firms through the acquisition of capital equipment, facility maintenance, and staff expansion. The SBIR program is valuable in that it provides an accessible revenue stream. It provides research support to move technology into core commercial directions.

From several perspectives, the SBIR program is lauded as providing numerous opportunities for young and less-established firms. First, it offers a means to expand a firm's technical understanding of its research areas. Second, it encourages the firm to develop commercially viable applications of its research. Third, it fills a funding gap difficult to bridge with conventional private investment. In several cases, the SBIR program provided the direct impetus for the founding of a company. There are a host of indirect but beneficial spillover effects related to the SBIR program. The SBIR experience creates relationships that continue to benefit participating firms. These include contacts with suppliers, manufacturers, and potential partners.

The SBIR program is used by many companies as a marketing tool. Receiving an award can illustrate a potential application for a technology, and provide the means to attract a partner. The SBIR award does not in and of itself certify the technology or the company on its own, but it gives the firm an advantage as it competes for the attraction of investors. Private investors are not attracted to the firm solely because of its SBIR award, but the SBIR award serves as an indication that the firm's technology may be promising because it has been recognized by a government agency.

Firms in all three categories appeared to use their SBIR awards as marketing tools, but in distinct ways. For the scientific firms, such as Lipitek and Radiant Research, the presence of an SBIR award confirms the importance of their research agendas. For the contractor firms, such as TPL Inc., Mission Research Corp., and SPEC, their ongoing participation in the program demonstrates that they can deliver projects and maintain the in-house expertise sought by government agencies. The technologist firms highlight their SBIR awards as indicative of innovation. Outside investors may still evaluate the technology of these firms on their own terms, but the award helps to distinguish their work from others competing for the interest of investors and venture capitalists.

Many successful SBIR contracts do not result directly in a product, and thus appear to fail when judged according to a narrow return on investment criteria. Yet the SBIR program can still lead to success in the long run by expanding a firm's understanding of its technology area, generating intellectual property rights, and supporting a young firm with promise. Often, commercial success is realized when a product is developed after a succession of SBIR awards. This has been the case with several of the larger firms classified as contractor firms. In the case of TPL Inc., SPEC, and Coherent Technologies, commercial opportunities were identified after a series of contracts were executed.

These firms and others classified as contractor firms appear to be committed to innovative R&D work, and have come to rely on the SBIR program for a large portion of their revenue. These firms believe that they are good at what they do and are using the SBIR program to respond to articulated government needs. These firms appear to be filling a niche market that consists of performing basic and applied research that responds to the need of sponsoring agencies. The basic research provides the building blocks for future product development, and thus is located earlier in the innovation and product cycle. The applied research often responds to the agenda of contracting agencies, and thus may be distinct from more commercial opportunities.

The larger amount of work that these firms undertake means that there are some applications for which the private market is interested. All of the four firms classified as contractors (Coherent Technologies, Mission Research Corporation, SPEC, and TPL, Inc.) have explored commercial opportunities, often in partnership with much larger and more established firms. However, in each of these cases, government R&D contracts continue to represent a significantly higher portion of their annual revenues.

Other firms are relying even more heavily on the SBIR program for operating revenue. These firms are often small firms concentrating on basic research. These cases describe two of the firms classified as scientific (Radiant Research and TRAC). Firms that appear interested in basic research are not as attractive to capital investors interested in a rate of return. Innovative research, by its nature, is the foundation upon which product development is based. Therefore, solid, innovative research is not readily commercializable. In this sense, the SBIR program's more recent focus on commercialization undermines one of the program's strengths: the SBIR program's ability to support the innovative work of small or young companies.

THE FAST TRACK PROGRAM

The Fast Track program offers a valuable tool to firms looking to attract investors because they can offer prospective investors a means to leverage their investment. The Fast Track program does give an advantage to firms with a higher potential for commercialization and an interest in producing for the market. Firms that are interested in commercialization prior to their application for an SBIR award appear most likely to succeed as Fast Track firms. This may be because these firms are more likely to have their sights set on pursing technological innovations that will attract market attention. Thus, they will be closer to marketing their technology than firms that are focusing on more conventional R&D activities. Of the four Fast Track projects in this study, three were from firms classified as technologists. Picolight, AvPro, and Chorum Technology have articulated business plans that address how their work will respond to unmet market needs. Each of these firms has identified not only its innovation, but the clients and firm clusters that their products will serve. Two of these firms, Picolight and Chorum Technologies, have generated additional investment that has far exceeded the SBIR contract amount and the third-party Fast track match. The fourth Fast Track firm, TRAC, has not had a successful Fast Track experience, and in fact lost its matching investment midway through the project. The TRAC case suggests that some scientific firms, being far away from commercialization, may not benefit—and may even be hurt—by a Fast Track designation. In TRAC's case, its small staff, limited administrative capacity, and undefined sense of its target market created difficulties after it received its Fast Track award.

Depending on the technology area, the time between developing a product for market and the basic R&D stage varies, making Fast Track firms not as viable for some firms. For example, TPL Inc. is primarily a materials company that projects a longer time horizon for product development than would a software or electronics firm. Even when they feel as though they have identified an important innovation, they recognize that it will take many years to develop a technology, streamline a manufacturing system, and market to other firms. Although they may be competitive in their field, this extended timetable may not be attractive to investors requiring a quicker return.

The organizational characteristics of SBIR firms essentially determine whether a firm will be able to take advantage of the Fast Track program. Firms that are committed to applied research and developing products or applications for an identified market have a greater likelihood of attracting investment. The later stage of the product cycle is represented by several other characteristics, such as whether a prototype already exists, products are being tested, or production facilities are in place. These types of characteristics appear to be prerequisites for firms to succeed in SBIR's Fast Track. The BMDO co-investment program appears to have a different pattern for firm participation. In the cases included in this study, firms with a range of organizational characteristics are

TABLE 4 Fast Track and BMDO Co-Investment Projects by Firm Type

Project Status	Firm	Firm Type
Fast Track	AvPro	Technologist
Fast Track	Chorum Technologies	Technologist
Fast Track	Picolight	Technologist
Fast Track	TRAC	Scientist
BMDO Co-Investment	TPL Inc.	Contractor
BMDO Co-Investment	SPEC	Contractor
BMDO Co-Investment	Radiant Research	Scientist
BMDO Co-Investment	Picolight	Technologist

more readily able to take advantage of this program for several reasons. First, project matching funds may be in-kind rather than exclusively cash. Second, larger firms that have received multiple SBIR awards have existing relationships with other firms that they can draw upon in executing a project. Third, BMDO has prioritized the co-investment system, and so, any firm that wishes to receive an SBIR award from BMDO must explore the co-investment option.

Benefits and Costs

For firms participating in the Fast Track program, there is an advantage in making a quicker transition from Phase I to Phase II. This is especially important for young start-up or relatively small firms, which benefit from the short gap between award and arrival of the matching funds. Fast Track investment addresses one of the major obstacles of small businesses: cash flow. The infusion of capital that comes with a Fast Track award can help to alleviate many problems.

However important addressing cash flow issues is, the overwhelming benefit to the Fast Track firm is the ability to use government funds to attract outside investment. Fast Track represents a means for research companies to offer third-party investors the opportunity to leverage their investment. This makes investing in a young start-up firm significantly more attractive. Picolight and Chorum Technologies specifically highlighted this phenomenon when discussing their participation in the Fast Track program.

Fast Track participation also has some perils. The shortened timetable for submitting a Fast Track application can propel a project prematurely, limiting administrative oversight. In the one Fast Track case study with a poor outcome (TRAC), the principal assumed responsibility for this lack of diligence in project administration, but felt seduced by the prospect of matching funds. In this case, the structure of the Fast Track program gave the investor undue influence on the course of the project.

There is some concern about the trade-offs that accompany the attraction of investors into a small firm. Primarily, firms are worried about losing control to outside forces. Still, most of the firms that have attracted substantial outside investment, either through Fast Track or other means, are pleased with the results. Venture capitalists do bring valuable experience. In most cases, their presence on the board of directors is a vehicle for structuring input rather than taking control. Picolight and the spinoff firm from SPEC both have appreciated the expertise that venture capitalists have brought to their firms, increasing the likelihood of market success.

Success with the SBIR program can lead to choices that can affect the firm's mission. When a firm grows with support of the program, the firm must determine what kind of company it wants to be. It must decide whether it should pursue investors, pursue acquisition, or take the company public. These are all measures that change a company's profile significantly. For example, Chorum

Technologies was adamant that it was not interested in being acquired by a larger company. As a result of this decision, they have decided that they need to develop the internal capabilities for large-volume manufacturing.

A company may be willing to consider drastic measures because of the difficulties associated with commercialization. Most firms recognize that there is a major gap between completing a Phase II contract and successfully marketing a product. Although access to capital is a major component of this gap, money is not the only solution. More than a money gap, many companies experience a maturity gap. They can develop a product prototype but much more is required to successfully produce and distribute a product in the market. Often what is missing is an understanding of the markets. This is an expertise that a third-party investor may possess, and thus the Fast Track program provides a vehicle for firms to collaborate, each bringing its own expertise to a particular project.

Factors that Inhibit Fast Track Participation

The prime factor that inhibits a firm's participation in the Fast Track program is difficulty in attracting investors to meet the matching requirements of the program. This is because SBIR proposals that address issues of basic research represent work that occurs early in the product cycle.

Several firms that were multiple winners of SBIR awards, such as Mission Research Corp., Radiant Research, and TPL, Inc., noted that the matching requirements represented an obstacle to their participation in the Fast Track program. Instead of the four-to-one matching requirement imposed on first-time awardees, multiple recipients have to find a one-to-one match. However, the Fast Track program solves some problems for younger and smaller firms that do not afflict larger, better established firms. Two of the key advantages to Fast Track are the bridge financing and quicker timetable for funding. Contractor firms or those technologist firms that have attracted investment and expanded their activities do not have a financial need for financing between Phase I and Phase II projects; these firms already have access to external capital. Fast Track still offers a chance to leverage this external capital, which is of interest to firms of all sizes.

PROGRAM ADMINISTRATION

One central conclusion from this set of SBIR case studies is that the size of the firm affects the impact of the SBIR award. Larger firms with more staff resources have an easier time preparing applications, managing contracts, and performing the research. Firms with less staff cannot easily divide SBIR responsibilities. For example, the experience of being audited is particularly onerous for a small company with few staff resources. A small firm is also disadvantaged because administration of the contract takes time way from the people who would otherwise be "doing the science."

Small firms are particularly concerned when a cost-plus-fixed-fee system is employed in the disbursement of funds. Many firms, both small and large, spoke against this method of structuring SBIR contracts. Firms would rather be paid upon completion of work or the submission of results. Many costs that a firm might submit are ineligible under government regulations, but the firm does not know what these costs are and feels as though a big staff would be required to successfully understand all of the government regulations and nuances of the SBIR program. There is the sense that contracts are written to make management easiest on the program officer rather than considering the needs of the recipient firm.

Some technology companies also have a unique problem. Many technological advances are incremental; others represent breakthrough innovations that create products or processes that do things that have never been done before. Thus, firms must look for a home for the technology that is not readily apparent if it is a breakthrough. This is the reverse of the process that most commercial companies pursue, where they know the market and compete with products that cost less or outperform existing products. Technology products can do new things, and markets are less defined. This is an issue for companies such as Coherent Technologies, which produce innovative laser products with detection capabilities that far exceed today's industry standards. The challenge for the firm is not only to develop the technology but also to nurture and educate the market to increase the likelihood of adoption. In this sense, the SBIR program's emphasis on commercialization may benefit from finding ways to support companies as they pursue commercialization and increase their understanding of markets where they can compete. For example, Coherent Technologies may need support to think more clearly about the needs of potential clients and their own strategic advantage.

Many firms described how they would like to see additional support for Phase III of the SBIR program, where commercialization is emphasized. The development of a marketable technology does not lead easily to commercialization; there are many additional steps that must be taken for a firm to succeed. If DoD wants firms to achieve commercialization, perhaps more thought should go into what can be done in a supportive manner to get firms to this point.

The requirement of matching funds for the Fast Track and BMDO co-investment programs have motivated firms to more explicitly explore commercial opportunities created by R&D activities. In one strategy, a spinoff company is created. This facilitates the investment of outside capital, which can target its investment to a specific project or product. Although this might be viewed favorably by DoD, it creates additional administrative difficulties. There is increased work required to create the new company, including complicated legal and accounting services. Second, once the companies have been distinguished from each other, it has been difficult to get DoD to recognize the new firm and transfer the Phase II contracts accordingly. This was one of the main criticisms of the program from one of the spinoff firms. Even though the principals believed that they were doing what DoD wanted, they did not believe that the administrative procedures were in place to facilitate meeting these objectives logistically. Also, the spinoff is treated as a multiple SBIR recipient and thus cannot take advantage of the most favorable matching requirements.

The Fast Track program is a good idea in that it ensures cash and moves a project along a quicker timetable. In many technology areas, time is an important factor because long delays can have major market consequences. Accordingly, the speed with which programs are evaluated by DoD agencies is a major characteristic that sets them apart from other agencies. Small companies, in particular, may find it less desirable to work with agencies that have longer delays in evaluating proposals or transitioning a project from Phase I to Phase II. Additionally, DoD should recognize that developing commercial products takes different lengths of time, depending upon the technology area in question.

The interviewed firms claim that there are two ways to execute a contract. The first is a firm-fixed-price model, where deliverables are presented to the agency in exchange for incremental payments. The second is a cost-plus-fixed-price model, where a firm is reimbursed for eligible costs in additional to a contract price. This latter method creates a higher scrutiny for audit, and thus is less preferable for smaller firms. However, the choice of how each contract is structured appears to depend on the DoD program officer and is outside the influence of the executing firm. Regardless of the form of the contract, it is a large document that take much of the firm's time and energy to monitor.

CONCLUSIONS

The observations of this paper are based on a limited set of firms participating in the SBIR program. Yet this set appears to identify distinct patterns that may more broadly characterize the program. Firms that receive SBIR awards believe that they benefit greatly from their participation in the program. These benefits are not always expressed in commercial sales, but are related to expanding basic research, responding to government needs, and developing applications for technology.

The Fast Track program helps to promote the commercial success of participating firms in several important respects. The program helps firms move toward commercialization by offering a tool to attract outside investors and a means to bridge the financing gap that some firms experience between Phase I and Phase II of the SBIR program. Unfortunately, the relative newness of the Fast Track program prevents a full assessment of its impact on commercialization at this time. Yet it appears likely that Fast Track firms will be able to demonstrate a significant degree of commercial success.

The Fast Track program appears to work best for firms that are prepared to take a product or an innovation to market. It also appears likely that the Fast Track project may offer benefits to firms that are already well positioned to attract investment as a result of their technology and its position late in the product cycle. In some of these cases, the third-party investor benefits because their capital is leveraged with SBIR funds. This may be a concern if the third-party investment is not contingent on the SBIR award. However, the dynamics of leveraging capital works both ways because the public sector has its investment matched by the third party. Thus, the potential for positive spillovers and meeting government needs may justify the SBIR subsidy and the leveraging of capital that occurs.

Even when firms have not been able to take advantage of the Fast Track program, there has been a recognition of DoD interest in commercialization and firms have wanted to appear to be responsive to this SBIR objective. However, there is a concern on the part of firms that are having difficulty in pursuing commercialization. An overemphasis on commercialization can take away from one of the SBIR program's most beneficial qualities, which is the provision of an accessible revenue stream for firms engaged in basic research and innovative development activities. Balancing the importance of support for these firms and the goals of commercialization is a policy question that needs to be explicitly addressed.

One finding of this research is that there is a role for non-Fast Track projects. Since commercialization is not the only objective of the SBIR program, DoD should continue to support non-Fast Track projects. This is because many SBIR projects focus on basic research and R&D activities that respond to the needs of contracting government agencies. These projects may be successful even if they do not lead to immediate commercial sales. The firm's relationship to the product cycle is relevant to its commercial potential. The appropriateness of Fast Track seems questionable for products or technologies that are distant from actual markets. The technology may need further development or the market may need time to adapt to the product or innovation. The diverse objectives of the SBIR program appear to indicate that DoD should not exclusively pursue Fast Track projects because it would inhibit the involvement of several categories of firms that have successfully participated in the program in the past.

APPENDIX A

Selected SBIR Phase II Projects in the Southwest and Mountain Region

Company	Description of SBIR Phase II Project
AVPRO Inc. P.O. Box 1696 800 West Rock Creek Road, #109 Norman, OK 73070 (405) 360 4848	Low-Cost Curing and Repair Process for Composites
Bolder Technologies 4403 Table Mountain Drive Golden, CO 80403 (303) 215-7200	Development of a 30Ah Thin Metal Film Lead Acid Cell for High-Power Electric or Hybrid Vehicles
Chorum Technologies Inc. (Formerly Macro-Vision Communications, LLC) 1155 E. Collins Blvd., Suite 200 Richardson, TX 75081 (972) 238-1770	Self-Routing Wavelength Switch
Coherent Technologies, Inc. 655 Aspen Ridge Drive Lafayette, CO 80026 (303) 604-2000	Tunable UV LIDAR for Water Vapor Profiling and Ozone Monitoring
Lipitek International, Inc. Texas Research Park 14785 Omicron Dr San Antonio, TX 78245 (210) 677-6001	Innovative Design and Synthesis of Antiparasitic Agents
Mission Research Corporation 1720 Randolph Road, SE Albuquerque, NM 87106-4245 (505) 768-7600	A Novel Field-Programmable Gate Array for Space Applications
Picolight Incorporated 4665 Natilus Boulder, CO (303) 530-3189	Long-Wavelength Oxide Vertical-Cavity Surface Emitting Lasers
Radiant Research, Inc. 3006 Longhorn Blvd., Suite 105 Austin, TX 78758-7631 (512) 339-0500	Planarized Optical Clock Signal Distribution on Si

Appendix A—continued

Company	Description of SBIR Phase II Project
Systems and Processes Engineering Corporation 101 West 6th Street Austin, Texas, 78701 (512) 479-7732	Growth of SiC Using Seeded Supersonic Beams
Technology and Resource Assessment Corporation (TRAC) 3800 Arapahoe Avenue, Suite 225 Boulder, CO 80303 (303) 443-3700	Array Beam Imaging for High Resolution Stand-Off Mine Detection
TPL, Inc. 3921 Academy Parkway North NE Albuquerque, NM 89109-4416 (505) 344-6744	Ultra-High Dielectric Constant Dielectric Materials; Inorganic Conformal Coatings for SiC Packaging

APPENDIX B

SBIR Program Case Study Survey

I. Description of the Firm

1) History of firm: founding date, founder, founder's background.

2) Product or service: description of the firm's main good or service.

3) Technology: description of technology produced.

4) Projected market size and major competitors.

5) What is the firm's competitive advantage?

6) Reasons for location of the firm (e.g., labor pool, nearby research facilities or science parks, markets, access to capital).

II. SBIR Project Information

7) SBIR technology: description of technology for which SBIR award was given and technology's origin.

8) Number of SBIR awards won.

9) What is the relationship to the agency mission?

III. Impact of SBIR

A. SBIR & Firm Strategy

10) Role of SBIR award in company strategy: How important was the award to the firm's current position? What alternative sources of funding were considered and/or enhanced by the SBIR award? Do they anticipate applying for further SBIR awards? If the firm received earlier SBIR awards, how did these affect the current award?

11) Was the SBIR and/or Fast Track award used as a marketing tool? That is, did the firm use the presence of the award as leverage to attract additional outside capital? Did the firm use the Fast Track policy—i.e., the opportu-

nity for outside investors to obtain up to a 4:1 match on their investment—as leverage to attract outside capital?

B. Commercialization

12) Does a strategy exist for commercializing the product? For example, strategic alliances for production, such as a joint venture or licensing. If possible, please disclose names of firms.

13) Has the SBIR-funded technology generated any patents? Is it expected to generate patents? Have any scientific papers resulted?

13a) How was the commercialization strategy affected by participation in SBIR? By participation in Fast Track?

14) Has the firm sold a product resulting from the SBIR project? If not, when does it anticipate selling its first product? Does the firm have specific customers interested in the product? If so, who? Does the firm have anyone on its board of directors or in a management position that has built a successful company before and taken it public?

C. Financing and external partners

15) Does the SBIR awardee have external financing? If so, how much and will it identify the partner? Did the SBIR award play a role in the external partner's decision to provide funding. Where is the partner located? How was the relationship with the partner developed? Did the SBIR award play a role in securing other external investment? Does the SBIR awardee have internal financing? If so, how much?

D. Other Impacts

16) Has participation in the SBIR program generated other types of impacts, relationships, or opportunities?

IV. SBIR Program Administration

17) Did the requirements of SBIR in general and Fast Track in particular prove helpful or onerous in terms of delay or impact on external funding? For standard SBIR awards, did the delay between Phase I and Phase II increase time to market for the firm's product? Did the Fast Track award improve time to market for the firm or allow the firm to maintain continuity of effort?

18) How could the SBIR and especially the Fast Track program be improved?

19) How did the firm become aware of the SBIR program? Of the Fast Track element of SBIR?

20) Did the firm experience difficulties in preparing the SBIR application?

V. Perspectives of Third Party Investors (if relevant)

21) How did the third-party investor become aware of the project being funded by SBIR?

22) What prospects does the SBIR project hold for the third-party investor's organization?

23) Did the SBIR award influence the third-party investor's decision to invest in the project? Did the Fast Track policy (i.e., significantly higher chance of Phase II award for projects attracting outside investors) influence the decision by (a) enabling the third-party investor to leverage its investment in the company, (b) "certifying" the promise of the technology (through a government review of and implicit approval of the technology)?

VI. Cross-Cutting Research Questions:

24) Does the firm think that the DoD policy of giving a higher chance of Phase II award to companies that attract outside investors (per Fast Track) is (a) a useful way of focusing the SBIR program on companies with strong commercialization capabilities; (b) good public policy?) What factors influence a firm's decision to participate in Fast Track?

25) What factors inhibit a firm's participation in Fast Track?

26) What benefits do firms expect to gain from a Fast Track award not available through the regular Phase II process?

27) Does the Fast Track award affect the performance of firms? How (e.g., in terms of research capabilities or commercialization prospects)?

28) What specific effects does the presence of a third-party investor have on performance?

29) Are differential impacts of Fast Track observed by region, DoD funding agency, or firm characteristics?

30) What do participating firms see as strengths and weaknesses of the Fast Track program?

Does the Small Business Innovation Research Program Foster Entrepreneurial Behavior? Evidence from Indiana

David B. Audretsch, Juergen Weigand, and Claudia Weigand
Indiana University

EXECUTIVE SUMMARY

Purpose

This paper identifies the degree to which (1) recipients of Small Business Innovation Research (SBIR) awards have altered their career choices as a result of SBIR, particularly with respect to commercialization in the form of a new firm; and (2) their behavior has "spilled over" by inducing other colleagues to commercialize their knowledge in the form of starting a new firm. This identification provides insight to answering the question, "To what degree has the SBIR contributed to changing the behavior of knowledge workers and in creating a science-based entrepreneurial economy?"

The relevant information for this study came from a series of 12 case studies and the responses to a survey from a broader sample of firms. In particular, the case studies and survey helped to determine

- the career background of the firm founder,
- what led to the decision to commercialize knowledge,
- why commercialization took the form of a new firm,
- what would have happened in the absence of the SBIR program,
- specific ways in which the founder's career path has been altered by SBIR, and
- specific people who have been influenced by his/her experience and who have commercialized knowledge via a start-up firm.

Conclusions

The evidence provided here is of a preliminary nature, partially because of the smallness of the sample, which consisted of 12 case studies and 20 firms' responses to a survey instrument, but also because of the particular context within which SBIR operates in a state such as Indiana. The fact that a viable cluster of knowledge-based small firms has been lacking in Indiana has implications for the commercialization possibilities for scientists and engineers. Perhaps a more subtle impact is that it limits knowledge about commercialization possibilities and the existence of ancillary services and institutions facilitating commercialization.

The results suggest that the SBIR has influenced the career paths of scientists and engineers by facilitating the start-up of new firms. Furthermore, there are indications that the experience of scientists and engineers in commercialization via a small business has an externality by spilling over to influence the career trajectories of colleagues.

Both the survey and the case studies provide the following consistent evidence:

1. A significant number of the firms would not have been started in the absence of SBIR.
2. A significant number of the scientists and engineers would not have become involved in the commercialization process in the absence of SBIR.
3. A significant number of other firms are started because of the demonstration effect produced by the efforts of scientists to commercialize knowledge.
4. As a result of the demonstration effect by SBIR-funded commercialization, a number of other scientists alter their careers to include commercialization efforts.
5. Technology-based entrepreneurs start firms because they have ideas that they think are potentially valuable; they do not start firms and then search for useful ideas or products. This is reflected by the fact that not a single respondent on either the survey or from the case studies suggested that he or she would have tried to start the firm with a different idea in the absence of SBIR funding. However, once the firm exists, one-quarter of the respondents and one-sixth of the case studies indicated that they would have tried to continue the firm with a different idea in the absence of SBIR funding. These different results may suggest that the SBIR has a greater impact on potential entrepreneurs than on existing small firms in commercializing ideas that otherwise would not find their way into the market.

Recommendation

A large-scale study spanning a broad spectrum of SBIR awardees should be undertaken to confirm these preliminary findings. Incorporating greater variation in either the funding agency or the underlying science could help to identify how the impact of SBIR on influencing the entrepreneurial behavior of scientists differs across scientific fields and funding agencies.

INTRODUCTION

The magnitude of the Small Business Innovation Research (SBIR) program of around $1.2 billion annually has attracted the attention of both policy makers and scholars. However, only recently have studies begun to identify the impact of the SBIR program (U.S. General Accounting Office, 1998). The best of these studies has focused on the impact of the SBIR program in terms of the likelihood of survival and growth rates of firms selected for SBIR funding (Lerner, 1999). Two bothersome questions have been raised about measuring the success of SBIR in terms of growth and survival. The first involves selection bias in that the SBIR program may give awards to firms that already have the characteristics needed for a higher growth rate and likelihood of survival. The second argues that a number of SBIR recipients would have followed the same commercialization process even in the absence of an SBIR award.

Although enhanced firm growth and survival are important aspects of SBIR, they do not capture all of the benefits of the program. A very different way in which SBIR may benefit the economy is by changing the behavior of knowledge workers. An important finding by Audretsch and Stephan (1996), who trace the career paths of scientists starting biotechnology firms, is that the scientist deviates from an academic career path or a career with a large pharmaceutical corporation to start a new firm in a new industry. How to induce knowledge workers in general and scientists and engineers in particular to change their behavior to take advantage of commercialization opportunities is a focal point of the policy debate in European countries such as Germany and France.[1] It may be that policies such as SBIR have contributed more to the creation of an entrepreneurial economy, where the costs of commercializing knowledge are reduced, than is captured in studies simply focusing on the links between SBIR, survival, and growth.

There are at least two important ways that the behavior of knowledge workers may be influenced by SBIR. The first way is that it may induce some scientists and engineers, who otherwise never would have engaged in the commercialization process, to commercialize their knowledge by starting a firm. The second involves the demonstration effect when the examples of successful science-based entrepreneurs who received SBIR support influence the behavior of their colleagues by inducing subsequent commercialization. Although a large literature exists on the importance of learning, this literature typically focuses on firms' learning. This second aspect, by contrast, focuses on individual knowledge-workers learning by observing the choices and outcomes of their colleagues. For example, Audretsch and Stephan (1996) found that the clustering of scientists working with biotechnology firms in a particular location is attributable to the demonstration effect on their colleagues of scientists involved with commercialization. Thus, rather than focusing on the diffusion of particular processes, it focuses on the diffusion of behavior (Audretsch and Feldman, 1996). The third impact of SBIR may be to alter the type of science being undertaken.

---

[1]As the former Chair of the President's Council of Economic Advisors, Laura Tyson, points out, this is also a critical debate in Eastern and Central Europe (Tyson et al., 1994).

This project focuses on the benefits of SBIR. In particular, we propose identifying the degree to which (1) SBIR recipients have altered their career choices as a result of SBIR, particularly with respect to commercialization in the form of a new firm; and (2) their behavior has "spilled over" by inducing other colleagues to commercialize their knowledge by way of starting a new firm. This will enable us to shed light on the question, "To what degree has the SBIR contributed to changing the behavior of knowledge workers and in creating a science-based entrepreneurial economy?" To shed light on these issues, we undertook a series of 12 case studies and surveyed a broader sample of firms to determine

- the career background of the firm founder,
- what led to the decision to commercialize knowledge,
- why commercialization took the form of a new firm,
- what would have happened in the absence of the SBIR program,
- specific ways in which the founder's career path has been altered by SBIR, and
- specific people who have been influenced by his/her experience and who have commercialized knowledge via a start-up firm.

From the responses to these questions, we have been able to shed some light on the degree to which the SBIR program has altered the behavior of knowledge workers, especially scientists and engineers, in terms of commercialization. We also have been able to identify the role that the SBIR program plays in influencing the career trajectories of some scientists and engineers, in particular, the role it plays in the commercialization process.

The impact of SBIR on fostering science-based entrepreneurial behavior is likely to vary across industries, sciences, and technologies. Thus, we would not restrict ourselves to any particular industry or science, but rather include a broad spectrum to ascertain differences in the contribution of SBIR in fostering science-based entrepreneurial behavior.

The impact of programs such as SBIR on sparking entrepreneurial activity may be of particular interest in states such as Indiana. Systematic evidence suggests that Indiana does not generate much innovative activity, measured in terms of research and development (R&D), patented innovations, or the introduction of new product innovations (Audretsch and Feldman, 1996).

Scientists are mainly employed by universities and large corporations in Indiana. There has not been significant start-up activity of technology-based firms. This may imply that scientists and other knowledge workers do not have low-cost access to information indicating (1) that commercialization through starting a new firm is a feasible and profitable action and (2) how a new firm can be started and maintained. Thus, the SBIR program may actually have a greater impact in states such as Indiana, where greater market imperfections exist in terms of linking entrepreneurial activities to scientific talent. In addition, small firms tend to face brighter prospects in regions where a rich cluster of small firms already exists. The lack of an existing cluster of knowledge-based small firms may deter the start-up of technology-based firms. The SBIR program can play an important role in overcoming this barrier to start-up.

The second section of this paper explains how the 12 case studies were selected. The third section explains how a survey instrument was devised and sent out to all SBIR firms in Indiana. The fourth section includes a description of the 12 firms included in the case studies. The fifth section describes the technologies involved in the SBIR projects. The sixth section describes the impact of the SBIR program on the firms. The seventh section identifies suggestions by the firms for improving SBIR program administration. The eighth section focuses on the perspectives of third-party investors, and the ninth section on crosscutting research questions. A summary and conclusions are provided in the final section. The principal finding from the case studies and the survey is that there is evidence suggesting that in some cases the SBIR program has altered the career trajectory of scientists. In the absence of the SBIR program, at least some of the scientists and engineers contacted in this project would not have become involved in the commercialization process. These preliminary results based a limited sample size and context—Indiana—suggest that a larger-scale project should be undertaken to identify the impact that the SBIR has on changing entrepreneurial behavior.

SELECTION OF THE CASE STUDIES

The firms interviewed were selected on the basis of (1) inclusion in the 1999 *Directory of Small Research and Development Companies of Indiana,* published by the Indiana University Industrial Research Liaison Program; and (2) the recommendation of Ben Dulaski and Sid Johnson of the Indiana University Industrial Research Liaison Program. The Industrial Research Liaison Program was founded in 1986 to provide R&D assistance as well as information services to Indiana's business and industrial communities. The Industrial Liaison Office is responsible for implementing the SBIR program in Indiana. To accomplish this, the office serves as a link between specific SBIR funding opportunities and particular research interests and capabilities of scientists, engineers, and small firms in Indiana. The staff of the office spends considerable time in the field, getting to know the interests and capabilities of individual scientists, and then tries to recommend specific projects to potential scientists working for universities, firms, and the government. In addition, the office publishes a monthly newsletter, *R&D NOTES: SBIR and STTR*, in which it announces the solicitation dates of SBIR, as well as upcoming conferences and particular funding opportunities associated with SBIR. Recent SBIR awards in Indiana are also listed in the newsletter. In addition, the office sponsors a number of seminars where the purpose of SBIR is explained and potential entrepreneurs have an opportunity to meet scientists and engineers who have successfully obtained SBIR awards.

Mr. Dulaski and Mr. Johnson recommended several firms that they felt would be articulate in discussing their experience with SBIR. Although the *Directory of Small Research and Development Companies of Indiana* provides a comprehensive list of all firms receiving SBIR awards in Indiana, the recommendations of Mr. Dulaski and Mr. Johnson result in a more biased selection of firms. However, since the goal of this study is to uncover some of the different types of impacts that the SBIR has, we felt that the value of having articulate and willing participants more than outweighed the selection bias. However, an important qualification to be emphasized is that the firms selected were not statistically representative.

Table 1 provides a list of the firms that were interviewed. The interviews typically lasted between 45 minutes and 3 hours. In addition, each of these firms was sent a written questionnaire. Most of the firms were contacted on several occasions to ensure the consistency of information and to correct any misunderstandings that might have arisen in the interview process.

THE SURVEY INSTRUMENT

The survey listed in the Appendix was mailed to what were identified by the Industrial Liaison Office as SBIR firms. The survey was evaluated and approved by the Bloomington Campus Committee for the Protection of Human Subjects. After its first evaluation of the survey, a number of changes were recommended and the survey was modified. The survey was then mailed to all 84 SBIR firms identified by the Industrial Liaison Office. Of these 84 questionnaires, 20 were returned with answers. An additional 24 came back because the firm no longer existed.

TABLE 1 Firm Characteristics—Case Studies

Firm	Date Founded	Funding Agency[a]	Size (employees)
STAR Enterprises, Inc.	1985	NASA	5
Genetic Models, Inc.	1991	NIH	25
Ash Medical Systems, Inc.	1980	NIH	23
Batch Processing Technologies, Inc.	1983	NSF	2
Endotech, Inc.	1986	NIH	7
Agdia, Inc.	1981	USDA	23
Advanced Process Combinatorics	1993	NSF	10
Medical Decision Modeling, Inc.	1994	NIH	3
Beard Industries, Inc.	1965	DOE	50
Terronics Development Corporation	1985	USDA	16
Focus Surgery, Inc.	1996	NIH/NCI	10
Hard Coating, Inc.	1998	DOE	2

[a]DOE = U.S. Department of Energy, NASA = National Aeronautics and Space Administration, NCI = National Cancer Institute, NIH = National Insitutes of Health, NSF = National Science Foundation, USDA = U.S. Department of Agriculture.

DESCRIPTION OF THE FIRMS

Table 1 identifies the age, funding agency, and size of the 12 case study firms. Table 2 classifies the origins of the founder(s) and the other participants who work closely with the firms. Note that because of multiple founders in some of the firms the total number of founders exceeds 12. We did not actually interview all of the founders, and so, there are some founders associated with these firms that are not included in Table 2.

The most important point to be emphasized in Table 2 is that the most important career trajectory of the founders is from the university. However, a number of founders also came from large corporations. It is also important to note that only a few founders actually had experience with other small firms. A stylized fact in the literature on small business economics is that most founders of new firms already have experience in a small business. This could indicate that the SBIR may be a mechanism to compensate for the lack of experience with a small firm.

STAR Enterprises, Inc.

STAR Enterprises is located in Bloomington, Indiana. Jeff Alberts founded STAR on the basis of an SBIR award. The company currently has five employees and a Phase III SBIR Award to build hardware for animals to live in the Space Station. The SBIR award is from the National Aeronautics and Space Administration (NASA).

The origins of the company date back to 1985, when Jeff Alberts, who was a professor at Indiana University, was the first investigator in the United States to work with the Soviet scientists in a Soviet-launched spaceflight. His research involved experiments with live animals. Jeff wrote proposals for two years to obtain access to the Soviet spaceflights. He subsequently became the first U.S. scientist to work with the Soviets in doing his own experiment on board a Soviet spaceflight.

TABLE 2 Classification of the Firms from Case Studies

Founders	Number of Respondents	Other Participants	Number of Respondents?
From university	10	From university	3
From corporation	1	From corporation	0
From university and corporation	1	From university and corporation	3
From large corporation	7	From large corporation	2
Previous small business	2	From small business	1
Government	0	Government	2

Genetic Models, Inc.

Genetic Models was founded in 1991 by five individuals. It is located in Indianapolis. The firm employs 25 people and provides high-quality jobs that include medical, dental, disability, and retirement benefits.

In the 1970s, Joe Pesek began working for Union Carbide as a sales and marketing representative. After working for about 11 years, Pesek was asked by the president of Union Carbide to manage the Lindy Division, a subsidiary that specialized in producing special carbon linings for pipelines. In about four years, Pesek helped the division to grow from about $4 million to $12 million annually. The Lindy Division was a national company that licensed its technology abroad.

Around 1984, one of Joe Pesek's bosses asked him to help out a small cryogenics business in Indianapolis. The cryobusiness developed two products: (1) large tanks to house liquids and gases under pressure; and (2) technologies that used molecular sieves, such as oxygen tanks, oxygen masks, and cryogenic freezers. Pesek's goal was to turn the cryobusiness around and make it profitable.

In 1985, Union Carbide needed to raise money to help pay for the methyl isocyanate disaster at its plant in Bhopal, India. Joe Pesek was instructed to sell parts of the business, and he viewed it as a good divestment opportunity to purchase one of the product lines. Unfortunately, Pesek lost his bid for the molecular sieve technology division and thus proceeded to look for other investment opportunities.

While Joe Pesek was looking for other small business deals, he took a consulting job with American Monitor, which later became AM Diagnostics. AM Diagnostics manufactured clinical chemical analyzers and reagents. Pesek eventually became the CEO of the company. By updating the product, Pesek was able to achieve a $5 million sales backlog, but with no working capital

Finally, in 1990, after working four years with AM Diagnostics, Joe Pesek met Professor Dick Peterson of Indiana University. Dr. Peterson had developed a diabetic rat model, but needed small breeding rooms to develop genetic reagents from mice antibodies. Additionally, Dr. Peterson had a list of people that had used or expressed interest in the diabetic model and he needed to find a way to market his technology outside Indiana University. As the demand for the genetic model grew, Dr. Peterson could no longer supply all of the companies with his genetic model. Pesek recognized the market potential for the genetic model and performed some market research to find a building with a very low lease rate. Hence, Joe Pesek, Dr. Peterson, and three other investors founded Genetic Models on a single rat model.

In 1991, Joe Pesek and the other founders constructed their first laboratories for Genetic Models. Although the others helped on the weekends with the carpentry, Pesek was the only full-time employee. The first building was about 7,200 ft^2 (seven years later the company would have a second building that is about 18,000 ft^2). Genetic Models became incorporated in November 1991 and bought its first breeding stock in January 1992. One of the company's first customers was Eli Lilly, who bought the animals for use in drug testing. The company has experienced 50 percent growth every year since its inception.

The core product of Genetic Models consists of inbred animal models (mostly rats and mice). The animals are bred with specific diseases, such as obese male rats with diabetes. These animals are then sold to research laboratories seeking cures for particular diseases.

Genetic Models combines the science of the model (e.g., physiology) with the application of the model for the customer. So far, the combination of science and testing the model for customers has been favorable. Genetic Models has found that crossbreeding two models produces a very interesting end result of gene combinations. One of the models they use is a model for congestive heart failure developed by researchers at Ohio State University. However, the main model used by Genetic Models is the Zucker Diabetic Fatty (ZDF) model developed by Walter Shaw, a former employee of Eli Lilly. While at Eli Lilly, Dr. Shaw conducted diabetic research and isolated the first colony of Zucker strains from male rats. During his research, he saw one fat male rat become spontaneously diabetic.

Building on Dr. Shaw's research, Dr. Peterson and Dr. Julia Clark, also of Indiana University, refined and developed the ZDF model into a reliable and inbred model that ensures that all obese male rats will acquire diabetes between 7 and 12 weeks in their development. The Indiana researchers also learned that they could feed rats certain foods that would lead to hypertension in the rats. Joe Pesek and Dr. Peterson believe that their company produces the "premier advanced aging model" because their rat specimens are bred with many of the same pathologic conditions (high triglyceride and cholesterol levels, heart hypertrophy [enlarged heart]) that occur in older humans in America. Further, the main cause of mortality for these rat models is end-stage kidney disease, not congestive heart failure. Thus, Joe believes that the ZDF model could become the most reliable kidney failure model for pharmaceutical companies that need animal specimens to test their drugs.

Ash Medical Systems, Inc.

Ash Medical Systems was started in 1980 by Steve Ash, who is also a professor at Purdue University. The firm's main product is plasmatherapy. Dr. Ash had the idea for starting the firm while he was doing research at Purdue. He has the only company producing an artificial liver.

Batch Processing Technologies, Inc.

Batch Processing Technologies was established in 1983. The firm currently employs two people and grosses about $150,000 annually in sales revenues (this estimate of sales revenues has not really changed since 1997). The focus of Batch

Processing Technologies is to develop simulation software for chemical engineering firms.

In 1981, Girish Joglekar earned his Ph.D. in computer science from Syracuse University. After receiving his Ph.D., Dr. Joglekar began conducting his post-doctoral research at Purdue University. While at Purdue, Girish and two other professors came up with the ideas for starting Batch Processing Technologies.

Endotech, Inc.

Endotech was founded in 1986 and specialized in developing human cell culture technologies.

After earning his B.A. in Biology from Indiana University in 1971, Anthony Hubbard worked for medical research firms in Indianapolis, at both the staff and executive levels. In 1984, Anthony earned his M.B.A. from Indiana University and continued working for a medical research firm in Indianapolis. During this time, he met a physician who had invented a human cell culture technology. Recognizing the market potential of the physician's technology, Anthony persuaded the primary inventor to fund part of the start-up costs of producing and marketing the cell culture technology. Thus, in 1986, Anthony Hubbard founded Endotech, Inc., on the basis of a human cell culture technology.

By 1992, Endotech had gone out of business because of a lack of consistent revenues. From his experience with Endotech, Anthony learned that the SBIR award was a viable source of "additional" funding, but not a reliable "primary" source of funding.

Agdia, Inc.

Agdia develops nucleic acid hybridization techniques for commercial use. The privately held company currently employs 23 people. Last year, sales revenues for the company increased by 20 percent compared to the previous year.

In 1959, R. Henn received his Ph.D. in physical chemistry from Iowa State University. Before founding Agdia in 1981, Dr. Henn conducted biochemical research for Miles Labs in Ames County, Iowa, and Ortho Diagnostics, a subsidiary of Johnson and Johnson. Through working at a high position in a major corporation, Dr. Henn learned from customers and professional collaborators about some innovative technologies in the immunochemistry field. On the basis of these technologies and his own corporate experience, Dr. Henn decided to pursue his own technical business ideas and found Agdia.

Advanced Process Combinatorics

Established in 1993, Advanced Process Combinatorics commercializes mathematical optimization technologies. Currently, the firm employs 10 people, 3 of whom are the company founders. All three founders have earned their doctoral degrees and were formerly employed in one of the following careers: a professor at Purdue University, a research scientist at DuPont, and a postdoc at Purdue University. Last year, Advanced Process Combinatorics experienced a 70 percent increase in sales revenues compared to the previous year.

The ideas for starting Advanced Process Combinatorics came from industrial and university cooperative research programs. Timing and business climate for mathematical optimization technologies also influenced the founders' decision to start the firm.

Medical Decision Modeling, Inc.

Established in 1994, Medical Decision Modeling disseminates information about the outcomes of medical studies through the Internet. The firm is a Subchapter S corporation that currently employs three people; in the previous year, the firm employed only one person. Harry Smolen, the 32-year-old founder of Medical Decision Modeling, earned his B.S. in electrical engineering from the University of Southern California in 1989, and his M.S. in industrial engineering from Purdue University in 1994.

While working on his Master's thesis at the Indiana University School of Medicine, Harry Smolen learned bout the SBIR program from one of the leaders of his research group. This same person also influenced Smolen to start his own company and commercialize the outcomes of medical studies using the Internet.

Beard Industries, Inc.

Beard Industries is a Subchapter S corporation that was established in 1965. Currently, the firm employs 50 people and sells about $15 million annually. Last year, sales revenues decreased by 20 percent compared to the previous year.

Before founding Beard Industries, William Beard earned his B.S. in agriculture from Purdue University in 1949, and worked for farm partnerships in north central Indiana. The focus of Mr. Beard's firm is to manufacture computer controls of drain dryers.

Terronics Development Corporation

Established in 1985, Terronics Development Corporation manufactures electrostatics technologies. Currently, the firm employs 16 people and sells about $1.2 million annually. Last year, sales revenues decreased by 50 percent compared to the previous year.

Before founding Terronics Development Corporation, Eduardo Escallon earned his B.S.M.E. from Carnegie Tech in 1965, and worked from 1977 to 1984 as a principal engineer for Ball Corporation.

Focus Surgery, Inc.

Established in 1996, Focus Surgery develops ultrasound technologies for medical applications. Currently, the corporation employs 10 people and sells about $1 million annually. Last year, sales revenues increased by 20 percent compared to the previous year.

Before founding Focus Surgery, Naren Sanghvi earned his M.S.E.E. from Rose-Hulman and his Ph.D. from Purdue. While working as the director of the Indianapolis Center for Advanced Research at Indiana University-Purdue University Indianapolis from 1992 to 1997, Dr. Sanghvi also taught at Indiana University as an associate professor. During that time, Dr. Sanghvi met his business partners and, in 1996, they founded Focus Surgery.

Hard Coating, Inc.

Established in January 1998, Hard Coating currently employs two people and sells $60,000 annually. The Subchapter S corporation specializes in physical vapor deposition (PVD) of superhard nanolayer composite coatings.

Prior to founding Hard Coating, Robert Oglesby earned his B.S. in chemical engineering from Tennessee Tech University in 1962. From 1972 to 1990, Mr. Oglesby managed the product finishing division of Faultless Caster, earning an annual salary of $35,000.

Table 3 classifies the origins of the founders of the firms responding to the survey as well as the other participants who work closely with the firms.

The results from the survey are generally consistent with those from the case studies. Most of the founders come from universities, followed by large corporations. Only a handful of founders had experience in small businesses.

PROJECT INFORMATION

Table 4 presents the technology or basic science on which each firm's SBIR award is based.

TABLE 3 History of the Firms from the Survey

Founders	Number of Respondents	Other Participants	Number of Respondents
From university	13	From university	5
From corporation	2	From corporation	0
From university and corporation	2	From university and corporation	6
From large corporation	10	From large corporation	4
Previous small business	2	From small business	2
Government	2	Government	3

TABLE 4 SBIR Technologies—Case Studies

Firm	Technology
STAR Enterprises, Inc.	Animal experiment hardware for spaceflight
Genetic Models, Inc.	Inbred animals with specific diseases
Ash Medical Systems, Inc.	Plasmatherapy
Batch Processing Technologies, Inc.	Simulation technology for chemical engineering
Endotech, Inc.	Human cell culture
Agdia, Inc.	Nucleic acid hybridization
Advanced Process Combinatorics	Mathematical optimization technology
Medical Decision Modeling, Inc.	Outcomes of medical studies via the Internet
Beard Industries, Inc.	Computer controls of drain dryers
Terronics Development Corporation	Electrostatics technologies
Focus Surgery, Inc.	Ultrasound technologies
Hard Coating, Inc.	Physical vapor deposition (PVD) of superhard nanolayer composite coatings

STAR Enterprises, Inc.

When NASA started doing similar types of experiments, Jeff Alberts became a consultant to NASA. This led to doing experiments on the U.S. shuttle as a principal investigator for NASA, through the university. His contacts at NASA suggested that his work overlapped with that of a small firm, SHOT, that had already been awarded several SBIRs. At a conference, Jeff had breakfast with the founders of SHOT. Together they determined that SHOT needed Jeff to apply for a Phase III SBIR award. SHOT had already successfully obtained several Phase I and Phase II SBIR awards.

In July 1998, Jeff Alberts signed a contract for STAR to work with SHOT in a Phase III award. The first part of the award involves $25 million to build hardware for animal storage and experiments aboard the Space Shuttle. The second phase involves $40 million. STAR is subcontracting some of the production to SHOT.

Jeff Alberts is under the impression that STAR has had more SBIR awards than any other firm in the state of Indiana. It has received a total of four Phase I SBIR awards and four Phase II SBIR awards.

Genetic Models, Inc.

As one of five cofounders of Genetic Models, Joe Pesek's original goal was to receive the SBIR award to provide money to produce an economic and feasible product. Their Phase I proposal requested funding for one genetic model in the first year. The authors of the proposal emphasized the great value in developing a diabetic female rat model. In 1992, Genetic Models received $75,000 to conduct a feasibility study for one year. NIH was the funding agency.

In 1993, Genetic Models won a Phase II award for $500,000 over the next two years. By this time, the company had developed a successful rat model. The company also had learned that a special rat diet was causing diabetic conditions in their specimens. However, the company still has not been able to determine the exact cause of diabetes in the diet.

Since they received their last SBIR award in 1993, Genetic Models has applied for more SBIR awards. The company's focus has been on awards that will fund new "inbred hybrid" models. Unfortunately, the latest SBIR proposals were rejected in Phase I.

Genetic Models has received one SBIR Phase I award and one SBIR Phase II award. In 1992, the firm received a Phase I Award from NIH for $75,000. Although the firm got spectacular results, the application for a Phase II SBIR was denied. When it was resubmitted, it was approved for $500,000. The firm plans on submitting future SBIR proposals to develop hybrid models.

In 1990, Dr. Peterson contacted Joe Pesek on the recommendation of the Business Modernization and Technology (BMT) program in Indiana. Peterson was and continues to be a professor in the Anatomy Department of the Medical School at Indiana University-Purdue University Indianapolis. After he was put in charge of the dog laboratory, he became involved in diabetes research. The technicians at the laboratory thought that there was the potential for developing a diabetic model, where animals were inbred to produce the condition. As word circulated that the laboratory was producing such animals, demand started to grow. The laboratory could no longer keep up with requests. The university turned down requests to expand the operations because it was "not in business to sell animals." Peterson perceived a potential commercial opportunity and was directed by the BMT to Joe Pesek.

Ash Medical Systems, Inc.

Ash Medical Systems has been awarded 15 SBIR grants. The company was just awarded a Fast Track grant from NIH to develop a diabetic product. The firm has previously won SBIR awards for urology, hepatology, and bioengineering, all from NIH.

Batch Processing Technologies

Shortly after founding Batch Processing Technologies in 1983, Girish Joglekar applied for a Phase I SBIR award to develop simulation software for chemical engineering firms. NSF accepted Joglekar's proposal and awarded him $50,000 in 1984. Unfortunately, the Phase I award was the only SBIR funding granted to Batch Processing Technologies. The company's Phase II proposal was rejected the following year, thus forcing Mr. Joglekar to look for an alternative source of funding.

Endotech, Inc.

In 1989, Anthony Hubbard's proposal for a Phase I SBIR award was accepted and Endotech was granted $50,000 by NIH. The next two years proved to be financially difficult for Hubbard and Endotech because their Phase II proposals were rejected.

Agdia, Inc.

In 1998, USDA granted Agdia a Phase I award for $65,000. Agdia works closely with other firms and institutions such as Animal and Plant Health Inspection Service, the USDA Agricultural Research Service (many locations), Purdue University, Michigan State University, University of Wisconsin, Iowa State, and many more universities around the world.

Advanced Process Combinatorics

Advanced Process Combinatorics has received three Phase I SBIR awards thus far. The first two Phase I awards of $60,000 each were granted in 1996. The third Phase I award also totaled $60,000 and was granted in 1997. All three SBIR awards were sponsored by NSF.

The company works closely with Purdue University in developing mathematical optimization technologies.

Medical Decision Modeling, Inc.

Medical Decision Modeling received a Phase I SBIR award for $90,000 in 1996, and a Phase II SBIR award for $750,000 in 1999. NIH sponsored both SBIR awards.

Harry Smolen, the company founder, used his own savings as the start-up capital for Medical Decision Modeling. Before receiving his SBIR awards, he was rejected twice in 1995 for his Phase I applications and once in 1997 for his Phase II application. In all cases, Smolen did not provide enough detail in the SBIR applications about the technology his company was developing.

While working on his Master's thesis at the Indiana University School of Medicine in 1994, Smolen learned about the SBIR program from one of the leaders of his research group.

Beard Industries, Inc.

Beard Industries received a total of $150,000 for its Phase I and Phase II applications in 1983. The DOE sponsored the SBIR award.

Terronics Development Corporation

Terronics Development Corporation has received three SBIR awards thus far. Two Phase I awards of $50,000 each were granted in 1986 and 1987. One Phase II award of $160,000 was granted in 1989. All three SBIR awards were sponsored by USDA.

Mr. Escallon's ideas to found Terronics in 1984 came from his own observations and understanding of the market. While working at his previous job, Mr. Escallon saw ways to improve electrostatics technologies. He was also aware of the market need or perceived market need for electrostatics. Because 1984 was a very good business year, Mr. Escallon knew he could do something useful technologically. Hence, Terronics Development Corporation was born.

Focus Surgery, Inc.

Focus Surgery recently received two Phase I SBIR awards for its commercialization of high-intensity ultrasound for prostate treatment. The first Phase I award of $95,340 was granted by NIH in 1998. The second Phase I award, of $99,331, was granted by NCI in 1999.

Dr. Sanghvi gained his ideas while working at his previous jobs, and working with other universities and medical schools. The main influences on Dr. Sanghvi for startng Focus Surgery were the basic science and technology of ultrasound, the market, financial support from his partners, and his own belief in himself.

Hard Coating, Inc.

Hard Coating received a Phase I SBIR award of $75,000 in June 1998 for its commercialization of PVD of superhard nanolayer composite coatings. The Phase I award was granted by the DOE. Hard Coating is submitting a Phase II application for $750,000 to DOE in April 1999. If the firm receives the Phase II grant, it expects to try for Phase III matching funds of $2 million from DOE. Depending on the status of the SBIR awards, Mr. Oglesby will use either his company's bank line of credit or cash flow from other corporations as needed.

As Mr. Oglesby's interest in high-tech coatings grew, he learned about the PVD process through contact with a doctoral student at Northwestern University and a doctoral student in Canada. By January 1998, Mr. Oglesby had enough of his own capital from his existing firm, along with his personal savings, to finance the start-up costs of Hard Coating.

IMPACT OF THE SBIR PROGRAM

Table 5 provides a perspective of the influence of the SBIR award on the entrepreneurial behavior of the 12 firms that participated in the case studies. Specifically, the table shows the firms' responses (yes or no) to whether the SBIR

TABLE 5 Influence of the SBIR Award on Entrepreneurial Behavior—Case Studies

Influence of SBIR Awards	Number of Respondents	
	Yes	No
Did the SBIR award influence the decision to start the firm or continue with the firm?	6	6
If the firm had not been awarded the SBIR, the company founder(s):		
Would not have started the firm	2	10
Would not have continued the firm	5	7
Would have started the firm with money from an alternative source	2	10
Would have continued the firm with money from an alternative source	6	6
Would have commercialized the idea through an existing firm	4	8
Would have abandoned the idea	2	10
Would have tried to start the firm with a different idea	0	12
Would have tried to continue the firm with a different idea	2	10
Has your SBIR experience with starting a firm influenced the activities of any of your colleagues?	4	8

award influenced the start-up or continuation of their firms, what the firms would have done if they had not received the SBIR awards, and whether the firms' SBIR experiences had influenced any of their colleagues.

In one-half of the cases, the SBIR award influenced the decision to start the firm. In the absence of the SBIR, over one-half of the firms would not have been started or continued. Only two of the firms would have been started using finance from an alternative source. An additional four would have pursued commercialization through an existing firm. In two of the cases, the idea would have been abandoned entirely.

In addition, one-third of the firms were able to name at least one instance of someone else who had started a firm or commercialized his or her knowledge as a result of the example of the SBIR firm. Thus, there is at least some evidence to suggest that the entrepreneurial behavior exhibited by the SBIR firm has an externality in that the career paths of other scientists are influenced toward commercialization.

STAR Enterprises, Inc.

Jeff Alberts, the founder of STAR Enterprises, emphasizes that he would not have been involved in commercialization at all had it not been for the SBIR program. The SBIR award made it possible for him to develop and then build products for the space shuttle under NASA. The SBIR then brought STAR and SHOT together to build products for the Space Station. Jeff thinks that, in the absence of the SBIR program, all of his energy and efforts would have been allocated toward basic research in the university laboratory and his administrative duties as Director of the Research and Development Office at Indiana University. He never would have considered commercialization had it not been for the encouragement of people at NASA first to develop a prototype and later actually produce products for the space shuttles and space stations. In responding to NASA's encouragement to develop the prototype and later produce viable products, Alberts started STAR.

Mr. Alberts also sees a large potential commercial market coming from his SBIR awards. This involves his research on controlling odors from the mice used in space experiments. Because the astronauts do not like the odor of mice urine, and NASA is sensitive to the well-being of the astronauts, Alberts has been awarded SBIR grants to develop methods to reduce the odor. STAR has funded a university lab to identify the determinants of mouse urine odor. Once the determinants of the odor have been identified, STAR will undertake research about how to mitigate the odor. Alberts has already had numerous inquiries from university labs, mouse facilities, hospitals, and other institutions housing mice, rats, and other animals about purchasing products to mitigate the odor.

Genetic Models, Inc.

Both Joe Pesek and Dr. Peterson, cofounders of Genetic Models, believe that, without the SBIR award, their company probably would not exist today. Both agree that they surely would not have the innovative reputation without the funding from the SBIR award. By having received the SBIR award, Genetic Models has been able to

- fully develop a female diabetic model,
- develop the technology and technical proficiency,
- acquire equipment for Phase II contract research,
- hire and train new employees,
- develop additional data to be able to examine additional female models (future markets), and
- develop a Web site that aids researchers in selecting a genetic model.

Joe Pesek emphasized the importance of Genetic Models by saying, "You won't find a drug company to support research on raising animals." Genetic Models had no credit with which to obtain a loan in the second year. Could they have found a venture capitalist? "Probably not," according to Joe Pesek, mostly because of conditions in financial markets; at the time Genetic Models considered an initial public offering (IPO), it was simply a difficult time to raise funds through an IPO. Pesek sees his company as a profitable niche market that will not gross more than $10 million to $20 million. The SBIR program helps to carry forth ideas that will benefit humankind.

The benefit of the SBIR program to Joe Pesek is unequivocal: "Without SBIR this company would not exist." The SBIR has made it possible to develop a commercially viable product–the fat male rat model—and is enabling the development of the female model. In addition, the SBIR enabled the company to acquire (lease) equipment that is the mainstay of its current research. With this equipment, the company can expand its model to new applications. The SBIR program has also enabled the firm to "hire people, train people, and create a unique Web site." The Web site goes way beyond simply marketing. It includes the results of specific tests of products. This has led to international sales.

Dr. Peterson says that this product is unique because "drug companies won't support this research. They don't want a niche for a model producing $1 million of revenues."

A spillover benefit of the SBIR award has been the influence of the firm and Peterson on other colleagues at the university. Peterson observes that a number of his colleagues in the anatomy department have increased their involvement in commercialization activities as a result of his success. One particular example involves a firm that has developed a drug delivery mechanism, which allows for a slow release of the drug to the patient over time.

The company sees itself as occupying a strategic niche in bringing new models for the study of disease. The biggest competitors are Harlam ($75 million of revenues) and Charles River. They think that the size of the world market of this niche is about $1.5 billion.

Ash Medial Systems, Inc.

The biggest problem for Ash Medical Systems has been a high turnover of personnel. The Fast Track award presumably will help that by enabling the firm to maintain high-quality trained personnel between Phases I and II.

Dr. Ash, the company founder, acknowledges that without the SBIR grants, "we would have to have had a lot more dilution." In particular, "The SBIR saved us in the early days."

Batch Processing Technologies, Inc.

According to Girish Joglekar, company founder, the SBIR award was not beneficial for Batch Processing Technologies. The SBIR award neither influenced his decision to start the firm nor continue with the firm. Batch Processing Technologies was able to continue operating with funds from other companies that were potential users of the software. Furthermore, Joglekar believes that his SBIR experience has had no influence on the activities of his colleagues. "Small businesses need to find their own way to continue their firms because the SBIR program is unreliable."

Endotech, Inc.

According to Anthony Hubbard, founder of Endotech, the SBIR award influenced his decision to continue with the firm. If Anthony had not received his Phase I award in 1989, he would have tried to commercialize the idea through an existing private investor. Anthony also believes that his SBIR experience has led about 20 colleagues to express interest in starting their own firms. However, Anthony feels (and has shared with his colleagues) that the SBIR award is an unreliable source of funding. "We felt we had a very strong proposal for Phase II. We had Phase I results with all of our data, but our Phase II proposal was still rejected."

Agdia, Inc.

Although the SBIR award did not influence Dr. Henn to start or continue Agdia, Henn's SBIR experience has led two colleagues to express interest in starting their own firms. Dr. Henn would have tried to commercialize his latest idea through an existing firm if he had not been awarded the Phase I SBIR in 1998.

Advanced Process Combinatorics

According to the company founders, the SBIR award did not influence them to start or continue Advanced Process Combinatorics, nor has their SBIR experience led any colleagues to express interest in starting their own firms. Depending on the availability of funds, the company founders would have done one of the following if they had not been awarded the SBIR: (1) they would not have continued the firm, (2) they would have started the firm with money from an alternative source, or (3) they would have continued the firm with money from an alternative source.

Medical Decision Modeling, Inc.

According to Harry Smolen, the SBIR award did influence his decision to continue Medical Decision Modeling, but the SBIR experience did not lead any colleagues to express interest in starting their own firms. If he had not been awarded the SBIR funds, he would have continued the firm with money from an alternative source.

The company works closely with the Indiana University School of Medicine and the Vanderbilt University School of Medicine.

Beard Industries, Inc.

According to William Beard, the SBIR award helped Beard Industries sustain its engineering staff during difficult times in the early 1980s. However, Beard does not think his SBIR experience has influenced the activities of any of his colleagues. If he had not been awarded the SBIR funds, Beard would not have continued the R&D portion of his engineering department.

Terronics Development Corporation

According to Eduardo Escallon, the SBIR award did not influence his decision to start or continue with Terronics Development Corporation, nor did his SBIR experience influence the activities of any of his colleagues. If he had not been awarded the SBIR funds, Mr. Escallon would have done one or more of the following: (1) he would not have started the firm; (2) he would have continued the firm with money from an alternative source; or (3) he would have tried to continue the firm with a different idea.

Terronics Development Corporation works closely with four Indiana suppliers: CPC in Yorktown, Airmotive in Elwood, Versitile Welding in Frankton, and Neel Tool in Muncie. The firm also works with Purdue University and Ohio State University.

Focus Surgery, Inc.

According to Dr. Naren Sanghvi, the SBIR award did not influence his decision to start or continue with Focus Surgery, nor did his SBIR experience influence the activities of any of his colleagues. If he had not been awarded the SBIR funds, Dr. Sanghvi would have either continued the firm with money from an alternative source or would have commercialized the idea through an existing firm.

Focus Surgery works closely with other firms and research institutions, including Hitachi of Japan, Indiana University School of Medicine, Takai Medical School of Japan, University of Illinois, and University of Washington.

Hard Coating, Inc.

According to Mr. Oglesby, the SBIR award did influence his decision to start and continue Hard Coating. However, he does not think his SBIR experience has influenced the activities of any of his colleagues. If Mr. Oglesby had not been awarded the SBIR Phase I award, he would have either continued the firm with money from an alternative source or would have commercialized the idea through an existing firm.

Mr. Oglesby used cash flow (stocks and a money market account) from his existing firm to finance the initial $250,000 start-up costs of Hard Coating. Then the firm received the SBIR Phase I award of $75,000 in mid-1998. Using its $250,000 collateral, the firm was able to borrow $100,000 from Citizens Bank to establish a line of credit. Hard Coating now has a $2 million line of credit. Mr.

TABLE 6 Influence of the SBIR Award on Entrepreneurial Behavior—Survey

	Number of Respondents	
Influence of SBIR Awards	Yes	No
Did the SBIR award influence the decision to start the firm or continue with the firm?	11	9
If the firm had not been awarded the SBIR, the company founder(s):		
Would not have started the firm	3	17
Would not have continued the firm	5	15
Would have started the firm with money from an alternative source	3	17
Would have continued the firm with money from an alternative source	11	9
Would have commercialized the idea through an existing firm	4	16
Would have abandoned the idea	2	18
Would have tried to start the firm with a different idea	0	20
Would have tried to continue the firm with a different idea	5	15
Has your SBIR experience with starting a firm influenced the activities of any of your colleagues?	5	15

Oglesby hopes to continue his relationship with Citizens Bank on an as-needed basis: If his company receives a Phase II award for $750,000 in 1999, it will try for a Phase III matching award of $2 million from DOE. During this SBIR application process, Hard Coating will use its bank line of credit or cash flow from other corporations as needed.

Hard Coating works closely with the ACT Group at Northwestern University and the U.S. DOE Oak Ridge National Laboratory in Tennessee.

Table 6 probes whether the SBIR award influenced the start-up or continuation of the firm, what the firm would have done if it had not received the SBIR award, and whether the firm's SBIR experience had influenced any of its founder's colleagues.

The influence of the SBIR award on entrepreneurial behavior of the 20 firms responding to the survey is shown in Table 6. The responses to the survey generally mirror those from the case studies. Over one-half of the respondents indicated that the SBIR award influenced their decision to start the firm. In the absence of the SBIR, 40 percent would not have started or continued the firm. Fifteen percent still would have started the firm, using an alternative source of finance; an additional 20 percent would have pursued commercialization of their knowledge through an existing firm. Ten percent of the respondents would have entirely abandoned the idea without SBIR funding.

There is also evidence of spillovers from SBIR recipients in their influence on the entrepreneurial behavior of colleagues. One-quarter of the respondents indicated that the commercialization activities of colleagues have been influenced by learning about the commercialization activities of the SBIR firm.

TABLE 7 Overall Impression of SBIR Program Administration—Case Studies

Impression	Number of Respondents
Favorable	11
Unfavorable	1

SBIR PROGRAM ADMINISTRATION

Table 7 provides the overall impression of the SBIR program by the 12 firms that participated in the case studies.

Table 8 presents the suggestions for improving the SBIR program given by the 12 firms that participated in the case studies.

STAR Enterprises, Inc.

In the early 1990s, there was no budget in NASA for animal experiments. Jeff Alberts, founder of STAR Enterprises, emphasizes what he considers to be a built-in deterrent to commercialization in the SBIR program. At least 50 percent of his time must be spent working for the company under Phases I and II. Because he did not want to terminate his position as professor at the university, he therefore stopped his involvement in the SBIR program for a while. He feels strongly that this condition should be dropped from the program requirements or at least modified. At least in his case, it deterred rather than promoted commercialization.

Jeff Albert's major two suggestions to improve the SBIR are:

1. Remove the 50 percent time requirement.
2. Require the SBIR program to codify conflicts of interest.

TABLE 8 Suggestions for Improvement in SBIR Program Administration—Case Studies

Area of Improvement	Number of Respondents
Review process	3
Time between Phase I and Phase II	2
Clarification of conflict of interest	1
Increase award amount	1
Transparency	1
Communication	2

Genetic Models, Inc.

Dr. Peterson and Joe Pesek, cofounders of Genetic Models, strongly suggest improving the review process for SBIR award proposals. They feel that the process needs more scientific scrutiny with a greater understanding of the topics being proposed. "Most companies can't afford an in-house expert to understand a kidney failure model." Both Pesek and Dr. Peterson propose an integrated review board that comprises both scientists and economists.

Their most recent experience with the SBIR review process was negative. Dr. Peterson's understanding of the SBIR review panel was that the panel consisted of two outside reviewers, basically an ad hoc independent review, with no meeting of a single review board.

Dr. Peterson and Joe Pesek shared the following suggestions for improving the SBIR award program:

- Build better communication between the review board, independent reviewers, and candidates.
- Change the review board's mentality from a "University RO1 Grant" mentality (which only looks at genetic research proposals that focus on discovering the mechanism) to an "innovative, open-minded" mentality (which recognizes the value of a broad product on which different pharmaceutical companies can test different drugs—the purpose of the SBIR award is to develop a product that is different). "There is no mechanism at the university to develop/fund the animal model. Molecular biology has perturbed the animal-model genetic mechanism."
- Use performance measures to evaluate the benefits of a potential product; that is, the review board should not look only at the number of rats sold or millions of dollars earned. Rather, it should consider the extra contract research, employee development, and benefits to other companies. "Banks wouldn't invest in our product because the present value of the start-up costs are too high compared to the long-term benefits."
- "If value is found in intangible benefits, are there practical and prudent ways to prevent pork-barrel projects where people just throw our money?"

The biggest concern that Peterson and Pesek have about the SBIR program is the review process. They feel that "the reviewer doesn't understand the science." In their experience, the reviewers have made false assumptions or have not taken the time to understand the underlying science.

Peterson and Pesek argue that the review process needs a business perspective as well as that from university scientists. They feel that almost all of the reviewers have imposed a basic science standard on proposals that are oriented toward developing a commercial product. They feel that reviewers need to be more sensitive to the fact that SBIR is not necessarily about progress for basic science. They also perceive that it has become increasingly difficult to obtain approval for an SBIR proposal. Peterson has a strong record at the university of obtaining academic grants and has considerable experience. He feels that the SBIR review process is too much like the RO1 university grant proposals, even though the goals for basic research and SBIR are different.

They suggest that the SBIR review process be changed from including only two outside scientific reviewers to including a broader group of reviewers.

They also suggest that a bridge program is needed between Phase I and Phase II to facilitate keeping the staff onboard. This is because the turnover of personnel is the biggest problem the firm faces.

They also suggest that the SBIR program would benefit from better communication between the agencies and the firms.

Ash Medical Systems, Inc.

Dr. Ash of Ash Medical Systems finds that the Fast Track instructions "are terrible and self-contradictory." For example, the instructions emphasize calling someone at the agency, yet there he found only confusion. This is due to a high turnover of personnel involved with the SBIR program.

Dr. Ash also finds the streamlined procedure for evaluating SBIR proposals to be inappropriate. This streamlining process involves tossing out the worst applications from the applicant pool in the first stage. Dr. Ash feels that this leaves applicants "at the mercy of two reviewers who may or may not know what they are doing." In addition, the reviewers may know what they are doing "but may have a personal or vested interest in the opposite approach." This leaves him "disappointed and frustrated." He feels that several of his Phase I and Phase II applications that have been rejected were better than those that had been approved.

Dr. Ash also criticizes the requirement that a Phase I be completed before a Phase II. He suggests that if the critiques for a rejected application would arrive a week earlier they could be revised and resubmitted to make the next deadline.

Dr. Ash was just notified of Fast Track approval. He says that originally he submitted one application but was told that even with Fast Track you need to submit a Phase I and Phase II application. He subsequently divided his submissions into two separate applications. He feels that this administration wastes a lot of time on needless paperwork.

Dr. Ash does say that "Fast Track is great. It enables us to keep key personnel." He greatly values the reduction in uncertainty between Phases I and II that comes with Fast Track approval.

Batch Processing Technologies, Inc.

The review process for the Phase II SBIR award was frustrating for Girish Joglekar, founder of Batch Processing Technologies, because the SBIR review board did not provide him with any comments as to why they rejected his proposals.

When asked how he would improve the SBIR program, Joglekar offered the following suggestion:

- The SBIR Review Board needs to provide feedback in the form of comments to a firm applying for an SBIR award, especially when the review board rejects the firm's proposal.

Endotech, Inc.

Anthony Hubbard's strong feeling that the SBIR awards are an unreliable source of funding is based primarily on his Phase II experience. Hubbard felt that Endotech had a very strong proposal for Phase II, which included all Phase I data and results. However, Hubbard felt that the comments on his Phase II proposal were significantly different from the Phase I comments he had received one year earlier. "There was no continuity of reviewers between our Phase I and Phase II proposals. It was like the Phase II review board ignored our Phase I results and overlooked that we had met our Phase I goals."

Hubbard offered the following suggestions for improving the SBIR program:

- Before reviewing a Phase II proposal, the SBIR review board needs to track a firm's success in meeting its Phase I goals.
- The SBIR program needs to establish continuity of reviewers between Phase I and Phase II submittals. This way, a firm submitting a proposal will be assured of a more consistent review in the SBIR process.

Focus Surgery, Inc.

Before Focus Surgery received its ultimate start-up funds, some of the firm's SBIR applications for financing were rejected by venture capitalists from Indiana and venture capitalist firms outside Indiana. The firm's first SBIR application for $3 million to $5 million was rejected in 1997 because the venture capital firms from Indiana "were not interested and thought the project was too high tech." In 1999, venture capital firms outside Indiana were ready to invest if the company would move out of Indiana. Thus, the firm's second SBIR application for $15 million was rejected because the venture capital firms thought that (1) the firm needed a new CEO who had raised money before, and (2) Indiana is not a competitive place for the medical device business.

PERSPECTIVES ABOUT THIRD-PARTY INVESTORS

STAR Enterprises, Inc.

To date, STAR Enterprises has had no third-party investors. Jeff Alberts, company founder, feels that it is essential to the product development of STAR to maintain control of the firm and not to lose any element of control to a venture capitalist.

Genetic Models, Inc.

Genetic Models has never had external financing. This is because when the firm was founded, "We had no credit to get a loan." The firm has generally leased equipment and so has required only a minimal of financing, which has come from the SBIR program. The firm tried to get venture capital type of funding for a while, but "stagnated" every time. It has a clear bias against venture capital. The owners do not want to lose control of the company.

Joe Pesek and Dr. Peterson, cofounders of Genetic Models, also argue that it would not be possible to obtain financing from traditional financial institutions because the scientific content eludes the capabilities of evaluators in traditional financial institutions.

CROSSCUTTING RESEARCH QUESTIONS

STAR Enterprises, Inc.

The founder of STAR Enterprises, Jeff Alberts, does not feel that Fast Track would have really benefited his firm very much. This may be because he does not have too much investment sunk in key personnel and he has ready access to qualified personnel at the university.

Jeff Alberts feels that the usual large corporate contractors for NASA would never have been able to deliver the same quality of product for the low cost. A typical rival contractor would have been Lockheed. According to Alberts, "Lockheed would screw it up" because they would put the engineering first and the rats second. In his experience with Lockheed, Alberts feels that Lockheed "would follow the book" in the both the prototype design and the actual construction. Although they have first-class engineering capabilities, "they don't know the rats the way I do." Alberts observed that, "At STAR, rats are part of the design team." This has led to a product that is more effective, and undoubtedly costs less. In fact, NASA originally turned to Alberts because they had a bid for the product by Lockheed, which Alberts reviewed. The bid involved lower technological standards at twice the cost. Lockheed and McDonnell Douglas were each funded $5 million a year to develop designs. Jeff Alberts had written the product requirements for NASA and evaluated the original proposals by Lockheed and McDonnell Douglas. They each had a budget of $80 million for "a much less interesting design."

Jeff Alberts observes that a benefit of SBIR is that it provides the agencies with a superior quality product at lower cost: "The corporate structures of the traditional contractors evolved in an era when they couldn't afford to build simple things."

STAR avoids employing university students. Jeff Alberts says this is to avoid any possible conflict of interest between the educational goals of the students and the commercial interests of firms. However, he emphasizes that there are many spillovers from STAR. Although he is building hardware for the space station, including hardware for experiments, he anticipates being awarded the opportunity to undertake some scientific experiments in the space shuttle. This will benefit his university research and the research of his students.

Some of his doctoral students have been influenced by STAR's success and have pursued careers involving the commercialization of science. One example involves a former female student who became a professor at the University of Vermont. As a result of Jeff's example, she has started her own firm and is now the lead habitat person for the space station.

Genetic Models, Inc.

Peterson and Pesek, cofounders of Genetic Models, have not applied for Fast Track. They do not feel that Fast Track would have helped their firm. At the same time, they see the advantages of Fast Track. They think that the biggest advantage is that it enables a firm to maintain employment of skilled, trained technical personnel. "Keeping people and equipment employed" is critical. They also pointed out that Fast Track would be important if there are other partners involved, such as distributors or networks.

Ash Medical Systems, Inc.

Dr. Ash of Ash Medical Systems was just notified that he has a score of 249 on a Fast Track application. He anticipates Fast Track approval for Phases I and II.

CONCLUSIONS

Although it is important to analyze the impact of a government promotion program such as the SBIR on the performance of firms, such as their ability to survive and grow, there may be an even more fundamental impact on whether the scientists and engineers start the firms in the first place (Audretsch, 1995). The project has attempted to shed some light on the influence that the SBIR program has had on altering the career paths of scientists and engineers by facilitating the commercialization process by either starting a new firm or becoming involved in an existing small firm. The evidence provided here is of a preliminary nature, partially because of the small sample size consisting of 12 case studies and 20 firms responding to a survey instrument, but also because of the particular context within which the SBIR program operates in a state such as Indiana. Not only does the relatively low amount of private R&D and innovative activity limit commercialization opportunities for scientists and engineers, but perhaps a more subtle impact is that it limits knowledge about commercialization possibilities and the existence of ancillary services and institutions facilitating commercialization. A stylized fact that has emerged in the literature on small business economics is that the propensity for people to start new firms is greater in the context of a viable cluster of dynamic small firms. In Indiana, such a cluster of knowledge-based small firms has been lacking.

With these qualifications in mind, there are indications that the SBIR program has influenced the career paths of scientists and engineers by facilitating the start-up of new firms. In the case studies, one-half of the scientists indicated that the SBIR award influenced their decision to start the firm. In the absence of the SBIR program, 20 percent of them would not have started the firm, and another 40 percent would not have continued the firm.

There are also indications that the experience of scientists and engineers in commercialization via a small business has an externality in that it spills over to influence the career trajectories of colleagues. One-quarter of the scientists interviewed in the case studies named specific examples of colleagues who were either starting a new firm or becoming involved in a small firm to commercialize their knowledge. This externality may, in theory, have a negative effect. If university researchers are lured away from the academic setting by commercial opportunities, then a core mission of universities—teaching students and priming the pipeline for future research—may be undermined. The evidence from Indiana suggests that SBIR grants are not creating such an effect. Moreover, the SBIR awards may serve to attract students interested in research projects known to have attracted public funding through the SBIR program.

The evidence from the broader survey generally confirms the findings from the case studies. Both the survey and the case studies provide consistent evidence of the following:

1. A significant number of the firms would not have been started in the absence of the SBIR program.
2. A significant number of the scientists and engineers would not have become involved in the commercialization process in the absence of the SBIR program.
3. A significant number of other firms are started because of the demonstration effect of the efforts of scientists to commercialize knowledge.
4. As a result of the demonstration effect of SBIR-funded commercialization, a number of other scientists alter their careers to include commercialization efforts.
5. Technology-based entrepreneurs start firms because they have ideas that they think are potentially valuable; they do not start firms and then search for useful ideas or products. This is reflected by the fact that not a single respondent on either the survey or from the case studies suggested that he would have tried to start the firm with a different idea in the absence of SBIR funding. However, once the firm exists, one-quarter of the respondents and one-sixth of the case studies indicated that they would have tried to continue the firm with a different idea in the absence of SBIR

funding. These different results suggest that the SBIR program has a greater impact on potential entrepreneurs than on existing small firms in commercializing ideas that otherwise would not find their way into the market.

It should be emphasized that these results are of a preliminary nature. They reflect one context in Indiana in which scientists and engineers make career decisions. It may be that the relatively low amounts of R&D and innovative activity in the state enhance the impact that the SBIR program has, both in terms of inducing scientists to start a new firm to commercialize knowledge and in terms of influencing other scientists, through the demonstration effect, to commercialize knowledge. In addition, there is not enough variation in either the funding agency or the underlying science to identify how the SBIR program's influence on the entrepreneurial behavior of scientists differs across scientific fields and funding agencies. Until the necessary large-scale study spanning a broad spectrum of technological and regional context is undertaken, these preliminary findings will remain conjectural. However, on the basis of this preliminary evidence on the impact of the SBIR program on the entrepreneurial behavior and career paths of scientists in Indiana, such a large-scale study is warranted.

ACKNOWLEGMENTS

We are grateful for the helpful comments of Al Link, John Scott, Chuck Wessner, and the participants at the Workshop on Evaluating the SBIR and the Fast Track, sponsored by the National Academy of Sciences, Washington, D.C., May 5, 1999.

REFERENCES

Audretsch, David B. 1995. *Innovation and Industry Evolution.* Cambridge, MA: MIT Press.

Audretsch, David B., and Maryann P. Feldman. 1996. "R&D spillovers and the geography of innovation and production," *American Economic Review* 86:630-640.

Audretsch, David B., and Paula Stephan. 1996. "Company-scientist locational links: The case of biotechnology," *American Economic Review* 86:641-652.

Lerner, Josh. 1999. "The government as venture capitalist: The long-run effects of the SBIR program," *Journal of Business.* July, v. 72 (3), pp. 285-297.

Tyson, Laura, Tea Petrin, and Halsey Rogers. 1994. "Promoting entrepreneurship in Eastern Europe," *Small Business Economics* 6:165-184.

U.S. General Accounting Office. 1998. *Federal Research: Observations on the Small Business Innovation Research Program.* Report to Congressional Committees, April.

APPENDIX

Questions

Part 1

- Please answer the following questions about your firm and about the SBIR-award.
- When answering the questions, please check the boxes, fill in the corresponding blank lines or tables, or circle your answer, as appropriate.
- You are not limited to one answer on many of the questions. Please check all answers that apply, as appropriate.

Firm Characteristics

1. When was the firm established? ______________

2. How many employees does the firm currently have? ______________

3. What are the current sales of the firm? $__________________

4. By what percentage have sales revenues or employees changed (increased/decreased) during last year compared to the previous year (Please circle)
   Sales revenues/employees; increase/decrease ________%

5. What technology or basic science is the firm based on? __________________

   ____

6. What is the firm's legal form?
   (a) Corporation ☐
   (b) Limited Liability Corporation ☐
   (c) Subchapter S Corporation ☐
   (d) Partnership ☐
   (e) Sole Proprietorship ☐
   (f) other ______________________________ ☐

Characteristics of the founder

1. What is the founder's age? Year of birth: ____________________

2. What is the founder's education level? Include degrees earned, dates, and educational institution.

I. Date	Degree	Educational Institution

3. Where and in what capacity were you employed in the 10 years before being fully self-employed?

Dates	Employer	Job Title	Salary

Questions about SBIR

1. From which agency/agencies did you receive your SBIR award(s)? ____________

____

2. List dates and phases of SBIR awards.

Date of Award	Dollar Amount	Phase

3. Did the SBIR award influence the decision to start the firm or continue with the firm? Yes / No

4. In regard to the firm or project-idea, what would you have done differently if you had not been awarded the SBIR? (More than one answer is possible.)

(a) I would not have **started** the firm. ¤

(b) I would not have **continued** the firm. ¤

(c) I would have **started** the firm with money from an alternative source. ¤

(d) I would have **continued** the firm with money from an alternative source. ¤

(e) I would have commercialized the idea through an existing firm. ¤

(f) I would have abandoned the idea. ¤

(g) I would have tried to **start** the firm with a different idea. ¤

(h) I would have tried to **continue** the firm with a different idea. ¤

5. Has your SBIR experience with starting a firm influenced the activities or possible activities of any of your colleagues?

   (a) It has had no influence on colleagues. ☐

   (b) It has led _______ colleagues to express interest in

   (c) starting their own firm. (Please fill in the number.) ☐

   (d) It has led _______ colleagues to start their own firm.

   (Please fill in the number.) ☐

Starting up the firm

1. What (and who) influenced your decision to start a firm?

____
____
____
____
____
____
____

2. Where did you get your ideas that are the basis of the firm?

____
____
____
____
____
____
____

Cooperation

1. Do you work closely with any other firms, individuals, or research institutions? If so, which ones?

____
____
____
____
____
____
____

An Assessment of the Small Business Innovation Research Fast Track Program in Southeastern States

Albert N. Link
University of North Carolina at Greensboro

EXECUTIVE SUMMARY

This paper presents descriptive findings from 12 case studies of Small Business Innovation Research (SBIR) award recipients in southeastern states. The focus of the case studies was to determine, to the extent possible, if the Fast Track Initiative encourages more rapid commercialization of research results through the acquisition of private investment capital, and if Fast Track projects progress more rapidly than standard SBIR awards. The key findings from the sample of 12 firms indicate that:

- Fast Track projects proceed to Phase II research faster than non-Fast Track projects;
- Fast Track projects develop a commercialization strategy sooner than non-Fast Track projects, but those Fast Track projects do not anticipate having commercial products sooner than non-Fast Track projects; and
- the post-Phase II funding expected to be needed to commercialize Fast Track projects is greater than is expected to commercialize non-Fast Track projects.

INTRODUCTION

The U.S. Department of Defense (DoD) requested that the National Academy of Sciences (NAS) review its Small Business Innovation Research (SBIR) Fast Track program to determine, to the extent possible,

- if the Fast Track Initiative encourages more rapid commercialization of research results through the acquisition of private investment capital, and
- if Fast Track projects progress more rapidly than do the standard SBIR awards.

To accomplish this, NAS undertook a multifaceted research strategy that included both a broad-based mail survey to a representative sample of SBIR awardees and focused regional case studies from that sample.

This descriptive paper presents the findings from 12 case studies of award recipients in southeastern states. It will join other researchers' papers that focus on various regions of the United States. In the second section, the overall NAS strategy for the collection of information related to the above two questions is described. Then, In the third section, the process for selecting these 12 southeastern firms is presented. In the fourth section, observations about the commercialization impacts realized to date from the Fast Track Initiative are offered. In the fifth section, observations about other project impacts are discussed. In the sixth section, estimates of the social benefits associated with the SBIR program, and the Fast Track Initiative in particular, are presented.[1] Concluding remarks are presented in the last section. The Appendix to the paper contains brief summaries of each of the 12 projects studied.

NAS STRATEGY FOR COLLECTION OF INFORMATION

NAS was asked by DoD to determine, to the extent possible,

- if the Fast Track Initiative encourages more rapid commercialization of research results through the acquisition of private investment capital, and
- if Fast Track projects progress more rapidly than do the standard SBIR awards.

Toward that end, a team of researchers was assembled, and each was assigned a different region of the country from which to identify a sample of Fast Track program awardees and non-Fast Track program awardees. Each researcher was given latitude with regard to how he/she approached the questions during the interview data collection process; however, certain crosscutting issues were common to each. These crosscutting issues related to information about the background of each firm being interviewed, information about how the SBIR award is affecting the firm's research and commercialization strategy, and each firm's general opinion about the administration of the SBIR awards program.

---

[1]A more detailed analysis is provided by Link and Scott, "Estimates of the Social Returns to Small Business Innovation Research Projects" in this volume.

SELECTION OF THE CASE STUDY FIRMS

As requested by NAS, the geographic focus for this paper is southeastern states. An inspection of background information provided by DoD and NAS shows that there have been 31 SBIR awards to firms in southeastern states over the period from 1993 to 1996.[2] Of these 31, six firms received Phase II awards through the Fast Track Initiative. These six firms are shown in the upper portion of Table 1. NAS requested that 12 case studies be conducted in southeastern states, six non-Fast Track firms were thus selected for the purpose of comparison.

Several factors were considered in the selection of the six non-Fast Track firms, including state, duration of the Phase II project, and number of employees in the firms. The six non-Fast Track firms are shown in the lower portion of Table 1. Also shown in Table 1 are selected characteristics of the firms and their projects. Because of the small size of the defined population of regional firms, comparability of firms and projects between the Fast Track and non-Fast Track groups is not defined on the basis of a statistical criterion.

DoD provided the name and telephone number of the principal investigator (PI) for each Phase II project in each of the firms in Table 1, based on information from the Phase II application. In most cases, the PI's name was correct, but more often than not the firm had moved or changed its telephone number. However, each noted PI was eventually located and contacted by telephone. During the initial conversation, the nature of the study was described, confidentiality issues were discussed, and an overiew of the type of information being requested was given. After that initial conversation, a subsequent telephone interview time was arranged. In three instances, the PI was not interested in participating in the telephone interview. In those three instances, the DoD technical monitor on the project was located and contacted; he intervened and reinforced to each of the three PIs the importance of the study to DoD and assuaged confidentiality concerns. Subsequently, each of these three individuals agreed to participate in a telephone interview, although two of the three (both Fast Track) agreed to answer only a limited number of telephone interview questions. The focus of these two limited interviews was the importance of the Fast Track program in closing the funding gap between Phase I research and Phase II research. Each of the other 10 full telephone interviews focused on this issue as well as other issues related to commercialization.

As seen from Table 1, the group of Fast Track projects and the group of non-Fast Track projects are similar in the following dimensions: Each Phase II project was proposed to last approximately 24 months; the average number of employees in the company that was specific to the Phase II project was nine; and the companies themselves had been in operation for approximately eight years prior to the Phase II research.

Each complete interview averaged just over 60 minutes.

---

[2]These firms are located in Alabama, Georgia, Florida, North Carolina, and Tennessee.

TABLE 1 Selected Characteristics of the Sample of Firms Interviewed

Firm	State	Fast Track (F)	Project Duration (months)[a]	Number of Employees[b]	Year Founded
Matis, Inc.	GA	F	24	5	1990
OPTS, Inc.	AL	F	24	5	1994
CG2, Inc.	AL	F	24	15	1995
Power Technology Services, Inc.	NC	F	12[c]	5	1984
Summitec Corporation	TN	F	30	19	1987
Bevilacqua Research Corporation	AL	F	24	7	1992
System Design and Analysis Corporation	AL		24	3	1996
Accurate Automation Corporation	TN		24	17	1989
MicroCoating Technologies[d]	GA		24	8	1993
Optimization Technology, Inc.	AL		24	21	1983
Optical E.T.C., Inc.	AL		24	5	1990
Intelligent Investments	NC		24	3	1995

[a]Proposed project duration.

[b]Number of full-time-equivalent employees at the time the project was funded; two part-time employees equal one full-time employee.

[c]As discussed in the Appendix, this company originally was funded for a basic Phase II but has received additional Phase II funding, thus extending the duration to 2 years.

[d]Formerly CCVD, Inc.

INDICATIONS OF COMMERCIAL IMPACTS ASSOCIATED WITH FAST TRACK PROJECTS

None of the firms interviewed has commercialized a product or process that was associated with the Phase II award under study. This absence of direct commercialization success was expected *a priori* because the Fast Track Initiative is a relatively young program and there are few firms that have even completed their Phase II study. In fact, from the 12 firms interviewed, only 2 are just now (in 1999) at the end of their Phase II research. However, other information was obtained during the interviews in an effort to glean some preliminary insight into the possible commercialization impacts associated with the Fast Track program. This other information is described in the Appendix. Because the sample size is small, the technologies differ across firms, and the research and commercialization expertise is unique to each firm, care should be exercised in generalizing beyond this sample of 12 firms/projects about the possible commercial impacts associated with the Fast Track program.

Before proceeding to discuss issues related to commercialization, it was the case in each of the 12 firms that the Phase II research was related to the research background of the PI or the firm. It was also the case in each of the 12 firms that this Phase II research would form the foundation for subsequent research—which may or may not be SBIR funded. In other words, there is no indication among the

TABLE 2 Average Funding Gap Between Phase I and Phase II Research

Status	Funding Gap (months)
Fast Track firms	< 1
Non-Fast Track firms	4.3

group of 12 firms that the Phase II research under study was sought in a one-time opportunistic fashion; it is part of each firm's long-term technology strategy.

Length of the Funding Gap

Each of the 12 interviewees was asked, as background information, if Fast Track facilitated continuous funding from Phase I into Phase II. That is, each was asked if there was a funding gap between these two milestones. Among the Fast Track projects, five firms reported no funding gap at all, and one firm reported a four-month funding gap. In this latter situation, the Phase II project was approved in a timely manner, although funding was not immediately available from DoD. Among the non-Fast Track projects, the average funding gap was 4.3 months, with a range from 0 months to 12 months. (See Table 2.)

If commercialization is enhanced by a reduction of time between Phase I and Phase II, then these descriptive findings, as summarized by the data in Table 2, are suggestive of one aspect of the benefits associated with the Fast Track Initiative.

Time to Commercialization

Each interviewee was asked how soon after the Phase II project's completion will the technology(ies) being developed be commercialized. Six of the Fast Track firms that responded to this question, expected the mean time period to be nine months. The six non-Fast Track firms also expected the mean time period to be nine months. See (Table 3.)

Based on the data in Table 3, it appears that the reduced funding gap associated with Fast Track firms is not related to the expected duration from the end of Phase II to commercialization. Hence, one cannot conclude that these Fast Track firms expect to commercialize faster than the non-Fast Track firms.

Commercialization Strategy

Four of the six Fast Track PIs stated that their commercialization strategy was currently in place as a result of working with their third-party private-sector investor. Each of the PIs in these four firms went on to say that the Fast Track Initiative was instrumental in their receiving third-party funding. One respondent

TABLE 3 Average Expected Duration from End of Phase II to Commercialization

Status	Time to Commercialization (months)
Fast Track firms	9
Non-Fast Track firms	9

referred to Fast Track as being "critical" for his obtaining an outside investor, and another respondent stated that SBIR's support gave his project "instant credibility" to the outside investor.[3] Two of the Fast Track firms are receiving third-party funding from other government sources and neither thought of their research as having a commercialization strategy.

All of the non-Fast Track firms eventually expect to obtain third-party funding. Two of these six firms reported that they do not yet have a commercialization strategy, but anticipate developing one when outside funding is obtained. These two firms anticipate finding outside funding through a joint venture arrangement with already identified but not yet contacted private-sector firms. A third firm has found a local investor, and a commercialization plan is being developed. The other three firms hope to be able to commercialize but are unsure of their ability to acquire additional private-sector funds and/or unsure of how to commercialize a product. No firm mentioned that its geographic location gave it any advantage in attracting third-party funding.

The above two findings are not necessarily at odds with one another. The fact that Fast Track firms develop a commercialization strategy sooner than non-Fast Track firms perhaps may say more about the expected success of their commercialization efforts than about the timing of their commercialization efforts.

Post-Phase II Precommercialization Funding Requirements

Each interviewee was asked the approximate level of additional funding that will be required to commercialize its technology(ies) between the end of Phase II research and the expected date of commercialization. Mean responses are in Table 4.

The expected funding needs in Table 4 are the averages for four Fast Track firms and for six non-Fast Track firms. Based on these statistics, Fast Track projects are expected to require more than twice the additional funds of non-Fast Track projects. To reemphasize, the type of project and the related technology differ between these two broad groups, as seen from the project summaries in the Appendix.

[3]Each of these respondents was emphatic about the confidentiality of the interview information, and each was very uncomfortable about a subsequent discussion with the third-party investor, although the names of all third-party investors are public information. Accordingly, no interviews were conducted with any third-party investors.

TABLE 4 Average Additional Non-SBIR Funding Expected for Commercialization

Status	Required Funding
Fast Track firms	$744,000
Non-Fast Track firms	$354,000

Total Research from Concept to Commercialization

Related to the additional non-SBIR funding expected to be needed for commercialization of the Phase II technology(ies), as discussed in the preceding section, Table 5 shows the average total cost to conduct Phase I research and Phase II research, from all sources, plus the additional funding expected to be needed for commercialization.

On the basis of the dollar amounts in Table 5, Fast Track projects will cost approximately 54 percent more than non-Fast Track projects, on average.

OBSERVATIONS ABOUT OTHER PROJECT IMPACTS

Other characteristics of the 12 Phase II research projects studied, as well as differences in those characteristics between the group of Fast Track projects and non-Fast Track projects, are described in this section.

Employment Growth

The number of employees in each firm at the time that the Phase II award was made is shown in Table 1 above.[4] Each PI was asked how many additional employees were added to the firm during the Phase II research. Using the algorithm that two part-time employees equals one full-time-equivalent employee, the average growth in employees among the six Fast Track firms and among the six non-Fast Track firms is shown in Table 6.

Three of the six non-Fast Track PIs reported that the number of new employees hired during Phase II would be reduced once the research was completed; in fact, two of the three PIs that reported a post-Phase II decline in employment are in the two firms with the greatest growth in numbers of employees (180 percent and 133 percent) during the Phase II research. None of the six Fast Track PIs made such a statement; in fact, all Fast Track PIs were of the opinion that employment growth due specifically to the Phase II project would be permanent.

[4]One should not generalize about the average employment size of an SBIR Fast Track firm compared to a non-Fast Track firm on the basis of the employment data in Table 1 because the comparable six non-Fast Track firms included in this study were selected, in part, on the basis of the number of employees.

TABLE 5 Average Total Research Cost from Concept to Expected Commercialization

Status	Research Costs
Fast Track firms	$1,894,000
Non-Fast Track firms	$1,233,000

Fast Track Versus Non-Fast Track Applications

Previous funding relationships between each of the 12 firms and SBIR is shown in Table 7. Both Fast Track and non-Fast Track firms have had previous award experience with the SBIR program.[5]

All six of the non-Fast Track PIs stated that they were aware of the Fast Track program, but each PI stated that his/her firm did not pursue that funding avenue primarily because of lack of time and experience in identifying a potential third-party investor.

SBIR Administration

Each PI was queried about his/her experience with the SBIR on this Phase II project. The six Fast Track firms reported complete satisfaction with the Fast Track process and had no suggestions for changes. Noteworthy is the fact that three of the six Fast Track PIs stated that they had previous investment dealings with the company that invested in their Phase II project. Two of the non-Fast Track firms did offer constructive suggestions. One PI recommended that SBIR provide assistance to firms, especially the very small ones, regardless of their previous relationship with SBIR, about how to market and how to commercialize products. Another PI noted that the six-month Phase I period is too short a time for a small firm with no or little commercialization experience to identify a potential investor, much less to establish a relationship and attract outside funding. All 12 interviewed firms expected to seek additional SBIR research support in the future. The six Fast Track firms anticipated applying again through the Fast Track program, and the six non-Fast Track firms were uncertain about their future use of Fast Track.

Intellectual Property Protection

None of the 12 interviewed PIs reported any patent activity related to their current Phase II project. Only one, a non-Fast Track firm, expected to file a

[5]Respondents also were asked if they have sought or expect to seek funding from the Advanced Technology Program at the National Institute of Standards and Technology. None knew about the program.

TABLE 6 Employment Growth During Phase II

Status	Employment Growth (%)
Fast Track firms	82
Non-Fast Track firms	74

patent at the completion of the Phase II research. The dominant reasons offered for the lack of patenting activity were the cost of filing and the cost of patent protection after the fact.

PRELIMINARY ESTIMATES OF THE SOCIAL RETURN TO SBIR FUNDING

As discussed in Link and Scott's contribution to this volume, as part of this NAS review of DoD's Fast Track Initiative, a model was formulated for estimating the social rate of return attributable to SBIR-sponsored research projects. Only the findings from the application of this model to the projects studied are described here.

All firms, Fast Track as well as non-Fast Track, reported that they would not have undertaken the entire research project absent SBIR support (one Fast Track PI reported that his firm would have undertaken a small portion of the research absent SBIR funding). For each project, two rates of return are shown in Table 8: a private rate of return absent SBIR support under the counterfactual situation in which the research was undertaken, and a lower bound on the social rate of return associated with the SBIR-sponsored project.

TABLE 7 Previous Funding Relationships Between Sample Firms and SBIR

Firm	Fast Track (F)	Previous SBIR Awards
Matis, Inc.	F	Phase 1
OPTS, Inc.	F	Phase 2
CG2, Inc.	F	none
Power Technology Services, Inc.	F	Phase 2
Summitec Corporation	F	Phase 2
Bevilacqua Research Corporation	F	Phase 2
System Design and Analysis Corporation		none
Accurate Automation Corporation		Phase 2
MicroCoating Technologies[a]		Phase 2
Optimization Technology, Inc.		Phase 2
Optical E.T.C., Inc.		Phase 1
Intelligent Investments		none

[a]Formerly CCVD, Inc.

TABLE 8 Private and Social Rates of Return

	Rate of Return (%)	
Status	Private	Social
Fast Track firms	28	132
Non-Fast Track firms	21	104

A more complete economic interpretation of the findings reported in Table 8, as well as cross-project differences in the net social rates of return are discussed by Link and Scott (1999). Also in Link and Scott (1999) is a detailed discussion of the elements of market failure associated with the projects studied and the related markets on which the technologies are focused. These aspects of market failure are noted, but are not discussed, in each project summary in the Appendix. It is sufficient for this descriptive paper simply to note that SBIR is providing a socially desirable role in funding both Fast Track and non-Fast Track projects; however, the estimated social rate of return (actually a lower bound on the social rate of return) for Fast Track projects is greater than for non-Fast Track projects. This finding is discussed in detail by Link and Scott (1999).

CONCLUDING OBSERVATIONS

The descriptive information presented in this paper indicates that in some dimensions Fast Track firms and programs are different from non-Fast Track firms and programs. However, the southeastern states sample studied and summarized herein is too small to infer more definite conclusions. A comparison of the case study results from other regions in the United States will confirm or not confirm the propositions stated in this paper. Also, and perhaps more importantly, a case comparison of Fast Track and non-Fast Track projects after Phase II research is completed and after sufficient time has passed to evaluate the commercialization results from the projects seems warranted.

ACKNOWLEDGMENTS

This paper has benefited from the comments and suggestions of David Audretsch and John Scott.

REFERENCE

Link, Albert N. and John T. Scott, 1999. "Estimates of the Social Returns to Small Business Innovation Research Projects," this volume.

APPENDIX

Research Summary of Phase II Projects

Matis, Inc., Atlanta, Georgia

COMPANY ORIGIN: Founded in 1990 in Georgia.

PROJECT: A Novel Computational System for Real-Time Analysis and Prediction of Antenna-to-Aircraft and Antenna-to-Antenna Interactions.

PRINCIPAL INVESTIGATOR: Vladimir Oliker, Vice President of the company; Engineer.

PROJECT SUMMARY: There is a major problem with communication systems in general, and with antenna systems in particular. If antennas have no obstructions, then signals are transmitted and received clearly. However, such an environment rarely exists. On aircraft and ships, there often are obstructions of one form or another. These obstructions could be communications hardware or parts of the vehicle on which the communications are mounted. It is therefore critical to the quality of the communication system that the antennas be in an optimal position to minimize interference. Matis is developing software to simulate the antenna's environment and to measure the communication quality of alternative antenna placements. Given simulated information on alternative placements, it is the responsibility of engineers to trade off communication efficiency with engineering feasibility. The technology to develop this software comes from previous research projects.

COMMENT: Absent SBIR funding, Matis likely would have taken on this project on a limited scale. Although the capital and labor costs to undertake this research are extraordinarily high, Matis has previous investment relationships with companies and could acquire partial funding.

OPTS, Inc., Huntsville, Alabama

COMPANY ORIGINS: Founded in 1994 in Alabama.

PROJECT: Imaging Automatic Gain Control for Target Acquisition, Automatic Target Recognition, and Tracking.

PRINCIPAL INVESTIGATOR: Charles Hester, President of the company; Industrial Scientist.

PROJECT SUMMARY: It is important for a missile to know what it is going to hit as opposed to where it is going to hit. For example, a missile might see two vehicles, a tank and a truck, at a predetermined location. To be most effective, the missile should be able to distinguish between vehicles and hit the most militarily important one. To be able to do this, the missile guidance system must be able to process infrared images into a pattern recognition program; however, there is tremendous noise in infrared imaging. Existing technology relies on what is called simple gain, meaning that the sited image is enlarged in all dimensions. OPTS is developing hardware to install on missiles that will enhance images selectively, or apply gain selectively, so as to improve recognition. This will take place in real time on the missile.

COMMENT: Absent the SBIR award, the firm would not have undertaken this research. The reason is that the capital costs are high and there are few investment sources. Also, there is a very limited commercial market for this technology; hence, finding a commercial investor was extremely difficult.

CG2, Inc., Huntsville, Alabama

COMPANY ORIGINS: Founded in 1995 in Alabama.

PROJECT: Virtual Reality Scene Generation by Means of Open Standards.

PRINCIPAL INVESTIGATOR: David King, President of the company; Engineer.

PROJECT SUMMARY: To test a missile, it has to be developed, tested under controlled conditions, and then fired. The model must be fired a significant number of times to verify its capabilities. The cost for each firing is between $10 million and $15 million. CG2 is investigating a lower-cost process for verifying the capabilities of a missile under development. The software that is being developed is designed to run a hardware-in-loop process. After a missile is launched once, all of the information from that launch is stored in a simulation computer. The simulation computer is then connected to the circuitry of a new missile, and to an image scene generator. Then the image scene generator is connected to the missile, completing the loop. The loop first repeats for the new missile the flight of the tested missile. Then, there is what is called a validated simulation. Once the simulation is validated, the missile can be tested in various environments that are created by the image scene generator. For example, the image scene generator can tell the missile that it is seeing various things (e.g., a mountain) and it will measure how the missile reacts. The missile's reaction is stored in the simulation computer. Once completed, this technology can save the DoD billions of dollars per year in unneeded missile firings.

COMMENT: Because of the high capital costs for this research and the lack of available funding sources, this research would not have been undertaken in the absence of the SBIR award. Outside investors would not have been interested because the market is so small, and the technology can be imitated quickly. Accordingly, CG2's outside investor is the government.

Power Technology Services (PTS), Inc., Raleigh, North Carolina

COMPANY ORIGINS: Founded in 1984 in North Carolina.

PROJECT: A New Dual-Gated Motion Control Technology for Hybrid Electric Power Systems.

PRINCIPAL INVESTIGATOR: John Driscoll, President of the company; Engineer.

PROJECT SUMMARY: Motion control motors use approximately 70 percent of all electricity in the United States. About half of them are used in industry. These motors currently are controlled by insulated gate bipolar transistor (IGBT) technology. These IGBT chips are very expensive and need to be imported from Japan. PTS is developing a double-sided flip chip that is smaller and will be less expensive than Japan's IGBT chip. In addition, these chips will be made domestically. The chips have an immediate use in military electric tanks. Driscoll was previously a scientist at General Electric, and he holds patents from that tenure. It is these patents that are forming the technological base for the research.

COMMENT: Absent the SBIR award, this research would not have taken place. The reason is the high capital cost of obtaining access to a fabrication facility to produce the chips. Efforts were made to acquire funds for this project before applying to SBIR, but no investors would incur the cost.

Summitec Corporation, Oak Ridge, Tennessee

COMPANY ORIGINS: Founded in 1987 in Tennessee.

PROJECT: Very-Low-Bit-Rate-Error-Resilient Video Communication.

PRINCIPAL INVESTIGATOR: Andrew Yin, President of the company; Researcher.

PROJECT SUMMARY: The lack of available bandwidth is the technical constraint on video imaging, especially wireless video imaging. With limited bandwidth, transmission of pictures is difficult and slow and video is nearly impossible. Summitec is developing a compression-like software that will select only the important pieces of information to transmit over a narrow bandwidth so that video images will be clear. As the technical monitor explained, this software is like getting 10 pounds of potatoes into a 5 pound bag. The primary use of the software is in surveillance. Video information can be transmitted to planes to assist them in locational bombing.

COMMENT: Outside investors are very difficult to locate because the commercial return to this technology will not occur quickly. The long-time to market is the hurdle that investors see.

Bevilacqua Research Corporation, Huntsville, Alabama

COMPANY ORIGINS: Founded in 1992 in Alabama.

PROJECT: A Dialectic Approach to Intelligence Data Fusion for Threat Identification.

PRINCIPAL INVESTIGATOR: Andy Bevilacqua, President of the company; Engineer.

PROJECT SUMMARY: The goal of this project is to produce a software architecture that will make computers think more like people think. DoD has a strong desire to be able to do intelligent programming. It has attempted this in the past through what was called "role-based expert systems." That technology worked fine in a FORTRAN world of "if this, then that." However, the needs of DoD are more complex, and alternative technology is needed. The software being developed will take systematic concepts and translate them into numbers so that the computer can process them. For example, when people think of a concept, they do so in terms of a vector of characteristics of the concept. However, if two concepts are combined, then the vector of characteristics of the combined concepts is not necessarily a linear combination of the individual concept vectors. Bevilacqua calls this architecture a "cognitive reasoning engine."

COMMENT: The company would not have undertaken this concept absent SBIR funding for two reasons. One, it did not have access to sufficient funding and the commercial applications of the technology would not have been readily understood by investors. Two, the architecture can be imitated quickly once commercialized. The third-party investor in this project is the government.

System Design and Analysis Corporation (SDAC), Huntsville, Alabama

COMPANY ORIGINS: Founded in 1996 in Alabama.

PROJECT: A Generic Ducted Rocket Test Facility Setup and Simulation Program.

PRINCIPAL INVESTIGATOR: Gary Kirkham, Partner in the company; Computer Science.

PROJECT SUMMARY: The company is developing a software tool specific to ducted rockets. When designing a facility to test rockets, there are cost reasons for being able to test the facility prior to testing the rockets. There are software packages on the market to assist in the design of test facilities, but they are generic in their abilities and thus are not sophisticated enough to meet the needs of ducted rocket facilities.

COMMENT: The company considered applying to the Fast Track program but could not find investors. The reason was that the research takes a long time to complete and commercialize. Hence, sources of capital were not available for such high-market-risk projects.

Accurate Automation Corporation, Chattanooga, Tennessee

COMPANY ORIGINS: Founded in 1989 in Tennessee.

PROJECT: Neural Network Figure of Merit Subsystem.

PRINCIPAL INVESTIGATOR: Alianna Maren, Researcher.

PROJECT SUMMARY: Many military vehicles use radar. Each vehicle can have multiple radar units, and radar can be located away from the vehicles as well. It is likely that radar in both locations might be tracking the same signal. Each radar will produce its own information but, because each is in a different position at different times, each will have periods of locational advantage, but there will be overlapping information. To be precise in tracking, what is needed is a total combined or "fused" picture of what is being tracked. Although it is technologically possible to fuse overlapping information, the question then becomes one of how good or accurate the fused picture is. The answer to that question required a definition of "good." This research project deals with the development of a proof of concept that algorithms can be developed to define good. The output of the Phase II research is a commercializable software package that incorporates the algorithms developed.

COMMENT: Commercial potential is far off and there could not be investors or borrowed money absent SBIR.

MicroCoating Technologies (formally CCVD, Inc.),Atlanta, Georgia

COMPANY ORIGINS: Founded in 1993 in Georgia.

PROJECT: Non-Chromate Combustion Chemical Vapor Deposition (CCVD) Coating for Naval Engine Components.

PRINCIPAL INVESTIGATOR: Andrew Hunt, President of the company; Materials Science.

PROJECT SUMMARY: Hexavalent chrome is widely used in the Navy as well as in industry. However, it is a known carcinogen, and thus its use creates a toxic waste problem. The U.S. Environmental Protection Agency (EPA) knows of the problems associated with hexavalent chrome, but it has not yet mandated that it cease being used because no replacement is yet available (EPA practice). Congress has given DoD an internal directive to find a replacement material, and so, MicroCoating is developing such a material. It is based on a thin-film oxide that can be applied to metal during a CCVD process. During that process, the thin film is sprayed on metal with a flame, and the residual gas contains a replacement molecular coating that performs like hexavalent chrome but has more environmentally friendly properties.

COMMENT: This project would not have occurred absent SBIR. In fact, the company would not be in business. Hunt did try to find venture funding but it was simply not available. Venture capitalists are not interested in materials science. Industry eventually will need to use this process, but until EPA mandates a replacement, industry will not invest. His process can very easily be imitated, but he holds a flame deposition patent that will protect it.

Optimization Technology, Inc., Huntsville, Alabama

COMPANY ORIGINS: Founded in 1983 in California. Branch located in Auburn and that is how McGraw became involved. California branch could not make it but the Auburn branch did.

PROJECT: C41 Distributed Performance Simulation Environment.

Principal Investigator: Chris McGraw, President of the company; Computer Science.

PROJECT SUMMARY: As computers have moved from a supercomputer mainframe environment to a "distributed paradigm" where desktop computers are networked, a problem has arisen as to how best to share the processing loads across components in the network so as to maximize the efficiency of the network. This is not an interoperability problem; it is an ex post workload problem. Powerful desktop computers are being purchased but, depending on where the software resides and how the processing load is spread across machines, it could well be that the whole operates less efficiently than the potential sum of the parts. It is impossible to predict a priori how best to distribute workload. One needs trial and error to see what network configuration degrades the system the least. C4D is a software package that simulates various office systems before systems are purchased. The user can determine how to distribute software and workload efficiently.

COMMENT: There seem to be two aspects of market failure here, according to McGraw: (1) The work is very theoretical and outside investors have a hard time understanding it; and (2) it is hard to identify investors. McGraw claims that it is not so much that the outside funding market is thin as it is that it is hard to know how to find investors.

Optical E.T.C., Inc., Huntsville, Alabama

COMPANY: Founded in 1990 in Alabama.

PROJECT: High-G Microelectromechanical Accelerometer.

Principal Investigator: Arthur Werkheiser, Vice President (partner) of the company; Engineer.

PROJECT SUMMARY: The Air Force uses light gas guns for testing the air dynamics of various devices. These guns are 300-500 ft long. Artillery shells or small missiles are shot through these guns and pictures are taken of events such as ionization, tumbling, and other aerodynamic properties. When these pictures are taken, the acceleration of the object must be known. Acceleration needs to be measured up to 120,000 Gs (humans black out at 8-10 Gs. A sensor was needed that could be mounted on the object to measure acceleration in such an environment. The company is building such a sensor on silicon chips, using infrared devices that can withstand the acceleration. The technology underlying this project is known as microelectromechanical systems.

COMMENT: The technical risk associated with this project is large because the sensor needs to be so small (0.33 in.^{3}). It also will be very hard to appropriate property rights once it is commercialized.

Intelligent Investments, Greensboro, North Carolina

COMPANY ORIGINS: Founded in 1995 in North Carolina.

PROJECT: Multiagent Tool for Effective Network-Based Training Systems.

PRINCIPAL INVESTIGATOR: David Goldstein, President of the company; Computer Science.

PROJECT SUMMARY: Software is being developed in an interactive way to develop Web pages. Multiple individuals in different locations can work on the development of the page simultaneously.

COMMENT: The software can be partially imitated, but not totally, since he claims to have greater technical experience than others.

Evaluation of the Department of Defense Small Business Innovation Research Program and Fast Track Initiative: A Balanced Approach*

Robert B. Archibald and David H. Finifter
The College of William and Mary

EXECUTIVE SUMMARY

The purpose of this paper is to evaluate the Department of Defense (DoD) Small Business Innovation Research (SBIR) program and the Fast Track initiative. The SBIR program has grown and changed in its brief history, and the Fast Track Initiative is a recent innovation in the program. The SBIR program has a dual nature. It is designed to invigorate federal R&D efforts by expanding the participation of small business in the largest federal research programs and to increase the commercialization of innovative products and services that result from federal R&D efforts. Given the dual nature of the program it is important that evaluation of the program consider both the research and the commercial outcomes.

The SBIR Program—The SBIR Program was started in 1982 as a set-aside program for small business. It diverts a portion of the extramural research or research and development budgets of eleven federal agencies to fund the awards. The SBIR program has grown over time. Currently, each agency must devote not less than 2.5 percent of its extramural research and development budget to its SBIR program, representing a considerable increase from the original 0.2 percent of this budget. The SBIR awards are divided into three types. Phase I awards are small and intended to determine the scientific and technical merit and feasibility of the ideas. These awards are very competitive; overall there is an average of one award for every six to seven Phase I proposals. Phase II awards are designed to

*Prepared for the May 5th SBIR Symposium sponsored by the Board on Science, Technology, and Economic Policy of the National Research Council, National Academy of Sciences (NAS).

enable firms to do the research proposed in Phase I. Current Phase I awards normally have a maximum of $100,000 and Phase II awards have a maximum of $750,000, although these ceilings may be exceeded at the discretion of the program managers. Phase III awards do not actually involve SBIR funds. The terminology is used to describe situations in which actual production for agency use is funded through the continuation of a successful SBIR project.

The rationale for aid to small businesses generally relies on claims of discrimination in capital markets. The discrimination in capital markets is frequently called statistical discrimination. The fact that statistics indicate small businesses typically have higher default rates than large businesses is used to deny the loan to a small business applicant. Statistical discrimination leads to social inefficiency because small firms have less access to capital than they would if information were perfect.

When it created the SBIR program, Congress was also concerned with discrimination in government procurement. The concern here is with research productivity, i.e., a budget that relies heavily on large business results in lower productivity for research and development than would a budget that increases small business participation. Much of the mission-essential research carried out by government agencies is in areas that are highly innovative. In the industries in which small firms may have an advantage, it is sensible technology policy for the government to target some of its R&D funds on small firms.

The Fast Track Initiative—The 1996 Fast Track Initiative of DoD represents a continuation of the shift in emphasis in the SBIR award process toward commercial success. Under Fast Track firms with Phase I contracts which can interest outside investors in committing funds to further the development of the project increase their chances of obtaining Phase II funding and are eligible for bridge funding between Phase I and Phase II. The increase in the importance of commercial success is clear. A firm that does a piece of research that increases knowledge of a government laboratory provides a useful service to the government by aiding the ongoing research of the laboratory. Such a firm will, however, not be able to attract outside investors, so it will not be eligible for Fast Track. Fast Track is reserved for firms that are likely to be commercial successes through producing a product or service that can be directly sold, or whose product or service holds sufficient commercial promise that outside investors are willing to invest in its further development.

Fast Track fits a particular model of small business. It is designed for a small business that has technical expertise and a desire to use that technical expertise to develop a product or service that it can sell, either in the commercial marketplace or as a government contractor, or both. This firm has no particular desire to be a small business. There are two other types of firms that participate in the SBIR program. First, some successful small businesses have no desire to be big businesses. These firms are not growth-oriented. Some of these were created started by refugees from laboratories in large businesses and others by researchers who got their start in university or government laboratories. The founders of these firms were frustrated by the bureaucracy in the large business or some part of their academic or government job and became convinced to start or join a small business. Second, there are growth-oriented firms with products that are not yet commercially viable. These firms would like to expand and become big businesses, but the product or service they are developing is too far from its final state to be interesting to outside investors.

Survey of Technical Monitors—Our independent contribution to the evaluation is measurement of the quality and usefulness to the federal government of the research conducted by the SBIR firm, i.e., we provide measures of research success. The individuals who are in the best position to assess the quality and usefulness of SBIR research to the government are government scientists and engineers who monitor the research progress. Our evaluation takes the form of a survey of the technical monitors of the SBIR Phase II contracts. The technical monitors, called either Contracting Officer's Technical Representative (COTR) or Technical Points of Contact (TPOC), are the DoD scientists or engineers who served as the liaison between the small business and the agency that awarded the contract.

One innovative aspect of the survey is that it was conducted via e-mail to make it practical and encourage a high response rate. Because it was designed for e-mail distribution, the survey was as short as possible while capturing key questions. We had a very high response rate for e-mails that were received by a COTR, although we were unable to find good e-mail addresses for a large number of COTRs. The database contains 379 SBIR Phase II contracts. The sample was drawn systematically. It includes all of the 1996 Fast Track projects and all of the BMDO Matching projects from 1992 to 1996 as well as a matched sample of regular Phase II projects. As such it does not represent a random sample of the entire DoD SBIR program. We have responses covering 195 (51.5%) of the contracts in the full sample. Our overall response rate for successful contacts, 78.9%, is very high.

The survey had two parts and covered five major areas. The first part of the survey focused on individual SBIR projects, measuring (1) research quality, (2) usefulness of the research, and (3) mission benefits of the research. The second part of the survey focused on the SBIR program, measuring (4) overall quality of SBIR proposals, and (5) impressions of Fast Track.

(1) **Quality of Research**—We asked two questions about the quality of research. One question asked for a rating of the quality of the research for the SBIR Phase II and the second asked for a rating of the quality of other research. For our sample, the mean value for the difference between the rating given to the research quality of the SBIR project and the rating given to other research is .025 with a standard deviation of 2.366.

(2) **Usefulness of the Research**—To discover whether or not the SBIR research was useful we ask survey respondents to indicate whether or not the research has affected the way that research is conducted in their unit/office. 25.94% indicated "No, the knowledge generated by this SBIR contract has had no impact on the other research we conduct or sponsor. The other 74.06% chose one of the responses that started with "Yes, this project produced results that have been useful to us,..."
(3) **Mission Benefits**—We asked the respondents to compare the mission benefits per dollar on the SBIR project with the average benefits per dollar on other research. 31.58% indicated that the SBIR project yielded more benefits per dollar, 42.11% indicated it yielded the same benefits per dollar, and 26.54% indicated it yielded less benefits per dollar.
(4) **Overall Quality of Proposals**—We asked each COTR about whether he or she was satisfied with the quality of the SBIR proposals his or her office received. 63.91% indicated that they had more good proposals than they could fund, 25.18% indicated that they had the same number of good proposals as they could fund, and 10.91% indicated that they had fewer good proposals than they could fund.
(5) **Fast Track**—We asked those COTRs who had been involved with a Fast Track about its effectiveness. 66.67% indicated that it was more effective than the normal program. 21.57% indicated that it was the same as the normal program, and 11.76% indicated that it was less effective than the normal program.

In summary, the DoD SBIR program received a very favorable evaluation from the technical monitors. The quality of the research in the SBIR program is indistinguishable from the quality of other research. Regarding usefulness of the research, over 74% of the responses indicated that the research was useful. Just over 73% of the responses indicated that a dollar spent on the SBIR project yielded the same or more benefits per dollar than other research efforts. Over 81% of the respondents said that they had the same, or more, good SBIR proposals than they could fund. And the Fast Track Initiative was rated as more effective than the normal SBIR program by two thirds of the respondents.

In the analysis of the survey results we divided the sample among the Fast Track projects, the BMDO Matching projects and the comparison group projects. Perhaps surprisingly, the Fast Track projects did very well in these comparisons. Despite the fact that the Fast Tract Initiative was designed to enhance the commercial success rate of its projects, according to our results the research outcomes of the Fast Track projects compared favorably to those in the other two samples.

Balanced Evaluation—The fundamental assumption of our approach to evaluation is that the two goals of the SBIR program should be given equal weights. Under this approach, projects that produce commercial success should count as successful SBIR projects, and projects that produce useful research for the contracting agency should count as successful SBIR projects. A proper evaluation of the program has to consider these two possible avenues to success for each project. This approach corrects a fundamental flaw in current evaluations, the fact that, exclusively, these evaluations have been separate evaluations of either commercial success or research success.

To accomplish a balanced evaluation we need information on both research outcomes and commercial outcomes. To obtain such data, we combine the results from our survey of COTRs with the results from a survey of firms conducted by Peter Cahill of BRTRC. The survey of firms was designed primarily to provide information on commercial outcomes. The linked survey responses therefore allow us to measure interactions between the research and commercial goals of the SBIR program. A majority of the projects for which we have information from technical monitors is contained in the data set of firm responses. There are 124 observations in the linked data set.

For given definitions of research success and commercial success, we can classify projects into one of four outcomes. Table 1 illustrates our evaluation strategy.

Table 1 is the basis for our evaluation of the SBIR program. There are several measures of research and commercial outcomes, and for some measures it is not obvious where one should draw the line between success and failure. For this reason, there are many ways one could classify projects. In the text we provide several examples using different definitions of the two types of success. Our major finding is that in almost all of the examples we considered, commercial and research success were statistically independent. This means we did not find a trade off between commercial success and research success. Also, we found that projects in the Fast Track Initiative generally did very well both in measures of research outcomes and measures of commercial outcomes.

Summary and Conclusions—There are three major conclusions in this paper.

- The balanced evaluation we outline is the appropriate way to evaluate the SBIR program. Almost all of the previous evaluations have only considered commercial success. The quality of the research in the SBIR program has received scant attention. This paper provides the first evaluation that looks simultaneously at the two goals of the program.

TABLE 1 Classifications of SBIR Projects

	Research Failure	Research Success
Commercial Failure	Group N—Neither a research nor a commercial success	Group R—Research successes that are not commercial successes
Commercial Success	Group C—Commercial successes that are not research successes	Group B—Both a commercial success and a research success

- Our survey of technical monitors indicates that the DoD SBIR produces high quality research that in useful to the overall research effort of the department. Also, this high quality research does not appear to conflict with the SBIR goal of increasing the commercialization of the results of government R&D.
- Contrary to our initial expectations, projects in the DoD Fast Track Initiative appear to produce research that is as good if not better than the research in the normal DoD SBIR program.

INTRODUCTION

The purpose of this paper is to evaluate the Department of Defense (DoD) Small Business Innovation Research (SBIR) program and the Fast Track Initiative. The SBIR program has grown and changed in its brief history, and the Fast Track Initiative is a recent innovation in the program. The results of surveys of technical monitors and SBIR awardees, which we report here in detail, indicate that these technical monitors have a strongly positive assessment of the Fast Track Initiative and the Fast Track firms have generally been very successful. The focus of this paper is on the appropriate evaluation of the SBIR program and the Fast Track Initiative.

The SBIR program has a dual nature. It is designed to invigorate federal research and development (R&D) efforts by expanding the participation of small business in the largest federal research programs and to increase the commercialization of innovative products and services that result from federal R&D efforts. Given the dual nature of the program, it is important that evaluation of the program consider both the research and the commercial outcomes.

Seven additional sections follow. In the second section, we give a brief description of the SBIR program and discuss its history. In the third section, we provide an outline of the theoretical rationale for the program. We introduce the Fast Track Initiative in the fourth section and explain its role in the context of the SBIR program. In the fifth section, we present evidence from a survey of technical monitors of SBIR contracts. In the sixth section, we provide an analysis of the results of the survey of technical monitors, focusing on research outcomes for the DoD SBIR program and the Fast Track Initiative. In the seventh section, we combine the survey of technical monitors with a survey of firms. The combined survey allows us to do a balanced evaluation of the DoD SBIR program and the Fast Track Initiative, including information on both goals of the SBIR program. The final section contains a summary and our conclusions.

DESCRIPTION OF THE SBIR PROGRAM

The SBIR program was started in 1982 as a set-aside program for small business. It diverts a portion of the extramural research or R&D budgets of 11 federal agencies to fund the awards. The SBIR program has grown over time. Currently, each agency must devote at least 2.5 percent of its extramural research R&D budget to its SBIR program, representing a considerable increase from the original 0.2 percent of this budget.

The SBIR awards are divided into three types. Phase I awards are small and intended to determine the scientific and technical merit and feasibility of the ideas. These awards are very competitive; overall there is an average of one award for every six to seven Phase I proposals. Phase II awards are designed to enable firms to do the research proposed in Phase I. They are granted to roughly one-third of the winners of Phase I awards. Phase III awards do not actually involve

SBIR funds, but this terminology is used to describe situations in which actual production for agency use is funded through the continuation of a successful SBIR project. On occasion, private commercial funding also is described as Phase III funding. Current Phase I awards normally have a maximum of $100,000 and Phase II awards have a maximum $750,000, although these ceilings may be exceeded at the discretion of the program managers.

Each agency establishes its own policies and priorities regarding the categories of projects funded by its SBIR program, receives and evaluates proposals, selects awardees for SBIR funds, and makes the appropriate payments. The Small Business Administration (SBA) sets the schedule for the solicitation of proposals and, along with the Office of Science and Technology Policy, receives annual reports from each agency that runs an SBIR program.

To evaluate the SBIR program, it is important to identify the legislative intent for and goals of the program. The Small Business Innovation Development Act of 1982 was the result of a recommendation of the first White House Conference on Small Business in January 1980. The delegates to this conference voted to recommend the bill that authorized the creation of the SBIR program. There were several reasons this piece of legislation found support in the conference and eventually in the Congress. First, evidence suggested that small businesses had been having difficulty obtaining funds in general and had a declining share of federal R&D contracts. Second, several well-publicized studies indicated that small businesses were a very important source of job growth[1], and the recessions of the early 1980s created a supportive climate for any proposal that could claim job creation potential. Third, a successful SBIR pilot project established by the National Science Foundation demonstrated that the program was feasible.

From the start, Congress had two goals for the SBIR program. A report of the Senate Committee on Small Business makes this clear.

> The purpose of the bill is twofold: to more effectively meet R&D needs brought on by the utilization of small innovative firms (which have been consistently shown to be the most prolific sources of new technologies) and to attract private capital to commercialize the results of the Federal research. (U.S. Congress, 1981, p. 1)

The two primary goals for the program are also clear from the legislation. The 1982 act that created the SBIR program listed the following objectives of the program:

> (1) to stimulate technological innovation;
> (2) to use small business to meet Federal research and development needs;
> (3) to foster and encourage participation by minority and disadvantaged persons in technological innovation; and
> (4) to increase private sector commercialization of innovations derived from federal research and development. (96 Stat. 217)

[1]See, for example, David Birch (1981).

The act was originally scheduled to expire on October 1, 1988, but during fiscal year 1986, Congress enacted legislation extending the law through September 1993 (99 Stat. 443). In 1992 the SBIR program was reauthorized. The Small Business Innovation Reauthorization Act of 1992 both raised the percentage of research expenditures dedicated to the SBIR program and increased the importance of the goal of commercializing SBIR projects. The goal of commercialization moved from being listed fourth in 1982 to second in 1992. This change in the ordering of the goals was purposeful and was reflected in important ways in the language describing the selection process after 1992. Specifically, the original language describing Phase I was:

> (A) a first phase for determining, insofar as possible, the scientific and technical merit and feasibility of ideas submitted pursuant to SBIR program solicitations: (96 Stat. 218)

This language was amended as follows (the added language is underlined):

> (A) a first phase for determining, insofar as possible, the scientific and technical merit and feasibility of ideas that appear to have commercial potential as described in subparagraph (B)(ii), submitted pursuant to SBIR program solicitations: (106 Stat. 4250)

For Phase II the change is more dramatic. The original language was:

> (B) a second phase to further develop the proposed ideas to meet the particular program needs, the awarding of which shall take into consideration the scientific and technical merit and feasibility evidenced by the first phase and where two or more proposals are evaluated as being of approximately equal scientific and technical merit and feasibility, special consideration shall be given to those proposals that have demonstrated third phase, non-Federal capital commitments; (96 Stat. 218)

This was changed to:

> (B) a second phase, to further develop proposals which meet particular program needs, in which awards shall be made based on the scientific and technical merit and feasibility of the proposals as evidenced by the first phase considering, among other things, the proposal's commercial potential, as evidenced by:
> (i) the small business concern's record of successfully commercializing SBIR or other research;
> (ii) the existence of second phase funding commitments from private sector or non-SBIR funding sources;
> (iii) the existence of third phase, follow-on commitments for the subject of the research; and
> (iv) the presence of other indicators of the commercial potential of the idea. (106 Stat. 4251)

These are clear mandates to change the selection processes by increasing the importance of commercial potential. Under the 1982 legislation, ties between projects deemed to be of equal scientific and technical merit could be broken in favor of projects that were more likely to be commercially successful. The likelihood of commercialization was clearly a secondary concern. This was changed with the 1992 legislation, which placed commercialization on an equal footing with scientific and technical merit.

The 1996 Fast Track Initiative of DoD represents a continuation of the shift in emphasis in the SBIR award process toward commercial success. Under Fast Track, firms with Phase I contracts that can interest outside investors in committing funds to further the development of the project increase their chances of obtaining Phase II funding and are eligible for bridge funding between Phase I and Phase II. The increase in the importance of commercial success is clear. A firm that does a piece of research that increases the knowledge base of a government laboratory provides a useful service to the government by aiding the ongoing research of the laboratory. However, such a firm will not be able to attract outside investors, and so it will not be eligible for Fast Track. Fast Track is reserved for firms that are likely to be commercial successes through production of a product or service that can be directly sold, or whose product or service holds sufficient commercial promise that outside investors are willing to invest in its further development.

RATIONALE FOR THE SBIR PROGRAM

An important feature of the SBIR program is that it is a set-aside. It does not result from new monies appropriated by Congress. It results from Congress mandating that agencies engaged in research target a portion of their existing funds for research projects carried out by small businesses. Because of this, the SBIR program represents a redirection of R&D spending, not an expansion. R&D provides a public good and, for that reason, it is a sensible public expenditure.[2] The question raised by the existence of the SBIR program is: Why does the government want to increase the involvement of small business in this activity?

We think of the SBIR program as addressing two failures: a market failure and a government procurement failure. First, the SBIR program is one of several programs designed to help small businesses. The loan and loan guarantee programs administered by the SBA provide other examples. These programs address a failure in credit markets. Second, the SBIR program seeks to correct deficiencies in federal procurement practices that lead to an excessive reliance on large businesses for federal R&D. Evaluations of the SBIR program should focus on how well the program corrects these two failures.

---

[2]See Griliches (1992) for a review of the literature on this point.

The SBIR Program as Aid to Small Businesses

The rationale for aid to small businesses generally relies on claims of discrimination in capital markets. Frequently called statistical discrimination,[3] it occurs when, for example, a lender decides to make a loan to a large business, not to a small business, simply on the basis of size; that is, the two businesses are otherwise equally creditworthy. The fact that statistics indicate that small businesses typically have higher default rates than large businesses is used to deny the loan to a small business applicant.[4] This kind of decision is cost-effective for the lender, but it is easy to see why government remedies are appealing. Statistical discrimination leads to social inefficiency because small firms have less access to capital than they would if information were perfect. This rationale for support of small business is generally recognized. One important question the SBIR program raises is: Is there anything about R&D activity or high-technology firms that exacerbates these problems?

The answer to this last question is probably "yes." R&D is a much riskier investment than are many other types of investments that a firm might undertake. Because of this riskiness, R&D investment requires a much higher rate of return in capital markets than financing an expansion in current capacity. As Bronwyn Hall points out "asymmetric information between firms and investors implies that, to fund projects about which they do not have full information, investors will demand a 'lemons' premium in the form of a higher rate of return." (1993, p. 290). This lemon effect for R&D investment applies to all firms, but it is very likely to be an even bigger problem for small firms. Josh Lerner (1999) explains how the SBIR program helps small firms to overcome these difficulties. He argues that a small firm that is successful in the competition for SBIR funds sends a signal of its capability to outside investors. In essence, to outside investors this certification indicates that winners of SBIR awards are less likely to produce R&D lemons.

One could argue that the special problems that small businesses have obtaining R&D funds might have been overcome by creating a special category of SBA loans or loan guarantees to support R&D. If the argument put forth by Lerner is correct, the current SBIR program is probably a superior alternative. The program gives small firms a chance to enter a competition in which federal technical experts judge their R&D proposals. A small firm's successful competition in this arena signals that it is likely to be better than other small businesses in the commercial arena as well.

---

[3]See Arrow (1973) for an early discussion of statistical discrimination.

[4]Stiglitz and Weiss (1981) demonstrate why financial institutions' response to this problem is to deny credit rather than simply to raise the interest rates they charge on loans to risky borrowers.

The SBIR Program as Aid to Federal Research Efforts

When it created the SBIR program, Congress was concerned with discrimination in government procurement. The 1981 Senate Report indicates that encouraging greater small business participation in federally sponsored R&D was an important part of the argument for the legislation:

> Numerous studies have shown that small businesses are our Nation's most efficient and fertile source of innovations. Yet only 3.5 to 4 percent of the Federal R&D dollar is spent with small firms. This underutilization of small business in Federal R&D programs is especially regrettable when considering the highly successful track record of small firms in generating jobs, tax revenue, and other economic and social benefits. (U.S. Congress, 1981, pp. 4-5.)

Such language clearly implies that the federal R&D procurement process had unwisely led to choices that limited the participation of small businesses.

The argument here is based on claims about research productivity; that is, a budget that relies heavily on large business results in lower productivity for R&D than would a budget that increases small business participation. Academic views about the relationship between firm size and innovation in the private sector have varied over time, but by the time the SBIR legislation was being considered, there was growing support for the notion that small firms might produce better research than large firms.[5] Clearly, there are disadvantages of large size. For one, it may be more difficult for a large firm to take on a risky project than for a small firm. In a large firm, many layers of control have to approve a particular project, and this can discourage innovation. There is considerable evidence that frustration with unimaginative management leads to the creation of small high-technology firms by talented refugees from the laboratories of large businesses.

As researchers started to do empirical work on the relationship between firm size and innovation, support for the views that large scale was necessary for innovation diminished.[6] Writing in 1970, F. M. Scherer summarized some of the evidence as follows:

> The weight of the available quantitative evidence favors a conclusion that among the largest 500 or so U.S. industrial corporations, increases in size do not as a rule contribute positively to the intensification of R & D inputs or inventive outputs, and in more cases than not, giant scale has a slight to moderate stultifying effect. (p. 361)

With the advantage of additional time, Chris Freeman and Luc Soete (1997) report a similar finding in the summary to their chapter on the relationship between firm size and innovation:

---

[5]As an example, a study by the Organization for Economic Cooperation and Development concluded that "Small firms have shown remarkable ability as purveyors of innovations, in particular in industries characterized by high growth rates and technical change" (OECD, 1982, p. 5).

[6]See, for example, Rubenstein (1958), Hamberg (1963), and Roberts (1968).

> But when all is said and done, there is still impressive support for the view that small firm innovations have genuinely increased their share of the total in the final quarter of the twentieth century. (p. 239)

Zoltan Acs and David Audretsch (1990) add some important detail in their careful study using data for innovations. They find that innovative activity is hindered in concentrated markets. They also find that "the innovative advantage of small firms is promoted in industries that are highly innovative and that utilize a high component of skilled labor." (p. 147)

Although not all of the arguments about R&D in the private sector necessarily carry over to the public sector, this last finding is important for the SBIR program. Much of the mission-essential research carried out by government agencies is in areas that are highly innovative. In the industries in which small firms may have an advantage, it is sensible technology policy for the government to target some of its R&D funds on small firms.

If it is the case that small firms are at least as likely to produce good R&D as are large firms, one might wonder why a government agency would spend its R&D budget predominantly at large firms. In some cases, government R&D efforts naturally by their very scale and complexity require large business; for example, few would want to fly a jet aircraft designed and engineered by a coalition of small businesses). Also, the reliance on large businesses might result from the same type of risk minimization behavior characterized by models of statistical discrimination. The perceived high variance of outcomes from small business might discourage procurement officials. Still, there are many types of government research that could be carried out by small business just as efficiently as by large business, and Congress was convinced that the optimal share for small business was larger than 3.5 to 4 percent.

THE FAST TRACK INITIATIVE

Components of Fast Track

The Fast Track Initiative has three components: First, a firm qualifies for Fast Track if it can link with outside investors who are willing to partially fund the Phase II research. The attraction for outside investors is that the SBIR program will match their investment. The willingness of outside investors signals the SBIR program that the project is very likely to lead to a commercial success. Also, the funds from the outside investors add to the research budget of the firm, further benefiting the government research effort. Second, a Fast Track firm receives bridge funds between Phase I and Phase II. These bridge funds allow the firm to maintain its research effort and research staff until the Phase II contract begins. Third, Fast Track contracts, as its name implies, have a smaller time gap between Phase I and Phase II.

Alternative Models of Small Businesses

Fast Track fits a particular model of small business. It is designed for a firm that has technical expertise and a desire to use that technical expertise to develop a product or service that it can sell, either in the commercial marketplace or as a government contractor, or both. This firm has no particular desire to be a small business. It views itself as on the way to becoming a big business, and it has developed its product or service to the point that it can convince outside investors to back the project. Such firms can be characterized as "growth-oriented with a commercially viable product."

Viewing the entire SBIR program as a program for growth-oriented firms with commercially viable products is prevalent. First, SBIR program managers speak of a successful SBIR firm as "graduating" when the firm hires worker number 501 and is no longer classified as a small business. On the basis of some of their statements, these program managers have the objective of graduating as many small businesses as they can. Second, this view is reflected in the emphasis placed on commercial success in evaluations of the SBIR program. A major General Accounting Office (GAO) report in 1992 was titled *Small Business Innovation Research Program Shows Success But Could Be Strengthened.* The report came to that conclusion based upon research that looked only at rates of commercialization. Presumably, if commercialization rates had been higher, the last clause would have been removed from the title.[7] Third, in the abstract of his paper evaluating the SBIR program, Josh Lerner (1999) explains that "This paper examines the largest U.S. public venture capital initiative, the Small Business Innovation Research (SBIR) program ...." Taking the view that the SBIR program represents "public venture capital" focuses entirely on the commercialization goal and implies that all SBIR firms are growth-oriented with commercially viable products.[8]

There are two other types of firms that participate in the SBIR program. First, some successful small businesses have no desire to be big businesses. These firms are not growth oriented. Some were started by refugees from laboratories of large businesses and others by researchers who got their start in university or government laboratories. The founders of these firms were frustrated by the bureaucracy in the large business or some part of their academic or government job and became convinced to start or join a small business.[9] Second, there are growth-

[7]The SBIR program was studied closely in 1992 in association with the congressional debates on its renewal. The other major study conducted by the SBA (1992) also focused exclusively on commercialization rates.

[8]The notion of public venture capital is inappropriate for the SBIR program in several other respects. In the SBIR program the government does not gain an equity holding in the firm, nor does it become involved in the management of the firm.

[9]Scherer (1970) puts the case for flight from big business clearly: "A direct consequence of this problem, which has been noted time and again in case histories and treatises on research management, is a bias away from really imaginative innovations in laboratories of large firms. But more important,

oriented firms with products that are not yet commercially viable. These firms would like to expand and become big businesses, but the product or service they are developing is too far from its final state to be interesting to outside investors.[10]

The 1992 testimony of James A. Block, President of Creare Inc. of Hanover, New Hampshire (U.S. Congress, 1992) provides an example of a firm that is not interested in growing. In discussions, DoD-SBIR program managers assured us that this is one of several cases we could have presented. Mr. Block outlined the commercialization philosophy of Creare Inc. as follows:

> We remain an engineering service company that commercializes its product opportunities through licensing or the creation of separate product companies. Ten such companies trace their roots to Creare. We spin off products for commercialization not because we are blindly following an objective written over 30 years ago, but because we have found it very difficult to optimize both products and services in a single organization. (p. 358)

The point of this example is that Creare, no matter how many commercial successes it initiates, will always want to be a small engineering service firm.[11]

There is clearly hostility to firms such as Creare from some parts of the SBIR program. The April 17, 1998 GAO report evaluating the SBIR program indicates 10 current concerns about the SBIR program (GAO, 1998). Concern number 5 is of interest:

> 5. the number of multiple-award recipients[12] and the extent of their project-related activity after receiving SBIR funding.

Discussions about the SBIR program often focus on "SBIR mills," that is, firms that are receiving a large number of SBIR awards and have no plans nor made

---

inability to get ideas approved by higher management drives many of the most creative individuals out of large corporation laboratories to go it alone in their own ventures. During the two decades following World War II thousands of research-based new enterprises were founded by frustrated fugitives from the laboratories of such U.S. giants as Sperry Rand, IBM, Western Electric, Hughes Aircraft, Raytheon, and many others." (p. 354)

[10]We are implicitly assuming that each small business is only working with one technology. This is a simplification because it is clearly possible for a small business to be involved in many technologies. The analysis follows if we drop firm and replace it with "division" or "technology group" or any other designation that indicates a portion of a firm's effort focused on a technology that has yet to yield commercially viable products or services.

[11]While discussing the effects of trends toward networking and the growth in complexity of technology to discourage concentration in high-technology industries, Freeman and Soete (1997) suggest that Creare might not be an isolated example. They say, "These trends may be reinforced by the preferences of many engineers and scientists to work in smaller and more intimate organizations." (p. 239)

[12]GAO defines multiple award recipients as Phase I awardees for that year who had received at least 15 Phase II awards in the preceding 5 years. There were 10 such firms in 1989 and 17 in 1996. The 17 companies in this category in 1996 had received between 15 and 61 SBIR Phase II awards in the preceding 5 years.

any effort to commercialize the results of their research. Although "multiple-award recipients" is a less pejorative term than "SBIR mills," the connotation is similar.

It is interesting to learn both why these firms are being singled out for special evaluation and how they are being evaluated. Commercialization, or the lack of it, is involved in the answer to both questions. The assumption of such an evaluation is that all small businesses are or should be growth oriented. If one views the SBIR program as a research and commercialization vehicle, then the goal is to carry out research, develop products, and eventually "graduate" small businesses, that is, help them to become big businesses. Under this interpretation, the existence of multiple-award recipients is a problem. These firms have received many awards but have not graduated. However, if one takes seriously the goal of the SBIR program to reduce the barriers to small firms as providers of federal R&D, and one believes that smaller may be better in some types of R&D, then the existence of multiple-award recipients is not a matter of concern. The model that relies on the graduation of firms ignores the fact that some commercialization takes place, as the Creare example illustrates, by firms spinning off new firms that concentrate on commercial development.

Attacks on multiple-award recipients take one of three approaches. First, some completely ignore the research component of the SBIR program. Second, some have the false impression that every small business would like to be a big business. Third, some argue that the multiple-award recipients are just good at writing proposals, but not good at producing research. This is possible, but only for a short time. There is enough continuity in the SBIR program for such a firm to be found out. Either the multiple-award recipients both write good proposals and do good research, or there is a serious problem with the integrity of those involved in the selection and monitoring of SBIR contracts. The intent of Congress is clear. As noted earlier, a firm's past commercial success, or lack thereof, should be taken into account in the selection of Phase II projects. Our survey results and our interviews with program managers indicate that this dimension of the program is fully understood.

As a final comment on multiple-award recipients, it is interesting that there is no concern over such firms among big businesses. There are many big businesses that engage in a large amount of government R&D. They are perhaps best known as long-term successful defense contractors. They are not termed "mills," and they get no special attention in GAO reports.

The third firm type, growth-oriented firms with not yet commercially viable products, represent riskier SBIR projects. In these firms, research is not sufficiently advanced to be able to demonstrate commercial potential to outside investors. The existence of these firms in the SBIR program raises a research policy question. Is the SBIR program incurring enough risk? If we view the SBIR program as a program entirely for growth-oriented firms with commercially viable products, the risk of research and commercial failures will be minimized.

TABLE 1 Types of Small Businesses

Types of Small Businesses	Desire to Grow	Desire to Become Producers	Reliance on Government Funding	Appeal to Outside Investors	Suitability for Fast Track
Growth-oriented commercially viable product	High	High	Low Short Term	High Short Term	High
Non-growth-oriented	Zero	Low	High Long Term	Low Long Term	Low
Growth-oriented not yet commercially viable product	High	Moderate or High	High Short Term	Low Short Term	Low

However, a research program that focuses too much on reducing the risks will not push the technological frontiers as far and as fast as one that is more willing to take risks.[13] The optimal amount of risk to include in a research program is difficult to determine, and we cannot pretend to know the answer, but this still is an important factor to consider when evaluating any aspect of the SBIR program. It is clear that the ultimate objective of R&D conducted by DoD is to enhance national defense. Ultimately, R&D should get a better weapon in the hands of the soldier, a better ship for the sailor, or a better plane for the pilot. The development process for the better weapon, ship, or plane can be very long. Research on the final stages of the process will more quickly be able to point to concrete benefits. In a world in which it is important to collect success stories to justify a program, it is very tempting to concentrate resources on projects that are currently commercially viable. It would be politically naive for those running the SBIR program to ignore the need for commercial success. Therefore, it is a challenge for agencies to balance short-run and long-run goals. There are small businesses doing good research on emerging technologies that should have a chance in the SBIR program. As they develop expertise with their technologies, it is possible that they will develop commercially viable products. It would be very shortsighted to crowd these firms out of the SBIR program.

Table 1 provides a summary of our discussion of the three types of small businesses.

---

[13]SBIR technical monitors recognize this trade-off. One of the comments we received in our survey of technical monitors illustrates this: "I expect half of my SBIRs to be outright bombs at the end, and no more than 10-20 percent to pan out into practical technologies. In contrast ... of this group probably has a 50-75 percent hit rate on practical products; however, his product are typically incremental improved versions, whereas mine are more likely to be novel."

Fast Track's Niche

The Fast Track Initiative was started as a limited experiment. Up to 5 percent of the DoD SBIR contracts could be in the Fast Track program. As Table 1 indicates, not all small R&D firms would find the Fast Track Initiative useful. Firms will not be suited for Fast Track either because of the nature of the firm or the nature of the research projects. In either case, these firms are contributing to the research goal of the SBIR program. They are also contributing to the commercialization goals of the program indirectly, either through spinning off new companies or commercializing by licensing, or by further developing technologies that have the potential to be commercialized.

The number and proportions of firms of each particular type is unlikely to be stable over time. One aspect of a program innovation such as Fast Track is that it is very likely to expand its niche. Fast Track provides powerful incentives for an SBIR participant to push its technology toward commercial viability. Also, Fast Track provides incentives for a particular type of firm to participate in the SBIR program. A small R&D firm with a product that is close to commercial viability has extra incentives to search for military applications for the product. This could expand participation in the SBIR program. In both instances the number of growth-oriented firms with commercially viable products in the SBIR program, Fast Track's niche, will grow. There are limits to the potential size of this niche. A balanced evaluation of the Fast Track Initiative, one that accounts for both commercial and research success, will tell us how successful the Fast Track Initiative has been in filling that niche.

As a final comment on the Fast Track Initiative, it is interesting that our interviews with DoD program managers uncovered several program innovations at the branch level that appear to be responses to Fast Track. Some of the features of Fast Track have been incorporated in the regular SBIR program. For example, one of the program managers sets some funds aside to use as bridge money between Phase I and Phase II. At the discretion of the program manager, these funds allow the research to continue in the time period between the end of Phase I and beginning of Phase II funding. Another program manager has initiated a system in which other components of the branch are asked to share with the SBIR office in the funding of some projects. This mimics the part of Fast Track that requires outside investors to commit funds for Phase II. In this way the program manager can see that other units in the same branch are very interested in the research. These two innovations show how, by its example rather than by its use, Fast Track has changed the SBIR program within DoD. Also, it shows that program managers have been given the freedom to innovate and that they have taken initiative.

SURVEY OF TECHNICAL MONITORS

The fundamental assumption of our approach to evaluation is that the two goals of the SBIR program should be given equal weight. Under this approach, projects that produce commercial success should count as successful SBIR projects, and projects that produce useful research for the contracting agency should count as successful SBIR projects. A proper evaluation of the program has to consider these two possible avenues to success for each project. This approach corrects a fundamental flaw in current evaluations, the fact that, exclusively, these evaluations have been separate evaluations of either commercial success or research success.

Our independent contribution to a balanced evaluation is measurement of the quality and usefulness to the federal government of the research conducted by the SBIR firm; that is, we provide measures of research success. In this section, we discuss the survey we conducted to measure the research outcomes of SBIR awards. In the next section, we describe the results that link this information with a separate survey of the commercial success of the firms awarded SBIR contracts for these projects.

Description of the Survey of Research Outcomes

The individuals who know about the quality and usefulness of SBIR research to the government are government scientists and engineers who monitor the research progress. Our evaluation takes the form of a survey of the technical monitors of the SBIR Phase II contracts. The technical monitors, called either Contracting Officer's Technical Representative (COTR) or Technical Points of Contact, are the DoD scientists or engineers who served as the liaison between the small business and the agency that awarded the contract. For simplicity, we refer to these individuals as the COTRs. Quite frequently, the COTR was involved in determining the topics for the SBIR competition, and in most cases the COTR had worked with the firm during Phase I as well as Phase II.

One potential flaw in relying on a survey of the technical monitors is that they may have too narrow a view of the innovation process or the SBIR program. It is quite possible that a particular SBIR project might not be deemed very high quality or very useful by a scientist in one laboratory, but it might have very important applications in some other area. Because the COTR might be unaware of the eventual success of a particular SBIR Phase II, we interviewed the program managers, that is, the individuals in charge of the SBIR program for the various military branches and research agencies in DoD. These individuals provided important perspectives on our survey results.

One innovative aspect of the survey is that it was conducted via e-mail to make it practical and encourage a high response rate. Because it was designed for e-mail distribution, the survey was as short as possible while capturing key questions. One clear advantage of the e-mail survey over a regular mail survey is that it dramatically shortens the time lag involved. The first day of the survey, we had many responses. Our response rate reveals the good news and the bad news with an e-mail survey. The good news is that we had a very high response rate for e-

TABLE 2 Response Rates

Military Branch/ Agency[a]	Sample Size	Successful Contacts	Completed Surveys	Percentage of Total	Percentage of Contacts
Air Force	66	45	32	48.5	71.1
Army	44	37	34	77.3	91.9
BMDO	167	89	68	41.3	76.4
DARPA	49	41	30	61.2	73.1
DSWA	4	4	4	100	100
NAVY	42	21	18	42.8	85.7
OSD	8	7	5	62.5	71.4
Total	380	244	191	50.3	78.3

[a]DARPA = Defense Advanced Research Projects Agency, DSWA = Defense Special Weapons Agency, OSD = Office of the Secretary of Defense.

mails that were received by a COTR. The bad news is that we have been unable to find good e-mail addresses for a large number of COTRs.

The database contains 380 SBIR Phase II contracts. The sample was drawn systematically and it includes all of the 1996 Fast Track projects and all of the Ballistic Missile Defense Office (BMDO) matching projects from 1992 to 1996 as well as a matched sample of regular Phase II projects. As such, it does not represent a random sample of the entire DoD SBIR program. Some of the analyses we present would be more appropriately performed with a random sample, and therefore the results should be interpreted with great care.

We have responses covering 191 (50.3%) of the contracts in the full sample. Table 2 accounts for our responses. Clearly, our biggest problem is making contact; our overall response rate for successful contacts, 78.3 %, is very high.[14] The low response rates for BMDO and Navy result primarily from difficulties caused by retirements and missing e-mail information.

The survey had two parts and covered five major areas. The first part of the survey focused on individual SBIR projects, measuring (1) research quality, (2) usefulness of the research, and (3) mission benefits of the research. The second part focused on the SBIR program, measuring (4) overall quality of SBIR proposals, and (5) impressions of Fast Track. Our 191 responses represent responses to the first part of the survey. In the cases in which a COTR was assigned to more than one project in the sample, we only asked the questions in the second part of the survey once. There are 132 responses to the second part of the survey. The question about Fast Track was asked only of those COTRs who had experiences with Fast Track. We have 51 responses to that question.

[14]We defined a successful contact as an e-mail message that was not returned to us by the mailing system.

Part 1: Responses about SBIR Phase II Projects

Area 1: Research Quality

One area of interest for any research program is the quality of the research. We asked two questions about the quality of research:

3. On a 1 to 10 scale, where 10 represents the best research ever produced in your research unit/office or for your research unit/office and 1 represents the worst research ever produced in your research unit/office or for your research unit/office, rate the quality of the research in this particular SBIR contract.

Responses (n = 189): Mean = 7.069, Std. Dev. = 1.865.

4. On the same 10-point scale, rate the average quality of the research projects conducted for your research unit/office from contracts other than SBIR contracts for the last three years.

Responses (n = 181): Mean = 7.064, Std. Dev. = 1.754.

We included Question 4 to provide a context for the responses to Question 3. Some people may be easy graders, giving high marks to any research project, while others are hard graders, giving low grades to any research project. The most interesting number is the difference between the responses to Question 3 and Question 4.

Previous studies contradict one another on the quality of SBIR research. As part of the assessment required by Congress, the GAO in 1989 studied the quality of the research in the SBIR program. The GAO report on the scientific quality of the SBIR program gave the program very high marks. The basic findings were based on a survey of 530 project officers who monitor SBIR research. The conclusion was:

- Overall, respondents assessed 29 percent of the SBIR projects as being higher quality than non-SBIR research and indicated that about half of the SBIR projects were similar in overall quality to other research.

This suggests that the overall quality of SBIR research is roughly comparable to the quality of other federal research. A 1996 article by Jeffrey Mervis in *Science* indicates that there are those who dispute this claim. Using data on evaluation scores for research proposals at the National Institutes of Health, opponents of the SBIR program have shown that funded SBIR awards have lower scores than do other research projects.

For our sample, the mean value for the difference between the rating given to the research quality of the SBIR project and the rating given to other research is 0.025 with a standard deviation of 2.366 (n = 180). Clearly, there is no statistically significant difference between these ratings. We conclude that the quality of SBIR research is roughly equal to that of other contract research in the DoD.

Area 2: Usefulness of the Research

To discover whether the SBIR research was useful, we asked survey respondents to indicate whether the research has affected the way that research is conducted in their unit/office.

5. Has the research conducted for this SBIR contract affected the way that your research unit/office conducts research or the type of research your research unit/office obtains in other contracts? List as many as apply.
   a. No, the knowledge generated by this SBIR contract has had no impact on the other research we conduct or sponsor.
   b. Yes, this project produced results that have been useful to us, and we have tried to follow up on the ideas initiated in this SBIR contract by encouraging the firm to apply for additional SBIR awards.
   c. Yes, this project produced results that have been useful to us, and we have tried to follow up on the ideas initiated in this SBIR contract in other research we conduct or sponsor.
   d. Yes, but this project found a blind alley, so we have not followed up on this line of inquiry.

Responses (n = 183)[15]: (a) = 25.96%, (b) = 19.13%, (c) = 49.18%, (d) = 5.74%.

The responses to Questions 5 are quite positive for the SBIR program. Combining the two unambiguously positive responses, (b) and (c), we find that 68.30% of the SBIR had results encouraging enough to suggest further research. The high percentage for the (c) response indicates that the research that the SBIR program sponsors frequently affects the general DoD research program. The response of (a) to Question 5 appears to indicate a research failure, that is, the research project generated no follow-up. It is interesting that we found a few (d) responses, projects that produce useful research by finding a blind alley. This indicates that the SBIR program funds some risky projects.[16]

Area 3: Mission Benefits

The next question in the survey focused on mission benefits:

6. In comparison to a dollar spent in your research unit/office on other R&D projects, did a dollar spent on this SBIR project:
   a. yield more benefits for your agency's mission than the average dollar spent on other contracts sponsored by your research unit/office.
   b. yield the same level of benefits for your agency's mission as the average dollar spent on other contracts sponsored by your research unit/office
   c. yield fewer benefits for your agency's mission than the average dollar spent on other contracts sponsored by your research unit/office

[15]We have distributed the joint responses, for example, (b) and (c), evenly so that the total responses = 100%.

[16]It is important to recognize that all of our projects are Phase II projects. The prevalence of blind alleys would be much higher for Phase I projects.

Responses (n = 186): (a) = 31.99%, (b) = 41.67%, (c) = 26.34%.

The responses to Question 6 also offer positive evidence for the SBIR program. We recognize that respondents were asked to make a difficult comparison, and some respondents pointed this out to us. They were asked to compare the benefits to the agency's mission per dollar from the SBIR project with the average benefits to the mission per dollar of other sponsored research. This is a complicated comparison. Still, the results are consistent with findings on other dimensions of research output. A vast majority of the SBIR projects, 73.66%, were rated as beneficial or more beneficial per dollar than the average of other R&D projects. This is a strong endorsement of the research conducted in the SBIR program.

Part 2: Responses by COTRs About SBIR and Fast Track

Area 4: Overall Quality of Proposals

We asked each COTR about the whether he or she was satisfied with the quality of the SBIR proposals his or her office had received.

7. In general, do you find that your research unit/office has had more (about the same, too few) good SBIR proposals than you can fund?

Responses (n = 131): more = 64.12%, same = 24.81%, too few = 11.07%.

These results reflect well on the DoD SBIR program. A large majority of the COTRs think that they have more good proposals than they can fund, indicating a very competitive program. The competitive nature of the SBIR program could indicate that the program is underfunded.

Area 5: Fast Track

We asked those COTRs who had been involved with a Fast Track to respond to the following question:

9. If you have experience with the Fast Track program, is it more effective, less effective, or not different from the normal SBIR program in advancing the research program of your unit?

Responses (n = 51): more = 66.67%, less = 11.76%, not different = 21.57%.

This result suggests that Fast Track has a good reputation among those technical monitors who have been involved with it. It is unfortunate that the sample size is so low for this question. There were only 53 respondents who had been involved in any of the Fast Track SBIR contracts, and 2 of them failed to respond to Question 9.

FURTHER ANALYSIS OF THE TECHNICAL MONITOR SURVEY RESPONSES

In this section we evaluate the DoD SBIR program and the Fast Track Initiative with a focus on measures of quality, usefulness, and mission benefits derived from our survey of technical monitors.

COTR "Ownership"

When we constructed our survey of the technical monitors, we were concerned that the responses from COTRs who had a stake in the evaluation of the project might be biased in favor of showing favorable results. For this reason, we asked both whether or not the COTR was involved in defining or generating the topic for the Phase II project and when the COTR became involved with the SBIR firm. COTRs who were involved in defining or generating the topic as well as having been involved with the firm from before the Phase I proposal might be suspected of being biased in favor of the project. They might have taken some ownership of the research. Our interviews with program managers indicate that they thought this was possible.

Table 3 gives the results of responses from COTRs who might have experienced an ownership effect and the remaining COTRs in the areas of Research Quality, Usefulness of Research, and Mission Benefits. There is little evidence of an effect. In fact, although the difference is not statistically significant, the average quality rating is lower for the projects for which the COTR might have exhibited an ownership effect. The COTRs in the ownership group do seem to rate the projects as more useful. This result is statistically significant.[17] Perhaps it not a surprise to see that the COTRs who had the most chance to shape the research found it most useful. The ratings of mission benefits indicate little difference at all. We interpret these results as saying that there is no bias in the ratings of research quality or mission benefits from an ownership effect.

Fast Track and BMDO Matching

The sample used in this study was designed to study Fast Track. It contained three types of SBIR projects: (1) Fast Track, (2) BMDO Matching contracts, and (3) a comparison group. The BMDO Matching sample was included because the BMDO SBIR program requires matching funds for Phase II as does Fast Track. BMDO Matching and Fast Track are not identical, but they do share the important feature that they require outside funds for participation in Phase II. Table 4 gives the results for these three samples.

---

[17]A chi-square test yields a statistic of 8.66, which exceeds the critical value of 7.815 for the 5% level.

TABLE 3 Analysis of Ownership Effect

Area of Interest	Ownership Group	Remaining COTRs
Research Quality (Q3-Q4)	−.1279	.1373
Usefulness of Research (Q5)		
a. no, not useful	17.9%	31.9%
b. yes, more SBIR	24.3%	15.2%
c. yes, general follow-up	55.1%	44.8%
d. yes, but blind alley	2.5%	8.1%
Mission Benefits (Q6)		
a. more than average	32.5%	31.6%
b. same as average	44.8%	40.0%
c. less than average	23.8%	28.3%

It is interesting to compare these results to those from our direct question about Fast Track. Two-thirds of the 51 COTRs who had direct experience with Fast Track rated it as "more effective than the normal SBIR program." Though none of the differences is statistically significant, generally the results in Table 4 are consistent with this assessment. The average Research Quality of Fast Track projects is highest and the percentage of "no, not useful" in the rating of the Usefulness of Research is lowest for the Fast Track projects.[18] The only assessment in which Fast Track projects are not rated most highly is for projects rated as having "more than average" Mission Benefits per dollar, though the Fast Track research does have the highest percentage of "more" or "the same" level of benefits. Our conclusion is that these findings corroborate the opinion of the COTRs who had experience with Fast Track projects.

The results in Tables 3 and 4 are for differences in means. It is possible that a closer look at the data, one that controls for other factors, will provide different results.

For each project, we have information on the following firm and program characteristics:

1. number of prior Phase II contracts for the firm;
2. whether the firm was woman or minority owned;
3. number of employees of the firm;
4. year of the award;
5. major topic area of the project:
   a. computers,
   b. electronics,

[18]One of the program managers we interviewed indicated that the Fast Track projects should be expected to have better results because they had higher budgets. These projects have large additions to their research budgets from a source or sources outside the SBIR program.

TABLE 4 Analysis of the Fast Track, BMDO Matching, and Comparison Group Samples

Area of Interest	Fast Track	BMDO Matching	Comparison Group
Research Quality (Q3-Q4)	*n* = 30	*n* = 51	*n* = 99
	.4333	.1764	–.1767
Usefulness of Research (Q5)	*n* = 32	*n* = 52	*n* = 99
a. no, not useful	15.6%	39.4%	22.2%
b. yes, more SBIR	21.9%	18.3%	18.7%
c. yes, general follow-up	59.4%	38.5%	51.5%
d. yes, but blind alley	3.1%	3.9%	7.5%
Mission Benefits (Q6)	*n* = 32	*n* = 53	*n* = 101
a. more than average	28.1%	26.4%	36.16%
b. same as average	46.9%	48.1%	36.6%
c. less than average	25.0%	25.5%	27.2%

c. materials,
d. mechanical performance of vehicles, weapons, facilities,
e. energy conservation and use,
f. environment and natural resources,
g. life sciences;
6. branch that awarded the contract;
7. whether the COTR might exhibit an ownership effect.

In Appendix A, we provide information on the three measures of research success by the program characteristics. Although there are some clear differences in the table in Appendix A, few of them are statistically significant.

Our research strategy is, holding other things constant, to search for relationships between these variables and the measures of quality, usefulness, and mission benefits. What type of results do we expect from this investigation? Consider, for example, the relationship between research quality and the number of prior Phase II contracts for the firm. If, holding other things constant, we found a statistically reliable positive relationship between these two measures, it would indicate that firms with longer track records in the SBIR program, on average, produced better research. One could infer from that finding that the overall quality of the SBIR program could be improved if such firms received a larger share of the contracts. On the other hand, if we found a statistically reliable negative relationship, the conclusion would be just the opposite. The overall quality of the SBIR program could be improved if such firms received a smaller share of the contracts. Either of these results would suggest that the DoD SBIR program needs to change its project selection to improve the overall quality of the research. If, however, there is not a statistically significant relationship, the overall quality of the projects cannot be improved by changing the project selection. In this case, the absence of statistically significant relationships indicates a well-run program.

Appendix B gives the results of our correlation and regression analyses. The correlation analysis indicates that there are few strong correlations between variables. The three measures of research success—(1) research quality, (2) whether the research was useful to the further research effort, and (3) whether the contract yielded more than average benefits to the mission—are related to one another. Each of these correlations is above .2. The correlations between these indicators of research success and the firm and program characteristic variables are, however, not high. None of these correlations is as high as .2, and most of them are below .1. Also, there is little evidence of correlation among the program and firm characteristics.

The regression analyses for quality indicated only one statistically significant relationship. The year of the project, with more recent projects being rated as being higher quality, was statistically significant. The lack of statistically reliable findings is, as we explained earlier, an indication that the DoD SBIR program is efficiently managed, at least in terms of the characteristics we could measure. There do not appear to be statistically reliable relationships that could be exploited to raise the average quality of the research.

The results for the logit regressions of the indicator variable that measured whether the research was useful to the further research effort (based on the responses to Question 5) indicate completely statistically insignificant results. Interestingly, the effect we found in the bivariate analysis between the ownership and the ratings of usefulness of the research disappears in the multivariate analysis. Again, as with research quality, the statistically insignificant findings suggest that the SBIR program cannot exploit existing information to enhance the usefulness of the projects.

The final regression, for the indicator variable for projects that provided more than average mission benefits (based on responses to Question 6), contains some statistically significant results. The coefficients for the indicator variables for the Air Force, the Army, and DSWA are all positive and statistically significant (the excluded variable for this group of indicator variables was BMDO). Also, the coefficient for projects in the Electronics topic category was positive and statistically significant (the omitted variable for this group of indicator variables was Materials).

In sum, these results provide a positive evaluation of the DoD SBIR program. Measures of research quality, the usefulness of the research, and the benefits to the agency mission are generally not statistically significantly related to information available at the time of the award of the contract. This indicates that there are no easy improvements in research performance.

These results do suggest a way of thinking about some issues surrounding the SBIR program. For example, the lack of a relationship between the measures of research success and the number of prior Phase II contracts provides no support for restrictions on multiple-award recipients. The results also provide no particular support for multiple-award recipients; that is, there is no evidence that they receive more SBIR contracts because they do better-than-average research. According to our results, the research performance of these firms is indistinguishable from that of firms with less experience in the program. Finally, these results suggest that Fast Track contracts and BMDO Matching contracts are similar to other SBIR contracts in terms of their research output. There is no evidence that the research outputs of the SBIR program have been compromised by the emphasis on commercial success from the Fast Track Initiative.

COMBINING MEASURES OF RESEARCH AND COMMERCIAL SUCCESS

In this section, we combine the results from our survey of COTRs with the results from a survey of firms conducted by Peter Cahill of BRTRC. The survey of firms was designed primarily to provide information on commercial outcomes. The linked survey responses therefore allow us to measure interactions between the research and commercial goals of the SBIR program. A majority of the projects for which we have information from technical monitors are contained in the data set of firm responses. There are 122 observations in the linked data set.

Table 5 gives a summary of the findings of the two surveys and the characteristics of the linked data set. We considered three measures of commercial success: (1) having actual sales, (2) having expected sales, and (3) having either actual or expected sales. The data indicate that a vast majority of the firms (or their licensees) have or expect to have sales from a product or service developed through SBIR. Generally, the linked data set is representative of the two larger data sets. Firms that have already had sales are slightly underrepresented in the linked data set. Interestingly, the linked sample includes projects that the technical monitors rated as slightly better for all three of the measures of research success than in the full COTR data, but the differences are very small.

Our analytical strategy is similar to the one we used for the analysis of the COTR survey. The major difference is that success now has two dimensions. Those involved in the SBIR program in many agencies have long struggled with defining success in the program. In our view, there is no single measure of success. Success in the SBIR program is inherently two-dimensional. For given definitions of commercial success and research success, we can place projects into four groups:

1. Group N—neither a research success nor a commercial success,
2. Group R—research successes that are not commercial successes,
3. Group C—commercial successes that are not research successes,
4. Group B—both a research success and a commercial success.

TABLE 5 Results of Both Surveys

Variable	Survey of COTRs (%)	Survey of Firms (%)	Linked Survey (%)
Firm survey questions			
8. Firm or licensee has had sales of products, process, services or other sales from the technology		42.6	36.9
11. Firm or licensee expected to have sales in the next three years but has yet to have sales		50.4	54.9
8. & 11. Firm or licensee has had or expects to have sales		93.0	91.87
COTR survey questions			
3 & 4. Quality of the SBIR research is greater than the average quality of non-SBIR research	46.6		48.4
5. The SBIR research has been useful to the government agency	74.0		74.5
6. The SBIR project yielded more benefits per dollar than average	32.0		33.9

These groups are then the focus for analysis.

We have several potential measures of research and commercial outcomes and, for some measures, it is not obvious where one should draw the line between success and failure. For this reason, there are many ways one could classify the projects. We provide two examples.

Example 1: Sales and Usefulness of Research

In this example we use Question 8 on the survey of firms:

> 8. Has your company and/or licensee realized any actual sales of products, process, services or other sales from the technology developed during this project?

For research success we will use Question 5 from the COTR survey:

> 5. Has the research conducted for this SBIR contract affected the way that your research unit/office conducts research or the type of research your research unit/office obtains in other contracts?

Combining these two definitions of success yields the results in Table 6.

Given these classifications, there are two interesting questions. First, are commercial and research success related? For these data, they are not. The chi-square statistic for the test of independence is 1.525, which is not sufficiently high to reject independence. Second, what are the correlates of assignment into

TABLE 6 Classification of Research and Commercial Success Using Usefulness of Research and Sales

	Research Success	
Commercial Success	Yes (74.5%)	No (24.5%)
Yes (37.9%)	Group B, 30 (25.8%)	Group C, 14 (12.1%)
No (62.1%)	Group R, 56.5 (48.7%)	Group N, 15.5 (13.4%)

the various classifications? Table 7 presents information on this question. The table shows the prevalence of projects with various characteristics in the linked sample and for the projects in the two extreme categories: Group B, both a research and a commercial success, and Group N, neither a research nor a commercial success. The entries in the table are percentages of projects in the particular classification that are part of a particular group. For example, 25.4% of the full sample are Fast Track projects, but 19.4% of the Category B projects are Fast Track projects.

Though none of the differences in proportions are statistically significant, the results for Fast Track are mixed. Fast Track projects are slightly underrepresented in Category B, which does not reflect well on Fast Track, and very underrepresented in Category N, which reflects well on Fast Track. If it is more important to avoid Category N than it is to hit Category B, as seems sensible, then these are essentially positive results for Fast Track. The results for BMDO Matching projects appear to be uniformly inferior. Overall the Comparison group firms seem better in this comparison. Firms with more than three prior Phase II awards appear to be inferior by these measures also. In contrast, having fewer than three prior Phase II awards results in a statistically significant lower proportion of firms in Category N. This is the only statistically significant difference for this table. Finally, there does not appear to be a clear pattern for results based upon the number of employees in the firm.

TABLE 7 Analysis of Example 1 Classifications by Groups

Variable	Full Sample (%)	Group B (%)	Group N (%)
Fast Track	25.4	19.4	6.7
BMDO Matching	25.4	16.1	53.3
Comparison	49.2	64.5	40.0
Prior Phase II ≤ 3	81.1	90.3	60.0
Prior Phase II > 3	18.9	9.7	40.0
Employees ≤ 10	58.0	54.8	50.0
Employees > 10	42.0	45.2	50.0

Example 2: Sales and Mission Benefits

In this example we keep the same definition of commercial success and change the definition of research success. We base research success on Question 6 in the COTR survey:

5. In comparison to a dollar spent in your research unit/office on other R&D projects, did a dollar spent on this SBIR project yield more/the same/less benefits for your agency's mission than the average dollar spent on other contracts sponsored by your research unit/office.

We define research success as yielding more benefits than the average on other contracts, a very strict definition of research success.

In this case, the test of the hypothesis of independence yields a chi-square statistic of 1.209, which is not statistically significant.

Table 9 presents the analysis of these categories by the groups we used above. In these comparisons, again Fast Track is underrepresented in Group B, but this time Fast Track is also overrepresented in Group N as well, the uniformly poor rating. The BMDO Matching sample has mixed results, overrepresented in Group B and slightly underrepresented in Group N. None of the results for Fast Track or the BMDO Matching is statistically significant. Prior Phase II experience is again not related to favorable outcomes according to these findings. Again, firms with three or fewer prior Phase II awards were significantly less likely statistically to be in Category N. The results for firm size indicate that small firms are more than proportionately related to being both a commercial and a research success. For firm size, the result that firms with 10 or fewer employees were more likely to be in Group B is statistically significant.

The two examples illustrate that this type of analysis depends critically on the way one defines success. In Table 6, we defined research success in terms of whether the research in the SBIR provided results that were useful to the agency—a minimal notion of success. In Table 8, the research success was reserved for projects that provided more benefits to the agency mission than the average contract—a much stricter notion of research success. In Table 8, with the lower hurdle for research success, over 86% of the topics registered some type of success. In Table 8, with the higher hurdle for research success, just over 56% of the projects registered some type of success.

TABLE 8 Classification of Research and Commercial Success Using Mission Benefits and Sales

	Research Success	
Commercial Success	Yes (33.9.5%)	No (66.1%)
Yes (38.1%)	Group B, 18 (15.3%)	Group C, 14 (22.9%)
No (61.9%)	Group R, 22 (18.6%)	Group N, 51 (43.2%)

TABLE 9 Analysis of Example 2 Classifications by Groups

Variable	Full Sample (%)	Group B (%)	Group N (%)
Fast Track	25.4	18.2	31.4
BMDO Matching	25.4	31.8	24.5
Comparison	49.2	50.0	55.9
Prior Phase II ≤ 3	81.1	95.4	72.5
Prior Phase II > 3	18.9	4.6	27.5
Employees ≤ 10	58.0	77.3	52.1
Employees > 10	42.0	22.7	47.9

SUMMARY AND CONCLUSIONS

Our summary and conclusions come under four headings: (1) conclusions about appropriate evaluation of the SBIR program, (2) results for the general DoD SBIR program, (3) results for the Fast Track Initiative, and (4) suggestions for future research.

Appropriate Evaluation of the SBIR Program

The SBIR program is difficult to evaluate. Most previous evaluations have focused on commercial success and have been based on surveys of firms. The quality of the research in the SBIR program has received little attention. This paper provides the first evaluation that looks simultaneously at the two goals of the program. There are three methodological points we want to emphasize.

1. The SBIR program has two goals: to increase the participation of small business in federal R&D and to increase the commercialization of innovations developed as a result of federal R&D. An appropriate evaluation has to account for both of the program goals.
2. E-mail is a good medium to evaluate outcomes of the research. The technical monitors—the DoD scientists and engineers who worked with the firms—were generally quite willing to respond to an e-mail survey.
3. Our analysis of the responses from the survey of technical monitors and the survey of firms illustrates the appropriate methodology for evaluating a program with two clear goals. One objective of any agency's SBIR program should be to minimize the projects that are neither a research nor a commercial success. Our methodology illustrates the first attempt to evaluate progress for this objective.

Results Regarding General DoD SBIR Program

Several of our results provide an evaluation of the performance of the overall DoD SBIR program as a research program. In general, the results are very positive. There are five findings we would like to highlight:

1. Based on our survey responses, the quality of the research conducted in the SBIR program is indistinguishable from other research conducted by or sponsored by DoD.
2. Based on our survey responses, a large majority (74.04 percent) of the SBIR projects were rated as having produced results that were useful for the ongoing research effort at DoD.
3. Based on our survey responses, a large majority (73.66 percent) of the SBIR projects were rated as having produced the same or more mission benefits per dollar than other R&D projects of DoD.
4. Based on our survey, a substantial majority (61.14 percent) of the technical monitors indicated that they had more good SBIR proposals than they could fund, and only a small proportion (11.07 percent) indicated that they had fewer good SBIR proposals than they could fund.
5. Using multivariate analysis, we found no clear evidence that the research success of the DoD SBIR program could be improved based upon information available to program managers prior to the award of the contract.

The SBIR program is being considered for reauthorization in the next Congress. One of the questions that will come up in that debate is whether it is sensible to expand the percentage of the R&D budget set aside for the SBIR program. This is a complex issue. Our results indicate that, at the current size, the quality of DoD SBIR research is high and there are more good proposals than can be funded. Although these results are not sufficient to suggest an expansion of the program, they probably represent necessary conditions.

Results Regarding the Fast Track Initiative

The sample we used was designed to study the Fast Track Initiative. We gathered information on Fast Track in two ways. We asked the technical monitors who had experience with the Fast Track Initiative for a direct evaluation and, in the analysis of our results, we separated the sample to see if performance by Fast Track projects differed from performance by the comparison groups. The Fast Track Initiative generally is viewed very favorably and the results for Fast Track projects are impressive.

- Two-thirds of the technical monitors who had experience with a Fast Track project rated Fast Track as more effective in advancing the research program in their unit. Only 11.76 percent of these respondents rated Fast Track as less effective in advancing the research program in their unit.
- Though the differences were not statistically significant, the quality rating for Fast Track projects indicated an advantage for Fast Track compared to other projects.
- Though the differences were not statistically significant, the percentage of Fast Track projects rated as being useful for the research effort was higher than for non-Fast Track projects.

These results are impressive because the Fast Track Initiative was designed to improve the commercial success, not the research success, of the DoD SBIR program. Given the current size of Fast Track—5 percent of the DoD SBIR program—there clearly is not a trade-off between commercial and research success. Although perhaps unexpected, the lack of any reduction in research success may result from the fact that Fast Track projects have larger research budgets. One of the clear advantages of the Fast Track Initiative is that it attracts funds to the federal R&D effort. Our results suggest this has been beneficial for research.

Issues Deserving Further Research

One limitation of the research reported in this paper is the sample design. A correct evaluation of the DoD SBIR program requires a random sample. The linked sample provides a very interesting view of outcomes in the SBIR program. The objective of program managers should be to reduce the number of failures, that is, projects that are neither research nor commercial successes. It would be very important to search for the correlates of projects that are classified in this category from a random sample of SBIR projects. The preliminary analysis with our sample uncovered some interesting findings. In particular, it appears that firms with three or fewer prior Phase II contracts are underrepresented in the failures whereas firms with more than three prior Phase II contracts are overrepresented. It would be very interesting to see if such findings hold for a random sample.

ACKNOWLEDGMENTS

We would like to thank Joe Heyman, Thayer Sheets, and Bob Yang of NASA-Langley Research Center for valuable discussions about the workings of the SBIR program, Charles Wessner and John Horrigan (NAS) for useful comments on an earlier version of this paper, Peter Cahill (BRTRC) for help with data, and public policy graduate students Stephen Bowman, Jennifer Jebo, Randy Rosso, and Keith Wandtke for valuable research assistance. The above-listed individuals are not responsible for any shortcomings of the paper. The views expressed are those of the authors and do not represent the view of the College of William and Mary. Please do not cite without permission.

REFERENCES

Acs, Zoltan J., and David B. Audretsch. 1990. *Innovation and Small Firms*. Cambridge, MA: MIT Press.

Arrow, Kenneth. 1973. "The theory of discrimination," pp. 3-31 in *Discrimination in Labor Markets*, Orley Ashenfelter and Albert Rees, eds. Princeton, NJ: Princeton University Press.

Birch, David. 1981. "Who creates jobs," *The Public Interest* 65(Fall):3-14.

Freeman, Chris, and Luc Soete. 1997. *The Economics of Industrial Innovation.* Cambridge, MA: MIT Press.

Griliches, Zvi. 1992. "The search for R&D spillovers," *Scandinavian Journal of Economics* 94:S29-S47.

Hall, Bronwyn H. 1993. "Industrial research during the 1980s: Did the rate of return fall?" *Brookings Papers: Microeconomics* 2:289-343.

Hamberg, Dan, 1963. "Invention in the industrial research laboratory," *Journal of Political Economy* (April):95-115.

Lerner, Josh. 1999. "The government as venture capitalist: The long-run effects of the SBIR program," *Journal of Business.* July, v. 72 3, pp. 285-297.

Mervis, Jeffrey D. 1996. "A $1 Billion 'Tax' on R&D Funds," *Science* 272:942-944.

OECD (Organization for Economic Cooperation and Development. 1982. *Innovation in Small and Medium Firms.* Paris: OECD.

Roberts, Edward B. 1968. "Entrepreneurship and technology," *Research Management* (July):249-266.

Rubenstein, A. H. 1958. *Problems Financing New Research-Based Enterprises in New England.* Boston MA: Federal Reserve Bank.

SBA (Small Business Administration). 1992. *Results of Three-Year Commercialization Study of the SBIR program.* Washington, D.C.: U.S. Government Printing Office.

Scherer, F. M. 1970. *Industrial Market Structure and Economic Performance.* New York: Rand McNally College Publishing.

Stiglitz, J. E., and A. Weiss. 1981. "Credit rationing in markets with incomplete information," *American Economic Review* 71:393-409.

U.S. Congress, House. 1992. *SBIR and Commercialization: Hearing Before the Subcommittee on Technology and Competitiveness of the House Committee on Science, Space, and Technology, on the Small Business Innovation Research [SBIR] Program* (testimony of James A. Block, President of Creare Inc.) pages 356-361.

U.S. Congress, Senate Committee on Small Business. 1981. Small Business Research Act of 1981. S.R. 194, 97th Congress.

U.S. General Accounting Office. 1989. *Federal Research: Assessment of the Small Business Innovation Research Program,* GAO/RCED-89-39, Washington, D.C.: GAO.

U.S. General Accounting Office 1992. *Small Business Innovation Research Program Shows Success But Can Be Strengthened,* RCED-92-32. Washington, D.C.: GAO.

U. S. General Accounting Office. 1998. *Federal Research: Observations on the Small Business Innovation Research Program*, RCED-98-132. Washington, D.C.: GAO.

APPENDIX A

Means of Variables

This table gives the means of the variables for various subsamples of the data. For simplicity these means are for the respondents who answered both Question 5 and Question 6. This causes some of the means to differ slightly from those reported in the text.

TABLE A1 COTR Survey Results

Subsample	*N*	Quality Q3-Q4	Useful Q5 % b,c,d	Mission Benefit Q6 % a	b
Total	184	.0511	73	32	42
0 Prior Phase II	86	.2500	79	33	42
1-10 Prior	78	–.0395	69	32	40
>10 Prior	20	–.4000	65	25	50
Women/Min.	47	.2727	66	38	36
1-9 Employees	82	.4494	74	40	33
10-19 Emp	38	.0429	68	26	47
20-50 Emp.	33	–.6613	73	18	52
>50 Emp.	31	–.2419	77	29	48
Fast Track	31	.5862	84	29	48
92 Award	16	–.5625	63	19	56
93 Award	21	.5000	86	33	43
94 Award	30	–.5000	60	20	27
95 Award	33	.0156	73	48	30
96 Award	63	.2756	77	31	49
Computers	31	–.1207	71	26	42
Electronics	83	.3063	70	36	42
Materials	12	–1.6667	58	8	58
Mechanical	9	.6111	78	44	22
Energy Cons.	15	–.2000	73	27	47
Environment	4	.0000	100	50	50
Life Sciences	1	–4.000	100	0	0
Air Force	28	.5357	79	43	32
Army	31	.3214	87	39	32
BMDO	67	.0308	63	25	49
DARPA	33	–.2419	76	33	39
DSWA	4	.6250	75	75	0
Navy	18	–.6470	72	17	50
OSD	3	–.3333	100	0	100
Ownership	77	–.0205	82	32	44

APPENDIX B

Multivariate Analyses

This appendix presents the correlation and regression results discussed in Section VI. The variable definitions are given below followed by the correlation matrix and the estimated regressions. Because there were several missing values for the topic categories, for each of the three dependent variables, SBIR Quality, Useful, and Benefits, there are two regressions, one using all the variables, and one excluding the variables for topic categories. The regressions for Useful and Benefits are logit regressions.

Variable	Definition
SBIR Quality	a rating on a scale of 1 to 10 for the quality of the SBIR project
Other Quality	a rating on the same scale of the average quality of non SBIR research conducted in the last three years
Useful	a categorical variable equal to one if the SBIR project's research results were useful
Benefits	a categorical variable equal to one if the SBIR project's benefits to the mission (per dollar) were greater than the benefits to the mission (per dollar) from other projects
AF	a categorical variable equal to one if the project was an Air Force funded project
ARMY	a categorical variable equal to one if the project was an Army funded project
BMDO	a categorical variable equal to one if the project was a BMDO funded project
DARPA	a categorical variable equal to one if the project was a DARPA funded project
DSWA	a categorical variable equal to one if the project was a DSWA funded project
NAVY	a categorical variable equal to one if the project was a Navy funded project
OSD	a categorical variable equal to one if the project was an OSD project
Fast Track	a categorical variable equal to one if the project was a Fast Track project
BMDO Match	a categorical variable equal to one if the project was a BMDO Matching project
AGE	age of the project computed as 97 minus the fiscal year for the project older projects will have larger values.
SIZE	size of the SBIR firm measured as the number of employees
Experience	experience in the SBIR program measured as the number of prior Phase II awards for the firm
Ownership	a categorical variable for those projects for which the COTR was involved with defining or generating the topic and was involved with the firm before the Phase I proposal
COMPUTER	a categorical variable equal to one if the topic area was Computers
ELECTRON	a categorical variable equal to one if the topic area was Electronics
MATERIAL	a categorical variable equal to one if the topic area was Materials
VEHICLES	a categorical variable equal to one if the topic area was Mechanical Performance of Vehicles, Weapons, or Facilities
ENERGY	a categorical variable equal to one if the topic area was Energy Conservation and Use
ENVIRON	a categorical variable equal to one if the topic area was Environment and Natural Resources
LIFE SCIENCES	a categorical variable equal to one if the topic area was Life Sciences

TABLE B1 Correlation Matrix

	(1)	(2)	(3)	(4)	(5)	(6)	(7)	(8)	(9)	(10)
(1) SBIR Quality	1.000									
(2) Other Quality	.1571	1.000								
(3) Useful	.3840	.0095	1.000							
(4) Benefits	.3731	–.1403	.2533	1.000						
(5) AF	.0762	–.0503	.0494	.1254	1.000					
(6) ARMY	.0639	.0487	.1487	.0606	–.1908	1.000				
(7) BMDO	–.1048	–.1424	–.1763	–.1108	–.3270	–.3270	1.000			
(8) DARPA	.0244	.1051	.0669	.0391	–.1986	–.i986	–.3403	1.000		
(9) DSWA	.0815	.0375	.0141	.1408	–.0657	–.0657	–.1125	–.0683	1.000	
(10) NAVY	–.0681	.0882	–.0296	–.1061	–.1452	–.1452	–.2488	–.1511	–.0500	1.000
(11) OSD	–.0116	–.0158	–.0397	–.1143	–.0736	–.0736	–.1262	–.0766	–.0253	–.0560
(12) Fast Track	.0967	–.0202	.1156	–.0043	–.1087	.3841	–.1700	. 0013	–.0657	–.0445
(13) BMDO Match	–.0124	–.0754	–.1998	–.0730	–.2699	–.2699	.8353	–.2809	–.0929	–.2053
(14) AGE	–.2291	–.1500	–.0493	–.0561	–.0871	–.3673	.4206	–.0236	.0057	–.0080
(15) SIZE	.0362	.0593	–.0315	.0487	.0017	.0842	–.1507	–.1068	.0868	.1263
(16) Experience	–.0054	.0030	–.0592	–.0293	.1310	.0019	–.0791	–.0993	.0400	.0400
(17) Ownership	.0187	.1119	.1432	.0105	.1978	.1978	–.3723	–.0402	.0210	.1209
(18) COMPUTER	–.0202	.0908	–.0837	–.0336	.1134	.0739	–.3167	.0584	–.0697	.1847
(19) ELECTRON	.0284	–.1153	–.0313	.0835	–.0599	–.1508	.1832	–.0552	.0159	–.0021
(20) MATERIAL	–.1122	.0511	–.1031	–.1425	–.0632	–.1215	.2378	–.0697	–.0418	–.0209
(21) VEHICLES	.1016	.0353	.0355	.0638	.1772	–.0999	–.0123	.0309	–.0344	–.0760
(22) ENERGY	–.0752	–.0388	.0173	–.0312	–.0220	–.1313	.1092	.0761	–.0452	–.0329
(23) ENVRION	.0304	.0375	.0967	.0599	.0368	–.0657	–.1125	.1312	–.0226	.0756
(24) LIFE SCIENCES	–.1266	.0399	.0480	–.0505	.1706	–.0326	–.0558	–.0339	–.0112	–.0248

(11)	(12)	(13)	(14)	(15)	(16)	(17)	(18)	(19)	(20)	(21)	(22)	(23)	(24)
1.000													
.0183	1.000												
−.1041	−.2699	1.000											
−.1504	−.3897	.3039	1.000										
.1858	−.0589	−.1164	−.0223	1.000									
.1191	−.1469	−.0474	.0233	.5007	1.000								
.0575	.1065	−.3380	−.1786	.1866	.0212	1.000							
.0986	.1134	−.2539	−.1656	−.0237	−.0493	.0354	1.000						
.0517	.1521	.1149	−.1332	−.0013	.0091	−.0876	−.4171	1.000					
−.0469	−.1215	.2110	.3569	−.0063	.1089	−.0260	−.1289	−.2504	1.000				
−.0386	−.0306	.0292	.0607	.0494	−.0138	.0062	−.1060	−.2059	−.0636	1.000			
−.0507	−.0767	.0832	.1345	−.0977	−.0476	.0622	−.1393	−.2705	−.0836	−.0688	1.000		
−.0253	−.0657	−.0929	.1176	−.0898	.0247	.0210	−.0697	−.1353	.0418	−.0344	−.0452	1.000	
−.0126	−.0326	−.0460	.0444	−.0342	−.0288	−.0649	−.0345	−.0671	−.0207	−.0170	−.0224	−.0112	1.000

TABLE B2 Regression Equations

	SBIR Quality		Useful		Benefits	
Constant	7.1470	6.2360	.5673	.4203	–1.3571	–1.5289
	(6.08)	(7.27)	(0.46)	(.061)	(1.13)	(1.98)
Other Quality	.1477	.1365				
	(1.49)	(1.58)				
AF	.9479	.8144	.7160	.6977	2.0056	1.6533
	(1.58)	(1.44)	(0.93)	(0.95)	(2.38)	(2.19)
ARMY	.5299	.2793	1.0123	.8918	1.6025	1.3328
	(0.78)	(0.50)	(0.84)	(1.14)	(1.46)	(1.71)
BMDO	–.5610	–.2555	–.1657	.0094	–.1606	.5494
	(0.71)	(0.37)	(0.18)	(0.01)	(0.14)	(0.57)
DARPA	.2247	.4794	.5610	.3082	1.1246	1.0867
	(0.34)	(0.80)	(0.75)	(.045)	(1.37)	(1.44)
DSWA	.1123	1.4200	–.7223	.2664	1.5889	2.7578
	(0.11)	(1.67)	(0.46)	(0.21)	(0.97)	(2.08)
OSD	.0204	–.0455				
	(0.02)	(0.05)				
Fast Track	.2108	.4113	.6739	.6526	–.3007	–.3426
	(0.43)	(1.15)	(0.77)	(0.97)	(0.38)	(0.64)
BMDO Match	.8140	.9040	–.4197	–.3575	.8590	.2153
	(1.29)	(1.61)	(0.56)	(0.51)	(0.95)	(0.29)
AGE	–.2690	–.2548	.0979	.0819	–.0357	–.0908
	(1.73)	(2.06)	(0.53)	(0.55)	(0.19)	(0.60)
SIZE	.0017	.0016	.0053	.0015	.0065	.0033
	(0.83)	(0.98)	(0.90)	(0.42)	(1.77)	(1.21)
Experience	–.0110	–.0047	–.0318	–.0114	–.0286	–.01239
	(1.10)	(0.61)	(1.12)	(0.62)	(1.13)	(0.59)
Ownership	–.5125	–.1524	.2736	.4000	–.8089	–.1729
	(1.46)	(0.51)	(0.57)	(0.97)	(1.69)	(0.46)
COMPUTER	–.9787		–.6018		–.5348	
	(1.38)		(0.61)		(0.62)	
ELECTRON	–.5125		–.1314		.2177	
	(0.89)		(0.14)		(0.28)	
MATERIAL	–.7202		–.3861		–1.364	
	(0.85)		(0.36)		(1.03)	
ENERGY	–.6909		.0728		–.0787	
	(0.83)		(0.07)		(0.08)	
ENVIRON	–.0416				.4701	
	(0.05)				(0.36)	
n	154	181	148	181	153	182
R^2	.141	.105	.0867	.0609	.1012	.0559

Role of the Department of Defense in Building Biotech Expertise*

Maryann P. Feldman
The Johns Hopkins University

EXECUTIVE SUMMARY

Typically, whenever we think about the source of funding for research and development in biotechnology, or bioscience more broadly, the National Institutes of Health (NIH) is the agency that comes to mind. Indeed, NIH is an important source of funding for research in biology, chemistry, medicine, molecular biology, genetic engineering, and the related fields that provide the scientific basis for this emerging sector. What often is overlooked is that the Department of Defense (DoD) has played an important, and largely unappreciated, role in funding and shaping the development of this important technology.

At first glance, the idea that DoD is a major funding agency for biotech research may conjure up sinister images of biological warfare or bionic warriors. In truth, however, DoD is a task-oriented agency, and biotechnology, as an emerging broad-based technological platform, offers novel solutions that enhance DoD's mission. These include applications related to disease prevention and mitigation, rapid emergency medical response and trauma management, environmental remediation, and advanced materials.

The role of DoD in the development of emerging industries such as microelectronics, software, and computers has been documented previously by other authors. This paper demonstrates that the DoD Small Business Innovation Research (SBIR) program has played a substantial role in financing bioscience research. This paper documents over $240 million in SBIR awards for bioscience-

---

*Prepared for the May 5th SBIR Symposium sponsored by the Board on Science, Technology, and Economic Policy of the National Research Council, National Academy of Sciences (NAS).

related research by small companies from Fiscal Year (FY) 1983 to FY 1997 based on an examination of biotech keywords contained in the project abstracts. This was allocated as $194 million in Phase II awards and $47 million in Phase I awards. This amount represents about 4 percent of DoD's annual SBIR budget.

In addition, through a series of structured case studies, this paper demonstrates the role that the DoD SBIR program has played in entrepreneurship and technological innovation in biotechnology through a structured series of case studies. Many of the DoD projects have obvious dual use in the civilian sector and DoD-SBIR recipient companies have used the awards to advance their scientific and commercial objectives. All of the DoD SBIR-funded companies that we interviewed have developed commercial products. Two companies that were interviewed, MedImmune and Martek, had a strong DoD legacy and the SBIR awards helped the companies to convert to commercial, civilian applications. The case studies further demonstrate that DoD and NIH funding are complementary for small start-up biotech companies. It appears that DoD has an interest in funding different applications than NIH, but it is common for firms that received DoD funding to subsequently apply to NIH.

INTRODUCTION

Typically, when we think about the source of funding for research and development (R&D) in biotechnology, the National Institutes of Health (NIH) is the agency that comes to mind. Indeed, NIH is an important source of funding for research in biology, chemistry, medicine, molecular biology, genetic engineering, and the related fields that provide the scientific basis for this sector. What often is overlooked is that the Department of Defense (DoD) has played an important, and largely unappreciated, role in funding and shaping the development of this important emerging technology.

DoD oversees the largest budget for R&D of all federal agencies. In contrast to other agencies that have a mission dedicated to a specific topic area, such as health or the environment, DoD's research is mission-oriented and encompasses a wide range of topics and applications. Both the size and the scope of DoD's R&D budget enable the agency to make significant contributions to technology development. The idea of coupling DoD and biotechnology may evoke images of biological warfare, the creation of bionic warriors, and other sinister applications. The truth is that military goals related to disease prevention and mitigation, rapid emergency response and trauma management, environmental remediation, and advanced materials are furthered by scientific advances in the biotechnology and its underlying disciplines. The role of DoD in the development of emerging industries such as microelectronics, software, and computers has been recognized (Tirman, 1984; Alic et al., 1992; National Research Council, 1999). Biotechnology generally is believed to be a similar type of enabling platform technology that has the potential to transform a variety of applications in medicine, agriculture, and the environment, and to produce a new generation of biochemical processes and synthetic materials. In this regard, many DoD projects have obvious dual use in the civilian sector. One of the findings of this pilot research project is that DoD provided over $240 million in Small Business Innovation Research (SBIR) awards for biotech-related research in small companies from Fiscal Year (FY) 1983 to FY 1997. This included $194 million in Phase II awards and $47 million in Phase I awards.[1]

This paper investigates only one aspect of DoD biotech funding: the SBIR program. In this way, it underestimates the role of DoD because no consideration is given to dedicated research facilities, such as the Walter Reed Army Institute for Research (WRAIR), sponsored research at universities or nonprofit institutes, or other initiatives and expenditures made by DoD divisions. This paper has two modest objectives: first, to document the dollar amount that DoD has invested in small start-up companies, and second, to demonstrate how DoD SBIR recipient companies have used the awards to advance their scientific and commercial objectives.

[1]Phase I data were available for FY 1990 forward. All amounts are reported in 1997 real dollars.

Biotechnology presents a unique opportunity to study the emergence and development of a radical new field that has a strong science base and great commercial potential. Indeed, we can date the beginning of the modern biotech industry with the Cohen-Boyer patent application in 1974. This patent provided a means to manipulate, or recombine, genetic material into useful, commercial products that are more naturally acceptable to the human body and its environment than synthetic chemical products. Most important, this patent provided a precedent that created propriety value for intellectual property and, in turn, enabled the formation of new firms. The majority of biotech firms are entrepreneurial start-ups. Powell and Brantley (1992) argue that the commercialization of biotechnology requires the formation of new firms because biotech originates from a radically new scientific knowledge base that does not fit with the existing technological practices of established firms. In this way, entrepreneurial start-ups in biotech are a vehicle to commercialize new ideas and to take radical scientific discoveries out of the laboratory and into the marketplace.

The next section provides data on DoD SBIR awards related to bioscience applications.[2] This analysis is based on a database constructed for this project. A comparison is made to NIH SBIR funding because NIH is the largest dedicated federal funder of biotech research. To understand how firms use DoD SBIR awards, how DoD SBIR funding differs from NIH funding, and the effect that DoD projects have on the company's development and progress in commercialization, this paper examines five case studies in detail. The paper concludes with some suggestive reflections and some ideas for further research on this topic.

DOD SBIR DATABASE

To estimate the financial contribution of DoD SBIR awards to biotech, we developed a systematic database of projects that were funded by DoD through the SBIR program.[3] Using a set of terms that define commercial biotechnology applications provided by the Institute for Biotechnology Information (IBI), we conducted a search of the titles, keywords, and abstracts in the database of all DoD SBIR Award Abstracts. This database provides information for all DoD Phase I Awards for FY90-FY97 and Phase II for FY83-FY97.[4] A project that contained a match to the IBI keywords is defined as a DoD biotech award for this analysis.

---

[2]Biotechnology may be narrowly defined as the use of recombinant DNA methods or broadly defined as anything related to life sciences. The definition used here is broadly inclusive of the spectrum of disciplines that utilize modern biology in their work. Thus, biotechnology is defined as any activity that substantially involves research, development, or manufacture of (1) biologically active molecules, (2) devices that employ or affect biological processes, or (3) devices and software for production or management of biological information.

[3]Details of the database construction are provided in the Appendix.

[4]The searchable database is available at http://www.sbirsttr.com/Awards/Default.asp.

Our analysis revealed 906 projects that could be classified as biotech applications. This is out of a total of 21,211, or about 4 percent, of the projects that were funded by DoD during this 14-year time period. There were 551 Phase I Awards over the seven-year period from 1990 to 1997 and 275 Phase II Awards. This is out of a total of 15,517 Phase I awards and 5,694 Phase II awards. Again, this accounted for approximately 4 percent of the total number of awards granted. The average award for Phase I, in 1997 dollars, was $78,403 while the average Phase II award was $628,024. The total amount spent by DoD on SBIR biotech research was $240,866,001, in 1997 dollars, over the 14-year period for which we have data.

Table 1 provides an overview of DoD funding to SBIR biotech projects from 1984 to 1997.[5] For every year, the number of awards, the total amount awarded, and the average amount of the award is listed, by Phase and for the total. Phase II data are available for the entire time period; Phase I data are available from 1990 onward. We were able to identify two biotech projects in 1984. The number of projects funded has grown steadily and 143 projects were funded in 1997. The total amount allocated to these projects also has increased. In 1990, $13 million was awarded to biotech projects. In 1997, the amount awarded had increased to $39 million.

The biotech projects were funded by a variety of agencies within DoD, as demonstrated in Table 2. The largest funding agency was the Army, which awarded $106,116,285 for Phase I and Phase II projects that related to biotech applications over the 14-year period. Many of the Army projects have funded medical applications.

Agencies such as the Ballistic Missile Defense Organization (BMDO) have funded a broader range of activity. Biotech, as a scientific knowledge base, has applications in other rapidly evolving technologies. For example, BMDO sponsored Phase I and Phase II awards for Astralux of Boulder, Colorado, to develop a biotechnology-based process to make nanostructures for semiconductors. The result was a technology that allowed for a uniform array of replicable silicon quantum boxes of identical dimensions that may be important to the next generation of optoelectronic products. There are other examples of this type of adaptation of biotech knowledge to a broad array of commercial applications. Some of these are explored later in the case studies.

One question of interest is how DoD SBIR biotech funding compares with SBIR funding from NIH. Table 3 provides a comparison of DoD biotech projects and the total projects funded by NIH. The implicit assumption here is that all NIH-funded projects have a biotech application, which we know is not very likely. Given the data available at this time, this is the best assumption. In this case, the

[5]The years reported refer to the DoD fiscal year.

TABLE 1 DoD SBIR Biotech Awards FY1984-FY1997 (1997 real dollars)

Year Awarded	Type of Award	Count	Total Amount Awarded	Average Award
1984	Phase II	2	1,291,690	645,845
	Total	2	1,291,690	
1985	Phase II	8	4,958,783	619,848
	Total	8	4,958,783	
1986	Phase II	9	5,520,467	613,385
	Total	9	5,520,467	
1987	Phase II	12	7,975,713	664,643
	Total	12	7,975,713	
1988	Phase II	14	8,929,105	637,793
	Total	14	8,929,105	
1989	Phase II	12	4,372,354	364,363
	Total	12	4,372,354	
1990	Phase I	46	3,103,910	67,476
	Phase II	20	10,028,279	501,414
	Total	66	13,132,190	
1991	Phase I	64	3,804,775	59,450
	Phase II	15	12,044,014	802,934
	Total	79	15,848,790	
1992	Phase I	43	2,503,535	58,222
	Phase II	21	12,543,755	597,322
	Total	64	15,047,290	
1993	Phase I	51	3,166,938	62,097
	Phase II	27	15,072,733	558,249
	Total	78	18,239,671	
1994	Phase I	102	8,863,993	86,902
	Phase II	26	15,692,876	603,572
	Total	128	24,556,869	
1995	Phase I	100	8,058,951	80,590
	Phase II	48	33,839,580	704,991
	Total	148	41,898,531	
1996	Phase I	92	8,139,534	88,473
	Phase II	51	32,258,781	632,525
	Total	143	40,398,314	
1997	Phase I	99	9,164,828	92,574
	Phase II	44	29,531,407	671,168
	Total	143	38,696,235	
Total	Phase I	597	46,806,464	78,403
	Phase II	309	194,059,537	628,024
	Total	906	240,866,001	

TABLE 2 Biotech Research as Funded by Different DoD Agencies (1997 real dollars)

Agency[a]	Type of Award	Count	Total Amount Awarded	Average Amount Awarded
Air Force	Phase I	139	10,874,782	78,236
	Phase II	71	49,578,088	698,283
	Total	210	60,452,870	
Army	Phase I	219	16,201,582	73,980
	Phase II	144	89,914,702	624,408
	Total	363	106,116,285	
BMDO	Phase I	34	2,287,882	67,291
	Phase II	10	6,071,791	607,179
	Total	44	8,359,674	
DARPA	Phase I	111	10,361,175	93,344
	Phase II	37	21,768,252	588,331
	Total	148	32,129,427	
DSWA	Phase I	8	767,699	95,962
	Total	8	767,699	
Navy	Phase I	76	5,320,411	70,005
	Phase II	45	25,222,782	560,506
	Total	121	30,543,193	
OSD	Phase I	8	823,795	102,974
	Phase II	2	1,503,921	751,961
	Total	10	2,327,716	
SOCOM	Phase I	2	169,137	84,569
	Total	2	169,137	
Total	Phase I	597	46,806,464	78,403
	Phase II	309	194,059,537	628,024
	Total	906	240,866,001	

[a]BMDO = Ballistic Missile Defense Organization; DARPA = Defense Advanced Research Projects Agency; DSWA = Defense Special Weapons Agency; OSD = Office of the Secretary of Defense; SOCOM = Special Operations Command.

numbers presented here overestimate the contribution of NIH to the development of biotechnology but we cannot be sure of the magnitude of the overestimation.

One difference between NIH awards and DoD awards lies in the range of application areas. Whereas NIH is charged to fund health applications, DoD's charge allows it to fund research in a greater variety of application areas.[6] Table 4 provides an overview of the applications that DoD awards supported. The

[6]Data on application area were not collected for NIH for this project. We can expect NIH funding to fall into the general/other area in addition to medical applications.

TABLE 3 Comparison of DoD Biotech and Total NIH Awards (1997 real dollars)

	DoD				NIH			
	Phase I Awards		Phase II Awards		Phase I Awards		Phase II Awards	
Year	Number	Total Amount	Number	Total Amount	Number	Total Amount	Number	Total Amount
1991	64	3,804,775	15	12,044,014	488	23,815,606	281	64,776,947
1992	43	2,503,535	21	12,543,755	561	27,605,144	297	69,663,613
1993	51	3,166,938	27	15,072,733	615	30,674,368	382	90,302,755
1994	102	8,863,993	26	15,692,876	563	41,768,637	358	86,063,359
1995	100	8,058,951	48	33,839,580	642	60,674,775	387	114,394,241
1996	92	8,139,534	51	32,258,781	554	53,596,678	403	131,256,640
1997	99	9,164,828	44	29,531,407	749	73,124,474	487	173,055,595
Total	551	43,702,554	232	150,983,146	4,172	311,259,682	2,595	729,513,150

TABLE 4 Distribution of Awards by Application Area

Application Area	Number of Awards	Percent of DoD Awards
Agricultural/food related	37	4.1%
Industrial	117	12.9%
Medical	401	44.3%
General/other	351	38.7%
Total	906	100.0%

awards are classified by the application associated with the keyword that matched our subject classification criteria.[7]

The largest number of DoD awards, 401 (44.3 percent), were for medical and human health applications. The next largest number of awards were for general and other applications. These are applications that are basic in nature or are more difficult to classify. There were 37 awards (4.1 percent), in the agricultural, crop, or food application area and 117 awards (12.9 percent) that were classified as industrial applications. Industrial applications include the development of biosensors, environmental applications, and waste remediation.

On average annually, in comparison with the NIH SBIR program, DoD funded 13 percent of the total number of Phase I projects that NIH funded. On average though, DoD awards were 5 percent higher. DoD funded 9 percent of the Phase II projects that NIH funded; however, on average, DoD Phase II SBIR awards were 57 percent higher. The probability of receiving a Phase II award from NIH after a Phase I award was 62 percent whereas it appears that DoD SBIR projects had a lower probability, 42 percent, of continuing with a Phase II award. In summary, DoD SBIR awards have funded a significant amount of biotech R&D. Of course these aggregate data do not reveal how individual companies use SBIR awards to develop technology. This is explored through detailed case studies in the following section; an overview of the case studies is presented in Table 5.

CASE STUDIES

To explore how companies have used DoD SBIR awards to develop new biotechnologies, we conducted five in-depth case studies. Most biotech commercialization is conducted by small, start-up companies and, in this regard, the SBIR program is especially beneficial. These case studies illustrate how the DoD awards helped to launch the new start-up companies. Two companies that we interviewed, MedImmune and Martek, had strong DoD legacies and the SBIR awards helped the companies to convert to commercial, civilian applications.

[7]See the Appendix for a description of the keywords.

TABLE 5 Overview of Biotech Case Studies

Company Name	DoD-Funded Application Area	Location
Phytotech	Industrial—environmental remediation	Monmouth Junction, NJ
Integrated Diagnostics	Medical—test kits for emerging diseases	Baltimore, MD
MedImmune	Medical—vaccines	Gaithersburg, MD
Martek	Industrial—products from micro-algae	Columbia, MD
HT Medical	Medical—virtual reality training systems	Gaithersburg, MD

Most of the case study companies are located in Maryland, which is the geographic home base of the author. These case studies, in this regard, were chosen for ease of gathering information and conducting interviews and in no way represent a random selection of companies.

Phytotech

Phytotech has the distinction of being the only biotech company, to date, that has received a DoD Fast Track Award. The award, made in 1997, was for phytoremediation—the use of plants to treat contaminated soil and water. Phytotech is a biotech firm that focuses on environmental remediation. Specifically, the SBIR award allowed Phytotech to develop technology to mitigate the metal accumulation from firing ranges. The company has developed an in-situ treatment that preserves topsoil, minimizes environmental disruption, and produces significantly less waste than other site remediation technologies—all at significantly lower cost compared to conventional technology.

Phytotech was started in April 1993 around research conducted at Rutgers University by Ilya Raskin, Professor at the Center for Agricultural Molecular Biology, and Laura Meagher, Associate Dean of Research at Cook College and Associate Director of the New Jersey Agricultural Experiment Station. Burt Ensley, a veteran of early employment at Amgen and previously Director of Scientific Affairs at Envirogen, was the third founder of the firm, complementing the scientific expertise of Raskin and Meagher with business experience. Ensley raised over $3 million in private-placement venture seed money, which allowed the company to hire essential expertise in agronomy, plant physiology, soil chemistry, engineering, and biochemistry. These funds were used for early-stage company financing and to fund a $1.3 million phytoremediation research project at Rutgers.

Phytotech works on the development of two types of phytoremediation: phytoextraction and rhizofiltration. Phytoextraction uses specially selected and engineered plants to treat soil and water contaminated with toxic metals such as lead and cadmium, as well as radionuclides from uranium. Phytotech also works on rhizofiltration, which is the use of plant roots to absorb, concentrate, and precipitate toxic metals from aqueous streams. The idea is that the plants are grown in situ on contaminated soil and harvested after toxic metals accumulate in the plant tissues. After the plants are harvested, the contaminant metals are disposed of. However, the amount of disposable biomass is a small fraction of the amount of soil treated. As a result, site cleanup costs are less than those associated with traditional technology and environmental disturbances are minimized. The company was catapulted into worldwide attention in 1996 when its sunflowers proved effective in reducing the level of radioactivity in the ponds of Chernobyl, Ukraine. The technology is appropriate for use in environmental remediation on firing ranges, which contain high concentrations of noxious materials.

As might be expected, the Army was very interested in Phytotech's technology and the company has benefited from Army SBIR Fast Track funding of Phase I and Phase II projects:

- 1996, Army, $111,404 (Phase I), for heavy metals phytoextraction and uranium radionuclides phytoremediation;
- 1997, Army, $560,000 (Phase II), for phytoremediation of uranium-contaminated soils.

The company used its own funds for the match or co-investment. As a first-time awardee, the company benefited from a federal match equal to four times its investment. It is uncertain whether the company would have pursued the project without the funding.

The project, which will be completed in August 1999, will result in a commercial service on which the company expects to realize sales within the next 3 to 6 months. Phytotech is currently negotiating a contract to license the technology that resulted from the project. The company attempted an initial public offering (IPO) in 1998 but withdrew because of an unfavorable market. Presently, the company is negotiating to merge into a currently traded public shell corporation.

The technology developed during the DoD SBIR project provided knowledge that has opened new commercial avenues. Recognizing that the technology used to remove and accumulate unwanted soil contaminants could be used to extract and concentrate nutritionally valuable minerals, Phytotech created a nutraceutical division in 1998. By concentrating pure forms of minerals into edible plants, Phytotech has developed and patented a unique and highly bioavailable form of mineral supplements that can be delivered as a nutraceutical in a capsule, tablet, drink powder, or sports bar and can be formulated to be applied as a cosmeceutical cream or lotion. For this technology, Phytotech is increasing production capacity to meet demand as well as identifying commercialization partners and appropriate distribution channels. It is possible that this division may be spun off in the near future.

Phytotech has established itself as a commercial leader in phytoremediation services. Burt Ensley estimates that the company has spent $8 million to date to develop its technology. This includes $1 million in venture capital, $600,000 from a private investor, and $1.5 million in company funds. The company received an award of $70,000 from the New Jersey Commission on Science and Technology. In addition to the two DoD SBIR awards, Phytotech has received a U.S. Department of Energy (DOE) Small Business Technology Transfer grant for work with the Pacific Northwest National Lab for $100,000, a DOE SBIR, a U.S. Department of Agriculture SBIR, and a National Science Foundation SBIR award.

Phytotech has expanded operations from the headquarters location in New Jersey and has offices in Pittsburgh, Pennsylvania, Denver, Colorado, and Boca Raton, Florida. The company conducts its own marketing and manufacturing and has negotiated licenses with other companies to allow them to use its technology.

Integrated Diagnostics

Best-selling nonfiction books, such as Preston's *The Hot Zone,* or popular movies, such as *Outbreak*, draw attention to emerging diseases as a real and serious global health threat. Integrated Diagnostics, or INDX®, is a Baltimore, Maryland, firm that provides a full range of emerging disease detection systems to serve both the United States and the international community. The diseases on which Integrated Diagnostics focuses include various forms of typhus, Lyme disease, dengue fever, and other, similar types of infectious diseases. The company designs tests for both human and veterinary applications and currently has over 70 products on the market.[8]

Integrated Diagnostics was formed in Helene Paxton's basement in 1981 while she was working in Maine. Paxton moved to Baltimore to take a job with Maryland Medical Labs in 1983 and kept operating the business on a part-time basis. When Maryland Medical Labs was acquired by Corning/Medpath in 1995, Paxton decided to pursue the business full time. At that time the company had sales of $500,000 for testing devices for Rocky Mountain spotted fever, rickettsial typhus, and other immunodeficiency diseases. The founders of Maryland Medical Labs were looking for an investment and became silent partners in Paxton's developing enterprise. Barbara Hansen, a virologist at the University of Maryland, joined the firm part-time and Paxton and Hansen started writing proposals for grants to fund more research and product development.

DoD has been active in funding research on emerging diseases because of the susceptibility of troops in a variety of environments. NIH, in large part, has not been interested in this topic. Integrated Diagnostics has had several Cooperative Research and Development Awards (CRADAs) with DoD affiliates. The first was awarded in 1989 with the Naval Medical Research Command (NMRC) to develop a dipstick for dengue. This successful project was a collaboration with Lt. Darrell Kelly and later was extended into another CRADA that also included Dr. Jeffrey Dasch and Dr. Curtis Hayes of NMRC. The company currently has

[8]See http://www.indxdi.com/ for a complete product listing.

one CRADA with WRAIR with products under development and one CRADA with the Center for Disease Control at Fort Collins, Colorado, that resulted in test kit for St. Louis encephalitis antigen in mosquito pools. The CRADAs have been an effective means for the small company to augment its research capability.

INDX® has received four Army SBIR awards to date:

- 1996, Army, $70,000 (Phase I), for advanced system for worldwide surveillance for rickettsial disease antibodies;
- 1997, Army, $750,000 (Phase II), for advanced system for worldwide surveillance for rickettsial disease antibodies;
- 1996, Army, $99,481 (Phase I), for development of a sensitive and specific antigen-detection system for strongyloides stercolis and hookworm infections; and
- 1998, Army, (Phase I), for development of a rapid, sensitive, and specific antibody detection system to facilitate diagnosis of ehrlichial and rickettsial diseases.

The company is planning to submit a Phase II application for the antibody detection system to facilitate diagnosis of ehrlichial and rickettsial diseases for the current competition in May 1999.

The advanced surveillance system for rickettsial disease antibodies adapted the company's civilian dipstick technology to screen for the presence of antibodies for multiple diseases such as Rocky Mountain spotted fever, typhus and scrub typhus, and Q fever. The dipsticks were tested in a variety of different environments and tailored for regional conditions to best address the needs of domestic and foreign health agencies. Paxton felt that the funding delay between the Phase I and Phase II was harmful for a small company. There was great momentum coming off Phase I and the company found it necessary but difficult to fund the research until the receipt of the award.

The Phase II research for an antigen detection system for strongyloides stercolis and hookworm infections was not awarded. Although the Phase I results were promising and the Phase II proposal received high marks, DoD financing that year was lean and the project was not funded. INDX was encouraged to reapply; however, the lack of commitment from DoD meant that the company had to abandon the research. This was unfortunate because there could be a strong international commercial market for a hookworm infection detection system and the small company was not able to develop the product on its own. The lack of funding meant that the company had to refocus its research efforts and, although the project was promising, picking it up again will be difficult.

Integrated Diagnostics is not interested in applying for a DoD Fast Track because of the match requirement. Because the company has received prior SBIR awards, there is not much incentive in terms of leveraging the match for increased federal funds. In this regard the penalty put in place to deter "mills" from applying to Fast Track acts to deter small companies that have received prior awards and have had successful commercialization results. The silent partners from Maryland Medical Lab are interested in selling their interest in the company and Paxton is currently negotiating to bring in new partners or to be acquired by another firm.

The 1997 Phase II award allowed the company to collect samples from a variety of locations around the world. These samples have been invaluable as a research tool and have helped the company in other product development efforts. Paxton mentioned that the company has found it difficult to sell products to the Army, specifically in negotiating the purchasing system. This is indeed ironic since the Army has contributed to the realization of these products that are benefiting customers around the world.

MedImmune

MedImmune, headquartered in Gaithersburg, Maryland, is the eighth largest dedicated biotechnology company in the world. The company has deep roots with DoD. MedImmune was formed in 1988 by Wayne Hockmeyer, a former chairman of the Department of Immunology at WRAIR, and Franklin H. Top, a physician and former director of WRAIR. In its first 10 years, the company has built a pipeline rich with products and drug development projects for infectious diseases, transplant medicine, cancer prevention therapy, and autoimmune disorders. The company currently has three commercial successes: CytoGam®, an intravenous immune globulin that prevents cytomegalovirus, a viral infection common after solid organ transplants; RespiGam™, also an intravenous immune globulin, which prevents respiratory syncytial virus (RSV), the leading cause of pneumonia and bronchiolitis in infants; and Synagis™, a sister drug to RespiGam™ but more potent and easier to use. DoD funding played a small role in the MedImmune story. The DoD research project allowed MedImmune to explore a risky research program that was one of several technological approaches explored in the early stages of the company's development. Although it was ultimately not a technology that the company pursued to commercialization, the project was knowledge creating. The DoD funding allowed Mark Collett to continue his research in synthetic peptide, which he ultimately pursued via the formation of another company.

MedImmune's first two years of operation were financed largely through government CRADAs and research grants. One of its first grants was a Phase I and Phase II Army SBIR for the development of vaccines based on synthetic peptides. The principal investigator was Marc S. Collett, Director, Virology & Antibody Engineering, and Director, Biochemical Virology at MedImmune, Inc. The project resulted in one scientific article but did not result in any commercial products. Although the research project was abandoned in 1993 when Phase II ended, it should not be considered a failure.

MedImmune had gone public in 1991, the same year that it began to market

CytoGam®. Going public gave the young company greater access to capital and a better public image. On the downside, however, it also meant dealing with disclosure risks and the potential for greater negative fallout in reaction to bad news. Despite the success that the company was enjoying in the stock market, CytoGam® was not faring well because of a license with a distributor who was not aggressively marketing the product. In 1992, MedImmune reacquired the marketing right to CytoGam® and launched an expanded marketing program through its own sales force. The strategy worked—resulting in a 30 percent compounded sales growth and building a presence within the transplant community.

MedImmune spent tremendous amounts of time and resources readying the next product—RespiGam™—for submission to the Food and Drug Administration (FDA). This drug was being developed to provide significant protection against RSV, a potentially life threatening infection that hospitalizes over 90,000 infants and kills 4,500 annually. Striking most frequently in the late fall, winter, and early spring, it is an especially serious risk for the smallest and most medically fragile infants, such as those born prematurely or with a chronic lung disease known as bronchopulmonary dysplasia.

When the FDA rejected the company's application to market RespiGam™ in 1993, just as the SBIR Phase II was ending, Hockmeyer faced some very tough decisions about the future of MedImmune and its people. One option would be to transform the company into an "R&D boutique," identifying a variety of promising drug candidates and then licensing them to other companies to develop and market. A second option was to rely on data from previous trials to support an application to market RespiGam™ only for infants with congenital heart disease, and downsize the company. The third possibility, and the strategy Hockmeyer chose, was to "bet the farm" and devote all of the company's resources to executing a new clinical trial of RespiGam™. It was an all or nothing strategy for MedImmune.

Hockmeyer's plan succeeded. The development team regrouped, scrutinizing every aspect of RespiGam™'s clinical studies. They faced tight time constraints, designing and commencing a new trial within 90 days. On the business development side, the company worked hard to raise over $30 million from sources such as Baxter Healthcare Corporation, who entered into an exclusive, royalty-bearing licensing agreement to commercialize RespiGam‘ outside North America. It also sealed codevelopment and copromotion agreements with American Home Products Corporation.

After two years of tireless dedication to RespiGam™ by the company, the FDA approved it for the prevention of RSV disease in certain high-risk infants. Once again, the company rode the wave of a bull market, experiencing a substantial increase in its share price. The following year—1996—proved to be a banner year for MedImmune. It saw a record level of sales for CytoGam®, the beginning of a Phase III clinical trial of Synagis™, the second-generation product for RSV disease, and the start of construction on a $50 million manufacturing facility. Along the way, the company raised another $125 million in capital.

At the end of 1997, MedImmune announced a comarketing agreement for Synagis™ with Abbott Laboratories and, once again, its stock price skyrocketed. In June 1998, MedImmune received FDA approval to market Synagis™, a monoclonal antibody product that was both easier to use and more potent than RespiGam™. Making the picture even rosier for MedImmune, the FDA approved the drug's usage for any pediatric patient at risk for the disease (such as infants with low birth weight or children with lung or heart problems), expanding the U.S. market to about 325,000 children annually. RespiGam™ was marketed only to prevent RSV disease in severely premature infants and infants with lung disease (about 100,000 annually). The approval of Synagis™ brought MedImmune a $15 million licensing payment from Abbott.

Many industry analysts predict Synagis™ will become a blockbuster drug, with global sales of $500 million or more once regulatory approvals in overseas markets are obtained. Strong revenues are anticipated because the drug is cheaper and easier to administer than RespiGam™, there are no competing products, and it provides a higher level of protection against RSV infection.

MedImmune has faced severe adversity and emerged a stronger, more focused company. It has proven sales and marketing capabilities in addition to its product development expertise, a rich product pipeline, strong commercial alliances, and near-term prospects for profitability and revenue growth. Although the future is never assured, at least for now, the red ink for MedImmune has turned black.

Marc Collett went on to be a founders of ViroPharma Incorporated,[9] a firm headquartered in Exton, Pennsylvania, near Philadelphia. The company was started in December 1994 and went public in 1996. The company received the 1997 Enterprise Award for being the best start-up company from the Eastern Pennsylvania Technology Council. Collet, Vice President for Discovery Research, has focused his research on RNA antiviral diseases such as viral meningitis, viral respiratory infection, pneumonia, hepatitis C, and influenza. ViroPharma currently has eight products in various phases of clinical trials. The MedImmune SBIR award was useful in evaluating the early-stage feasibility of the synthetic peptide technology. The technology did not work as intended and did not appear to have direct commercial potential. The project was useful, however, in providing research experience for Marc Collett to transfer to the formation of his new company, ViroPharma Incorporated, and for providing information as to where the company MedImmune might better focus its product development efforts.

Martek Biosciences

Martek is a recognized leader in the development of products for health and nutrition from microalgae, a diverse group of microplants that produce many

[9]http://www.viropharma.com/

different and unusual fats, sugars, proteins, and bioactive compounds of great potential value to humans. Starting with 5 scientists in 1985, Martek now employs 120 people. It has raised over $80 million in equity capital and obtained approximately $6 million from over 40 small business innovation grants, primarily from NIH. The first funding for this radical and experimental technology was from non-NIH sources, with DoD SBIR awards playing a role in the developing company.

Martek Biosciences Corporation started as a spinoff from the giant defense contractor Martin Marietta Corporation. In 1985, Martin Marietta, after successfully battling a takeover bid by Bendix Corporation, decided to focus on its primary defense business. The algae research group was one of the units that was divested. The research group was composed of five scientists working on the genetic engineering of algae. They decided to form their own company and the name Martek was chosen to represent their research focus on Marine Technology.[10]

Martek is unique in the application of biotechnology to algae. The Martin biosciences research group had begun studying microalgae under a National Aeronautics and Space Administration (NASA) contract to explore the use of microplants as a source of food and oxygen for astronauts in space in the 1980s. Martek, which incorporated in May 1985, started operations with $325,000 worth of contracts with DOE, DoD, and NASA.[11] In addition, Martin Marietta traded the group's specialized lab equipment for a 7 percent equity stake in the new company.

The DoD SBIR Phase I grant allowed Martek to further its expertise by developing a deuterated oil made by the fatty acids of microalgae. The deuterated oil was intended for use with industrial bearings that required a long-lasting lubricant. Martek's scientist also discovered other unusual fats made by microalgae that are identical to those found concentrated in the gray matter of human brains, the retina, the heart and nervous tissue, and basically wherever there is electrical activity in the body. Most critically, the breakthrough came with the realization that one of the acids, docosahexaenoic acid (DHA), is provided by human breast milk, but was not available in infant formula. Martek started development work on manufacturing technology to ferment mass quantities of algae for DHA production.

As the technology was developing, so was the business. Martek's earliest rounds of financing—a total of about $ 700,000—came from venture capital subsidiaries of three former local banks, D.C. National, Suburban, and American Security. In 1988, Martek hired Pete Linsert, one of three nonscientists in the company, who helped position the company for potential commercial success. Linsert, a venture capitalist, had invested in the company while he was head of the Suburban Bank's venture capital operation. He recognized the commercial

[10]Biosciences accounted for 7 percent of the corporate research budget, with the rest devoted to more aerospace-related work in areas such as computer science, semiconductor physics, and advanced materials.

[11]Martin Marietta subcontracted with the new company to continue its prior work.

potential of DHA, which experimental research had indicated is crucial to brain and eye development.

Martek developed DHA into the commercial product Formulaid®, an additive for infant formula. Formulaid® may help to close gaps that researchers have found exist between the cognitive development of breast-fed and bottle-fed infants, differences that a study published in *Lancet* found persist in 9-year-old children. The market potential is astounding: Infant formula has more than $2 billion a year in wholesale sales in this country and about $5 billion worldwide.

In 1992, Martek, which then had 50 employees, also was able to supplement $9 million raised from private investors with more than $5 million received from 32 federal SBIR grants from the NIH. In 1994, Martek, went public, selling 2.3 million shares at $7 a share to raise about $14.5 million after expenses. Typically, biotech companies eat up large amounts of cash before they can develop commercially viable products. Up until Martek had its public offering, the company had raised $8.8 million in four rounds of private financing. When it went public, the company had an accumulated deficit of $6.3 million.

In 1995, Martek purchased a fermentation plant for $10 million. The Winchester, Kentucky, plant has two 40,000-gallon fermentation tanks, which are used to make Formulaid® with a patented process. The ability to grow commercial quantities of microalgae allows Martek to license the use of Formulaid® to three of the world's top infant formula manufacturers. The three companies are Mead Johnson & Company (a subsidiary of Bristol-Myers Squibb Company), American Home Products Corporation, and Nutricia B.V. In the fall of 1994, Formulaid® was introduced to the market in Belgium, as an additive in baby formulas made for prematurely born babies and sold by Nutricia, the second leading formula maker in Europe. The agreement was that Martek would receive a flat fee to cover production costs, plus a royalty of about 3 percent of the sales of formula containing Formulaid®. Formulaid® is currently sold in formulas in 50 countries around the world.

Martek's nutritional oils have other uses. IAM Co., the pet food manufacturer, has agreed to begin including Martek's nutritional oils as an additive in its Eukanuba brand of puppy and kitten nutrition formulas for orphaned puppies and kittens and other young pets in need of supplemental nutrition. Neuromins®, a dietary supplement for DHA for adults, is sold over the counter and distributed by such retail giants as Rite Aid and GNC. For the future, the company believes that microalgae may prove useful as ingredients in new medicines that would fight drug-resistant bacteria.

HT Medical Systems, Inc.

HT Medical is a medical software company headquartered in Rockville, Maryland, that specializes in computer-assisted, virtual reality (VR) medical training systems. The company, which currently employs 50 people, focuses on developing VR technology to train medical practitioners. HT Medical produces two categories of products: (1) hardware platforms and (2) software programs that run on the hardware platforms. HT currently has four products on the market:

- T-Vox, a VR software development toolkit;
- CathSim, which trains health care workers to practice injecting needles for procedures such as giving vaccinations, drawing blood, and inserting intravenous catheters[12];
- PreOp Endoscopic Simulator, a computer-based VR system that trains doctors to perform endoscopic procedures; and
- PreOp Endovascular Simulator, a computer-based VR system that trains doctors to perform endovascular procedures such as inserting pacemakers.

In 1987, straight out of Western Maryland College, Gregory L. Merril founded HT Medical, then named High Techsplanations. Merril was interested in enhancing the communication of medical information. The company's early work was in medical videos and displays. A breakthrough came in 1992 when the pharmaceutical firm, Merck, hired High Techsplanations to develop a presentation that would guarantee a high volume of traffic through its kiosk at the American Urological Association convention. Merril suggested a VR simulator that would allow doctors to practice endoscopic surgery. The advantage was the ability to experiment on technique without the pressure of preforming on an actual patient. Merril developed a high-tech training system that would duplicate the look and feel of actual procedures. Merril was able to talk Sun Systems into providing workstations to the fledgling company. The presentation was a success. Afterward, Merril redirected his fledgling company as a high-tech R&D firm.

HT Medical has benefited from Navy SBIR funding[13] of Phase I and Phase II projects[14]:

- 1995, NAVY, $114,680 (Phase I), for virtual environment training for trauma management;
- 1996, NAVY, $748,521 (Phase II), for virtual environment training for trauma management;

---

[12]CathSim improves on prior training techniques in which medical practitioners practice on oranges, plastic models, or each other.

[13]Contract Number N00014-95-C-0098.

[14]In addition, HT Medical has benefited from four NIH SBIR grants, including:

- 1993, Department of Health and Human Services (DHHS), $49,919 (Phase I), for VR surgical simulation for hemicholectomy;
- 1994, DHHS, $500,000 (Phase II), for cholesterol education using novel interactive multimedia;
- 1995, DHHS, $80,244 (Phase I), for visualization and dissemination of embryonic data; and
- 1997, DHHS, $49,781 (Phase I), for benchmark VR innovative procedural tool.

The idea was to apply VR to the treatment of traumatic wounds in order to train medical personnel to manage battlefield injury. During the DoD SBIR project, HT Medical produced a Trauma Simulation Suite, which contained prototype simulators for central venous access, endotracheal tube placement, and chest tube placement. Surgical computer simulations—VR systems—allow practice on a computer model with the physical and physiological characteristics of a living human patient. The idea is similar to the flight simulators and battle simulators that have been used with great success by the airline industry and the military. HT Medical's idea was to extend this concept to creating realistic VR surgical simulations that would allow physicians to train for battlefield trauma.

Greg Merril is a master at building partnerships. To commercialize the VR technology that resulted from the NAVY SBIR project, the Maryland Health Care Product Development Corporation (MHCPDC) invested $400,000 in HT in 1996.[15] In return for the investment, MHCPDC negotiated royalties for 14 percent of HT's annual net profits. MHCPDC's total return is expected to be 25 percent per year. The MHCPDC investment required a private match of $400,000, which HT received from Cook, Inc., a medical device company in Bloomington, Indiana. HT Medical also received $250,000 from Maryland's Enterprise Investment Fund.[16] Additional state funding has come from the Maryland Industrial Partnerships (MIPs), which provided $35,000 to the University of Maryland to work with HT on the development of a tactical robotic arm.[17]

To further develop the medical simulator technology, HT Medical has received two awards from the Information Infrastructure for Healthcare Competition of the Advanced Technology Program (ATP) of the National Institute of Standards and Technology. The first award was in 1995 for the development of TELEOS, an authoring system for virtual reality surgical simulations. The amount of this award was $560,000. HT Medical's second ATP award, in 1997, was for $2,000,000 for the development of a Preoperative Decision Support System (PreOp). Under this project, HT Medical Systems is designing a VR system that will enable physicians to use patient-specific data for diagnosis, selection of medical devices, rehearsals for operations, and remote consultations.

Despite several attempts, HT has been unable to secure venture capital. However, it has raised funding through two private stock placements. The first round

[15]MHCPDC is a nonprofit public/private venture funded by Maryland's Department of Business and Economic Development and the DoD Technology Reinvestment Program. It provides equity funding for companies involved in developing technologies of importance to national security and the national economy. Private cofunding is required. It generally invests $300,000 to $500,000.

[16]Enterprise Investment Fund, Maryland Department of Business and Economic Development, makes direct equity investments of up to $500,000 in emerging high-technology businesses.

[17]The MIPs program, based at the University of Maryland at College Park, shares the cost with companies of having University of Maryland System faculty collaborate on research on new products or processes.

netted $1 million; most investors were angels—high-net-worth individuals—some of whom Merril contacted through Maryland's Private Investors Network (PIN).[18] The second round brought $2 million and was solicited through an investment bank. HT is currently arranging a third private placement, targeted at institutional investors. The company is waiting for a favorable opportunity for an IPO.

HT has entered joint ventures with universities and established companies to develop and market its VR technology. For example, to develop and market its Intravenous Training System, which runs on the CathSim platform, HT partnered with the State University of New York (SUNY) at Plattsburg and Beckton-Dickinson & Co. Nurses at SUNY's medical school, learning about HT's doctor-training VR software products, wanted a similar product to train nurses on how to catheterize a patient properly. The school contacted HT and agreed to develop the software, at a cost of $300,000. Under this project, HT Medical authored a prototype VR simulation of the placement of a central venous line in the subclavian vein. This simulation incorporates visual and tactile realism of the actual surgical procedure, providing a genuine scenario for teaching physicians central line placement. In return, SUNY receives a 5 percent royalty deal. Additionally, the U.S. Health Care Financing Administration is working in cooperation with HT Medical Systems to incorporate features into CathSim that allow hospital administrators to benchmark their nurses' performance in placing intravenous catheters. To sell the product, HT arranged a comarketing agreement with Beckton-Dickinson, a medical supply company that sells catheters. HT gets the marketing strength of Beckton Dickinson's sales force while the medical supply company hopes to benefit from the synergy of selling a product that requires the purchase of catheters. Finally, not unlike other emerging research-centered, high-technology companies, HT contracts out the manufacturing of its products.

Merril applied for Fast-Track funding at NIH. Fast Track allows a company to begin processing the application for a Phase II SBIR grant while it applies for Phase I. Securing Phase I practically guarantees the company will get Phase II, but HT's experience exposed a frustrating Catch-22: to get Phase II funding. HT had to provide technical specifications that it would obtain only after completing work funded by the Phase I grant, but NIH had not approved the Phase I grant because Phase II was held up. After several attempts to project what the required specifications would be, HT gave up. Merril knows of no company that has successfully used the Fast Track SBIR program at NIH.

---

[18]PIN, of the Baltimore-Washington Venture Group, brings together accredited investors and growing companies in the mid-Atlantic region. It currently services Maryland, Washington, D.C., Virginia (as far as Richmond), and Delaware (unfortunately at this time the network cannot consider deals outside of this region). The Baltimore-Washington Venture Group is part of the Michael D. Dingman Center for Entrepreneurship, University of Maryland at College Park.

REFLECTIVE CONCLUSIONS

Government R&D support may have many indirect, and often unintended, effects on technological innovation. Indeed, the idea that DoD makes a significant contribution to the nation's developing expertise in biotechnology may, at first glance, appear counterintuitive. The contribution of DoD to the development of biotech knowledge and the realization of commercial products is an aspect has received little attention in the analysis of science and technology policy. This pilot research effort on the role of the DoD SBIR awards in building the nation's biotech industry has revealed some patterns and conclusions.

First, and perhaps most significant, DoD has made a significant financial contribution to the funding of biotech R&D. Although NIH, because of its mission, is the premier agency in funding biotech, DoD, in seeking novel solutions to advance its task-oriented mission, has funded biotech applications. Both the size of the DoD R&D budget and the scope of the agency's mission have allowed the it to make significant contributions to the development of knowledge-intensive technologies. The findings of this paper indicate that biotech is another instance in which DoD funding has played an important role.

DoD has funded biotech R&D in a variety of applications that appear to have dual uses in the civilian sector. This was further demonstrated in the case studies. All of the DoD SBIR-funded companies that we interviewed have developed commercial products. Two companies that were interviewed for this paper, MedImmune and Martek, had a strong DoD legacy and the SBIR awards helped the companies to convert to commercial, civilian applications.

Finally, the case studies demonstrate that DoD and NIH funding are complementary for small start-up companies. Although DoD has different interests from NIH, it is common for firms that have received DoD funding to subsequently apply to NIH.

ACKNOWLEDGMENTS

I would like to thank Charles Wessner and John Horrigan (NAS) for useful comments on an earlier version of this paper, Peter Cahill and Christina Villa (BRTRC) for help with DoD SBIR data, and Robert F. Moore for help with the NIH SBIR data. I would also like to thank JoAnne Goodnight of the NIH and Morrie Ruffin of the Biotechnology Industry Organization for their comments. This project would not have been complete without public policy graduate student Mike Vincellette who provided technical assistance in constructing the database and Merry Perry who provided document preparation assistance. The views expressed are those of the author and do not represent the views of the Institute for Policy Studies of Johns Hopkins University.

REFERENCES

Alic, John A., Lewis Branscomb, Harvey Brooks, Ashton B. Carter, and Gerald L. Epstein. 1992. *Beyond Spinoff: Military and Commercial Technologies in a Changing World.* Boston: Harvard Business School Press.

National Research Council. 1999. *Funding a Revolution: Government Support for Computing Research.* Washington, D.C.: National Academy Press.

Powell, Walter W., and Peter Brantley. 1992. "Competitive cooperation in biotechnology: Learning through networks?" pp. 366-394 in: N. Nohria and R. G. Eccles (eds.), *Networks and Organizations: Structure, Form and Action.* Boston: Harvard Business School Press.

Tirman, John. 1984. *The Militarization of High Technology.* Cambridge, MA: Ballinger.

APPENDIX

Construction of the Database

We used subject terms from the Institute for Biotechnology Information (IBI), which is an independent research and consulting firm that provides comprehensive information on commercial biotechnology. The firm has been in existence since 1986 and is headed by Mark D. Dibner, who holds a Ph.D. in neurobiology and pharmacology and an MBA in strategic planning. Dibner has written over 75 articles and 6 books on commercial biotechnology. He has served on the boards of directors of three biotechnology companies: the Association of Biotechnology Companies, the Council of Biotechnology Centers, and the Emerging Companies Section of the Biotechnology Industry Organization (BIO). One of the services that IBI provides is the U.S. Companies Database, which provides, for our purposes, a listing of subject terms that define company research expertise and topic areas. We used this as an independent source to define the subject terms that would be used in biotechnology or, more broadly bioscience, research abstracts[19]. We eliminated common terms that IBI listed that are not specific to bioscience, such as food, cosmetics, energy, equipment, imaging, and services. In addition, we condensed some categories. For example, whereas IBI lists11 types of cancers, such as colon or skin cancer, our search used the main term cancer. Finally, we read the abstracts of all of the awards that matched our search terms in order to ensure that the projects were appropriate. We eliminated projects that did not substantially involve research, development, or manufacture of (1) biologically active molecules, (2) devices that employ or affect biological processes, or (3) devices and software for production or management of biological information. We excluded projects that focused on health information management systems, test batteries, and training methods.

TABLE A-1 Examples of the Most Common Terms from IBO

Human Health/ Medical	Agriculture	Industrial	General/ Other
Antidote	Additives	Biomass	Biotechnology
Blood	Flavors	Biosensors	Reagents
Leukemia	Food enzymes	Environmental testing	Combinatorial
Lymphomas	Food microbiology	Environmental treatment	Chemistry
Cardiovascular	Food safety testing	Bioremediation	Diagnostic tools
Heart Failure		Industrial enzymes	Genetic screening
Hypertension		Laboratory enzymes	DNA probes
Stroke		Fermentation	Imaging
Cytokines		Speciality chemicals	Drug abuse screening

[19]The listing of the subjects can be found at http://www.biotechinfo.com/company_wizard/subject_terms_list.htm.

Estimates of the Social Returns to Small Business Innovation Research Projects

Albert N. Link and John T. Scott***
*University of North Carolina at Greensboro**
*and Dartmouth College***

EXECUTIVE SUMMARY

This paper examines a fundamental rationale for the Small Business Innovation Research (SBIR) program—namely, that government support of private-sector research and development (R&D) through the program is justified because the social benefits associated with the funded research are greater than the social costs, yet without public support, the private costs would be greater than the private benefits. Hence, the socially valuable SBIR R&D would not occur absent the support of the program. Based on interview data collected during case studies of 44 awardees from throughout the United States, we conclude that

- the funded companies would not have undertaken the R&D without public support because the private return that they perceived they would earn would be less than the minimum accepted rate of return required for private financing of the projects,
- the estimated lower bound on the social benefits associated with the funded research is greater than the estimated private returns if there were no public support—in terms of the average rates of return on the investments, 84 percent for society compared to 25 percent for the private investors, and
- the magnitude of the difference between the social and private returns does not vary significantly, on average, between Fast Track and non-Fast Track projects.

INTRODUCTION

The Small Business Innovation Development Act of 1982 (P.L. 97-219) required that federal agencies provide special funds to support small business R&D that complemented the funding agency's mission. This is called the Small Business Innovation Research (SBIR) program. Based on the premises, as stated in the Act, that "small business is the principal source of significant innovation in the Nation," and small businesses are "among the most cost-effective performers of research and development and are particularly capable of developing research and development results into new products," the Act lists its purposes for, among other things, establishing the SBIR program:

1. to stimulate technological innovation,
2. to use small businesses to meet federal research and development (R&D) needs,
3. to foster and encourage participation by minority and disadvantaged persons in technological innovation, and
4. to increase private-sector commercialized innovations derived from federal R&D.

The Small Business Innovation Research Program Reauthorization Act of 1992 gave reauthorization to the SBIR program because the program has "effectively stimulated the commercialization of technology development through federal research and development, benefiting both the public and private sectors of the Nation."

In thinking about a program such as SBIR, economists and many policymakers see the generation of positive externalities (i.e., social benefits exceeding private ones) as an important rationale for the program (Lerner, 1999). The purpose of this paper is to examine that underlying premise. Is it in fact the case that the social benefits associated with SBIR-supported projects are greater than the private benefits? Stated alternatively, is there any empirical evidence to suggest that absent the SBIR program the private sector would underinvest in DoD-related technologies?

The remainder of this paper is as follows. In the second section, we present an economic rationale for public–private partnerships and generalize from it to posit a similar rationale for the selection of projects to be funded by SBIR program.[1] In the third section, we offer empirical evidence that the social returns from SBIR-supported projects are greater than the private returns, and that without the support of public funding, the socially valuable research would not have been undertaken. In the fourth section, we investigate, in a preliminary fashion, project-specific characteristics associated with the gap between the estimated social and private returns. Then, we offer concluding remarks.

[1]For a more general overview of U.S. public–private partnerships, see Link (1998).

ECONOMIC RATIONALE FOR PUBLIC–PRIVATE PARTNERSHIPS

Government's Role in Innovation

Many date the origin of a U.S. domestic science and technology policy with Vannevar Bush's *Science—the Endless Frontier* in 1945. Certainly, Bush's views about science and the role of universities in sustaining the nation's science base had a profound impact on the scientific community of his time, as evidenced by the founding soon thereafter of the National Science Foundation in 1950. Bush's legacy is one of policy focus, emphasizing clearly the importance of basic research in the innovation process. However important Bush's views, he was not articulate about the economic rationale for government's role in innovation, much less about addressing issues of public–private partnerships. Bush did articulate an intellectual rationale for public support of basic research and research related to issues of national security, industrial growth, and health and human welfare.

The first official policy statement about domestic technology policy, *U.S. Technology Policy*, was released by the Executive Office of the President in 1990, coincidentally during the Bush administration. As with any initial policy effort, it was an important general document. However, precedent aside, it failed to articulate a rationale or role for government's intervention into the private sector's innovation processes. Rather, much like *Science—the Endless Frontier,* it implicitly assumed that government had such a role, and then it set forth a rather general goal (1990, p. 2):

> The goal of U.S. technology policy is to make the best use of technology in achieving the national goals of improved quality of life for all Americans, continued economic growth, and national security.

President Clinton took a major step forward in his 1994 *Economic Report of the President* by articulating first principles about why the government had such a role in innovation and in the overall technological process (p. 191):

> Technological progress fuels economic growth .... The Administration's technology initiatives aim to promote the domestic development and diffusion of growth- and productivity-enhancing technologies. They seek to correct market failures that would otherwise generate too little investment in R&D .... The goal of technology policy is not to substitute the government's judgment for that of private industry in deciding which potential "winners" to back. Rather the point is to correct market failure ....

This role for government traces back at least to the writings of Bator (1958). The conceptual importance of identifying market failure for policy is also emphasized by the Office of Management and Budget (OMB, 1996) and summarized by the Organization for Economic Cooperation and Development (OECD, 1998). However, the *Economic Report* did not expand on how to correct for market failure, much less discuss appropriate policy mechanisms for doing so.

Risk, Barriers to Technology, and Market Failure

Risk, as well as closely related difficulties regarding the appropriability of returns, create barriers to technology investment and, as a result of these barriers, there may be market failure leading to an underinvestment in or underutilization of technology. Much of the market failure literature focuses on investments in the creation or production of technology (e.g., R&D). Equally relevant, although often overlooked, are investments for the use and application of others' technology (Tassey, 1997; Link and Scott, 1998a,b).

Risk measures the possibilities that actual outcomes will deviate from expected outcomes, and the shortfall of the private expected outcome from the expected return to society reflects appropriability problems. The technical and market results from technology may be very poor, or perhaps considerably better than the expected outcome. Thus, a firm is justifiably concerned about the risk that its R&D investment may fail, technically or for any other reason. Or, if technically successful, the R&D investment output may not pass the market test for profitability. Further, the firm's private expected return typically falls short of the social expected return.

The expected outcome is the measure of central tendency for a random variable's outcome. Risk is sometimes quantified as the variance of the probability distribution for a random variable's outcome—here, the technical outcome of R&D or the market outcome of the R&D output are the random variables—although other aspects of the probability distribution may affect risk as well. Thus, the contribution to a firm's overall exposure to risk associated with a particular investment will be different depending on the collection of projects in the portfolio. In that sense, a large firm, with a diversified portfolio of R&D projects, might find a particular project less risky than a small firm with a limited portfolio. Similarly, society faces less risk than the individual firm, large or small, because society has, in essence, a diversified portfolio of R&D projects and that diversification reduces risk that the decision makers in individual firms will consider because of bankruptcy costs or managers' firm-specific human capital. As risk to society is reduced, overall outcomes become more certain. Further, for each particular technological problem, society cares only that at least one firm solves the technical problems and that at least one is successful in introducing the innovation into the market. The individual firm pursuing the technical solution with R&D and then trying to market the result will of course face a greater risk of technical or market failure.

Facing high risk—both technical and market risk not faced by society—or simply because society has a longer time horizon than the decision makers of individual firms, a private firm discounts future returns at a higher rate than does society. Therefore, the private firm values future returns less and, from society's perspective, will invest too little in R&D. Put another way, the higher the risk the higher the hurdle rate, or required rate of return, will be for a project. Thus, when social risk is less than private risk, the private firm will use a hurdle rate that, from society's perspective, is too high. Socially useful projects accordingly will be rejected. Further, when the firm's expected return falls short of society's expected return, the firm has less future returns to value than society does, and again, underinvestment will result.

Market failure, resulting from risk and the closely related difficulties of appropriating returns to investments in technology—R&D specifically—will lead to a divergence between private and social benefits. The social rate of return will be greater than the private rate of return; there are, of course, expected, or ex ante, returns. This is illustrated in Figure 1, following Jaffe (1998). The purpose of this simple heuristic device is to characterize private-sector projects with returns not only less than the expected social returns but also less than the private hurdle rate for projects normally undertaken by the firm.

The social rate of return is measured on the vertical axis of Figure 1 along with society's hurdle rate on investments in R&D. The private rate of return is measured on the horizontal axis along with the private hurdle rate on investments in R&D. A 45-degree line (long dashed line) is imposed on the figure under the assumption that the social rate of return from an R&D investment will at least equal to the private rate of return from that same investment. The three illustrative projects discussed below are labeled as projects A, B, and C.

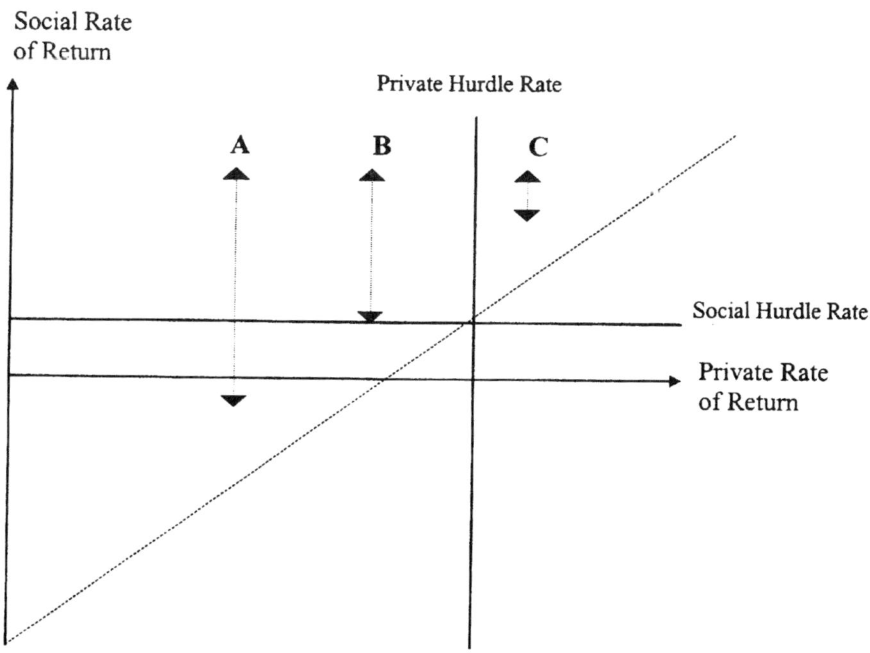

FIGURE 1 Gap between social and private rates of return to R&D projects.

For project C, the private rate of return exceeds the private hurdle rate, and the social rate of return exceeds the social hurdle rate. The gap (short vertical dashed line) between the social and private rates of return reflects the spillover benefits to society from the private investment. However, the inability of the private sector to appropriate all benefits from its investment is not so great as to prevent the project from being adequately funded by the private firm. In general, then, any R&D project with a private rate of return to the right of the private hurdle rate and on or above the 45-degree line is not a candidate for public support because, even in the presence of spillover benefits, the R&D project will be funded by the private firm.

Consider projects A and B. The gap between social and private returns is larger than in the case of project C; neither project will be adequately funded by the private firm. To address this market failure, the government has two alternative policy mechanisms. It can use a tax policy to address the private underinvestment in R&D or it can rely on public–private partnerships as a direct funding mechanism.

If the private return to project B is less than the private hurdle rate because of the risk and uncertainty associated with R&D in general, then tax policy may be the appropriate policy mechanism to overcome this underinvestment. Risk is inherent in a technology-based market, and there will be certain projects for which the rewards from successful innovation are too low for private investments to be justified. Tax policy, such as the research and experimentation tax credit, may in these situations reduce the private marginal cost of R&D sufficiently to provide an incentive for the project to be undertaken privately. For projects such as B, a tax credit may be sufficient to increase the expected return so that the firm views the post-tax-credit private return to be sufficient for the project to be funded.

However, for projects such as A, a tax credit may be insufficient to increase the expected return so as to induce the private firm to undertake the project. For example, a project expected to yield an innovative product that would be part of a larger system of products, even if technically successful, might not interoperate or be compatible with other emerging products. In such a case, direct funding rather than a tax credit may be the appropriate policy mechanism.

A priori, it is difficult to generalize about the way that any one firm's underfunded projects will be distributed in the area to the left of the firm's private hurdle rate. However, some generalizations can be made about the portfolio of private-sector firms' projects in general. For those R&D projects, like project B, for which the firm will appropriate some returns but for which the overall expected return is slightly too low, a tax credit may be sufficient to increase the expected return to the point that the expected return exceeds the private hurdle rate. Such projects may be of a product or process development nature and are likely to be a part of the firm's ongoing R&D portfolio of projects. For those R&D projects, like project A, for which the firm has little ability to appropriate returns even if the marginal cost of the project is reduced through an R&D tax credit, for example, the firm may not respond to such an R&D tax policy but may respond to a direct-funding mechanism. Such projects are likely to be of a generic technology nature, that is, technology from which subsequent market applications are derived and that enable downstream applied R&D to be undertaken successfully. Generic technology and the associated research process represent the organization of knowledge into the conceptual form of an eventual application and the laboratory testing of the concept. Generic technology draws on the science base but, unlike scientific knowledge, it has a functional focus.

Thus, the economic rationale for public–private partnerships is that such partnerships represent one direct-funding R&D policy appropriate to overcome market failure and they are more likely to be necessary, compared to fiscal tax incentives, when the R&D is generic in character.

Drawing upon the arguments set forth earlier, we maintain that a candidate project for SBIR awards is one like project A in Figure 1. That is, given that the proposed research aligns with the technology mission of DoD, SBIR should fund such projects for which there is a significant potential social benefit but also that are characterized by substantial downside risk such that the firm's expected private return is well below its private hurdle rate.

Case information reported by Link (1999) and Scott (1999) confirms for a small sample of SBIR-supported projects that not only are the firms' private returns less than their private hurdle rate but also that outside investors are unwilling to fully sponsor the research because of both technical and market risk. Hence, at the outset of an SBIR project, not only is a firm's private hurdle rate not expected to be met, neither is the required return for a third party. The case information provided by Link (1999) and Scott (1999) clearly indicates that there is a market failure.

PRELIMINARY ESTIMATES OF SOCIAL RETURNS

As part of this study, 44 SBIR award recipients were interviewed. Each company was interviewed toward the end of its Phase II award period. From each, we collected information that allowed us to calculate a lower-bound for the prospective expected social rate of return associated with each project and to compare that prospective expected social return to both the expected private return to the firm had it pursued the project in the absence of SBIR support and the expected private return expected by the firm with its SBIR support. Our analysis clearly indicates that the SBIR is funding projects like project A in Figure 1 and, given such funding, the project has become similar to project C.

Sample of Projects

The sample of projects studied is not intended to be representative of all projects funded by the SBIR program. Rather, because of the timing constraints associated with this study, the sample of projects was selected as follows: Inde-

TABLE 1 Characteristics of the Sample (N = 44)

Characteristic	Number
Region	
Northeast	17
Southeast	13
West	14
Funding Mechanism	
Fast Track	14
Non-Fast Track	30

pendent of this analysis, Link (1999) examined SBIR projects in the southeastern states and Scott (1999) examined SBIR projects in the northeastern states. Each of those two samples was selected for the purpose of comparing Fast Track projects to non-Fast Track projects in each region. In addition to those projects, other projects were identified on the basis of early responses to a broad-based mail survey conducted for the DoD by Peter Cahill, as discussed by Audretsch et al. (1999).

Given the practicality criterion that was imposed on the selection of the sample of 44 projects, no generalizations should be made about SBIR projects as a whole. However, the methodology that we implemented is sufficiently rich to be applied to other samples.

Table 1 describes selected characteristics of the sample of projects examined here.

Analytical Framework for Estimating Social Returns

Table 2 lists the variables required for implementation of our model. As noted in the table, data on selected variables were independently available from DoD project files, but all such information also was verified during the interview process and corrected when discrepancies were found.

Phase I and Phase II values for project duration (d), total cost (C), and SBIR funding (A) were combined into one value to cover both Phase I and Phase II of the project. That is, each project is viewed from the time that Phase I began, and expectations from that point forward are estimated. It is at that point in time that the market failure issues discussed earlier are especially relevant. The variables z and F refer to the additional period of time beyond the expected completion of Phase II until the research would be commercialized and the additional cost required during that period.

The variable v, the proportion of value appropriated, deserves some explanation. Firms cannot reasonably expect to appropriate all of the value created by

TABLE 2 Variables for Calculation of the Prospective Expected Social Rate of Return

Variable	Definition	Source
d	Duration of SBIR project	DoD files, verified and updated as necessary during interviews
C	Total cost of the SBIR project	DoD files, verified and updated as necessary during interviews
A	SBIR funding	DoD files, verified and updated as necessary during interviews
r	Private hurdle rate	Interview
z	Duration of the extra period of development beyond Phase II	Interview
F	Additional cost for the extra period of development	Interview
T	Life of the commercialized technology	Interview
v	Proportion of value appropriated	Interview
L	Lower bound for expected annual private return to the SBIR firm	Derived
U	Upper bound for expected annual private return to the SBIR firm	Derived

their research and subsequent innovations. First, the innovations will generate consumer surplus that no firm will appropriate, but that society will value. We ignore consumer surplus in our calculations of the prospective expected social returns, thus motivating our claim that our estimates are lower-bound estimates. Second, some of the profits generated by the innovations will be captured by other firms. Larger firms, for example, will observe the innovation and will successfully introduce imitations. As part of the interview process, each respondent was asked to estimate the proportion of the return generated by its anticipated innovation that it expected to capture. Then, in an extended conversation, other possible applications of the technology developed during the SBIR project were explored. Each respondent was asked to estimate the multiplier to get from the profit stream generated by the immediate applications of the SBIR projects to the stream of profits generated in the broader application markets that could reasonably be anticipated. Finally, each respondent estimated the proportion of the returns in those broader markets that it anticipated capturing. From these extended discussions, we estimated the proportion of value appropriated by the innovating SBIR awardee.

The lower bound for the annual private return to an SBIR-sponsored project is found by solving Eq. (1) for L, because that is the value that the private firm earns just to meet its private hurdle rate, or its required rate of return, on the portion of the total investment that the firm must finance. The firm would not invest in the SBIR project on its own unless it expected at least L for the annual private return on its investment.[2]

$$
\begin{aligned}
&-\int_0^d \left(\frac{C-A}{d}\right) e^{-rt}\,dt - \int_d^{d+z}\left(\frac{F}{z}\right)e^{-rt}\,dt + \int_{d+z}^{d+z+T} Le^{-rt}\,dt = 0 \\
&\Rightarrow -\left(\frac{C-A}{d}\right)\left(\frac{-1}{r}\right)e^{-rt}\Big|_0^d - \left(\frac{F}{z}\right)\left(\frac{-1}{r}\right)e^{-rt}\Big|_d^{d+z} + (L)\left(\frac{-1}{r}\right)e^{-rt}\Big|_{d+z}^{d+z+T} = 0 \\
&\Rightarrow \left(\frac{C-A}{dr}\right)(e^{-rd}-1) + \left(\frac{F}{zr}\right)(e^{-r(d+z)} - e^{-rd}) - \left(\frac{L}{r}\right)(e^{-r(d+z+T)} - e^{-r(d+z)}) = 0
\end{aligned} \tag{1}
$$

To determine the upper bound for the annual private return, U, Eq. (2) is solved for U. Any expected annual return greater than U would imply that the expected rate of return earned by the private firm would be greater than its hurdle rate in the absence of SBIR support, and therefore SBIR support would not be required for the project.

$$
\begin{aligned}
&-\int_0^d \left(\frac{C}{d}\right) e^{-rt}\,dt - \int_d^{d+z}\left(\frac{F}{z}\right)e^{-rt}\,dt + \int_{d+z}^{d+z+T} Ue^{-rt}\,dt = 0 \\
&\Rightarrow -\left(\frac{C}{d}\right)\left(\frac{-1}{r}\right)e^{-rt}\Big|_0^d - \left(\frac{F}{z}\right)\left(\frac{-1}{r}\right)e^{-rt}\Big|_d^{d+z} + (U)\left(\frac{-1}{r}\right)e^{-rt}\Big|_{d+z}^{d+z+T} = 0 \\
&\Rightarrow \left(\frac{C}{dr}\right)(e^{-rd}-1) + \left(\frac{F}{zr}\right)(e^{-r(d+z)} - e^{-rd}) - \left(\frac{U}{r}\right)(e^{-r(d+z+T)} - e^{-r(d+z)}) = 0
\end{aligned} \tag{2}
$$

Our estimate of the average expected annual private return to the firm is $[(L + U)/2]$. If the average expected annual private return is $[(L + U)/2)]$ and the portion of producer surplus that is appropriable, v, is known, then the total producer surplus equals $[(L + U)/2v]$ and hence this value is a lower bound for the expected annual social return. Again, it is a lower bound because consumer surplus has not been measured.

The expected private rate of return without SBIR support is the solution to i in Eq. (3), given solution values for L and U from Eqs. (1) and (2). The solution

[2]Equation (1) consists of three general terms. Each term represents the present value for a particular flow that is realized over a particular time period. The first term in the equation represents the present value of the negative cash flows that result to the firm from the cost of conducting the project, $C - A$, from its start to its expected completion, $t = 0$ to d. The second term is the present value of the future negative cash flows from the additional cost, F, of taking the generic technology from the project, at $t = d$, and commercializing it, at $t = d + z$. Finally, the third term is the present value of the expected net cash flows from the project, L, after it has been commercialized, at $t = d + z$, over its estimated life, to $t = d + z + T$. Note that the discount rate in Eq. (1) is the firm's hurdle rate, r. Therefore, the value for L that solves Eq. (1) is the value for which the private firm just earns its hurdle rate of return on the portion of the total investment that it must finance. The firm would not invest in the project unless it expected at least L for the average annual private return, so that its hurdle rate would be exactly met. Thus, L is a lower-bound estimate.

value of i in Eq. (3) represents the rate of return that just equates the present value of the expected annual private return to the firm to the present value of research and postresearch commercialization costs to the firm in the absence of SBIR funding.

$$-\int_0^d \left(\frac{C}{d}\right) e^{-it} dt - \int_d^{d+z} \left(\frac{F}{z}\right) e^{-it} dt + \int_{d+z}^{d+z+T} \left(\frac{L+U}{2}\right) e^{-it} dt = 0$$
$$\Rightarrow \left(\frac{C}{di}\right)(e^{-id} - 1) + \left(\frac{F}{zi}\right)(e^{-i(d+z)} - e^{-id}) - \left(\frac{L+U}{2i}\right)(e^{-i(d+z+T)} - e^{-i(d+z)}) = 0 \quad (3)$$

Finally, the lower bound on the social rate of return is found by solving Eq. (4) for i, given values for the other variables.[3]

$$-\int_0^d \left(\frac{C}{d}\right) e^{-it} dt - \int_d^{d+z} \left(\frac{F}{z}\right) e^{-it} dt + \int_{d+z}^{d+z+T} \left(\frac{L+U}{2v}\right) e^{-it} dt = 0$$
$$\Rightarrow \left(\frac{C}{di}\right)(e^{-id} - 1) + \left(\frac{F}{zi}\right)(e^{-i(d+z)} - e^{-id}) - \left(\frac{L+U}{2iv}\right)(e^{-i(d+z+T)} - e^{-i(d+z)}) = 0 \quad (4)$$

Equations (1) through (4) were estimated for each of the 44 SBIR-sponsored projects. Mean values of the two resulting important rates of return, averaged across the 44 projects, are shown in Table 3. There are two important points to be seen in Table 3: First, the average of the expected private rate of return in the absence of SBIR support is 25 percent, clearly less than the average self-reported private hurdle rate of 33 percent (see Table 4). Thus, in the absence of SBIR support. this sample of firms would not have undertaken this research and, in fact, they expressed this fact explicitly during the interviews. Second, the expected social rate of return (lower bound) associated with SBIR's funding of the these projects is at least 84 percent, and hence the projects are expected to be socially valuable.

We cannot conclude that a social rate of return of at least 84 percent is good or bad, or better or worse than expected. Those are nonaxiomatic conclusions. However, we can compare our estimate of the lower bound of the social rate of return to the opportunity cost of public funds. Following the guidelines set forth by the U.S. Office of Management and Budget to use a real discount of 7 percent for constant-dollar benefit-to-cost analyses of proposed investments and regulations, we find that, clearly, a nominal social rate of return of 84 percent is above that rate and thus is socially worthwhile.[4]

We also can calculate the expected private rate of return with SBIR support for each of the 44 projects as the solution to i in Eq. (5), given the values of d, C, A, z, F, and T and the derived values of L and U from Eqs. (1) and (2) using those values:

---

[3]Note that Eq. (4) is identical to Eq. (3) with the exception that the average expected annual private return, $[(L + U)/2]$, is replaced with the lower bound for the average expected annual social return, $[(L + U)/2v]$.

[4]Link and Scott (1998b) discuss the use of this guideline for National Institute of Standards and Technology economic impact assessments.

TABLE 3 Rates of Return for Average Project (N = 44)

Variable	Definition
$i_{private} = 0.25$	Expected private rate of return without SBIR funding
$i_{social} = 0.84$	Lower bound for expected social rate of return

$$-\int_0^d \left(\frac{C-A}{d}\right) e^{-it} dt - \int_d^{d+z} \left(\frac{F}{z}\right) e^{-it} dt + \int_{d+z}^{d+z+T} \left(\frac{L+U}{2}\right) e^{-it} dt = 0$$

$$\Rightarrow \left(\frac{C-A}{di}\right)(e^{-id} - 1) + \left(\frac{F}{zi}\right)(e^{-i(d+z)} - e^{-id}) - \left(\frac{L+U}{2i}\right)(e^{-i(d+z+T)} - e^{-i(d+z)}) = 0 \qquad (5)$$

The estimated private rate of return with SBIR support averages 76 percent for the 44 cases; this value is noticeably above the average private hurdle rate of 33 percent. However, there is no way for SBIR to have calculated the optimal level of funding for these 44 projects, or for any projects, unless, as part of the Phase I application, all relevant data, including hurdle rates, could have been assessed. In the absence of such information, which in practice would be difficult to obtain because of, if nothing else, self-serving reporting by proposers, the funding scheme that SBIR has implemented may be as close to optimal as possible.[5]

Figure 2 summarizes our estimated values for the average SBIR-sponsored project. Based on the sample of 44 projects, the average gap between the lower-bound social rate of return and the estimated private rate of return without SBIR funding support is 59 percent.

TABLE 4 Descriptive Statistics on Variables ($N = 44$)

Variable	Mean	Standard Deviation
d	2.68 years	0.36
C	$1,027,199	461,901
A	$782,000	127,371
r	0.33	0.08
z	1.30 years	1.07
F	$1,377,341	2,972,266
T	10.56 years	7.23
v	0.16	0.16
L	$902,738	1,228,850
U	$1,893,001	1,733,581

[5]Scott (1998) has proposed using a bidding mechanism that would result in the SBIR funding being just sufficient to ensure that the private participants earn just a normal rate of return. The proposal is a novel one, but it is as yet untried. Successful implementation would require additional development to make it practicable.

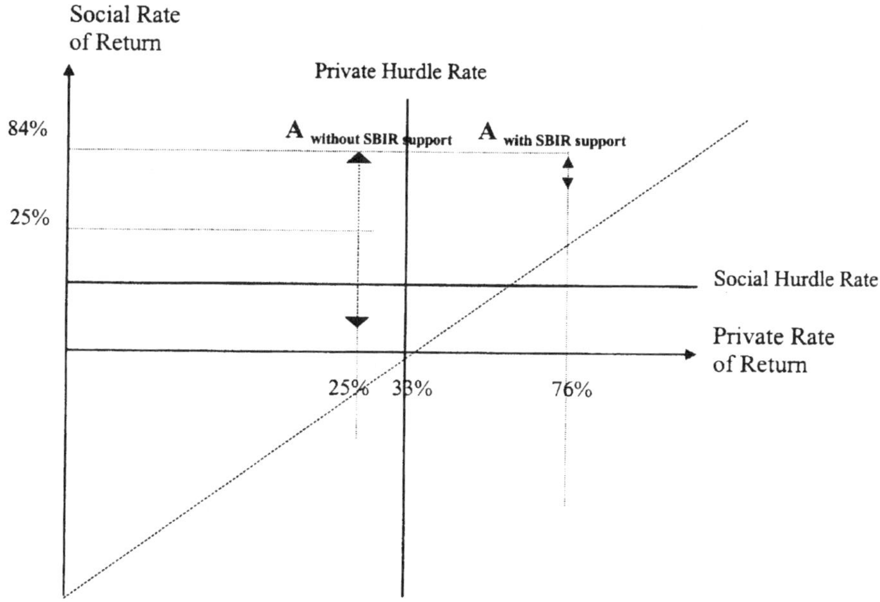

FIGURE 2 Gap between social and private rates of return to the average SBIR project ($N = 44$).

INTERPROJECT DIFFERENCES IN THE GAP BETWEEN SOCIAL AND PRIVATE RETURNS

On the basis of the estimation of the preceding equations, we calculated for each project the gap between the lower bound social rate of return and the private rate of return without SBIR funding support. In an exploratory fashion, we considered interproject differences in this spillover gap as a function of the geographic region of the company conducting the research, and the SBIR funding mechanism (Fast Track versus non-Fast Track).

Consider the following fixed-effects model:

$$Gap = \exists_0 + \exists_1 \, FT + \exists_2 \, NE + \exists_3 \, W + , \tag{6}$$

where *Gap* represents the difference between the lower-bound social rate of return and the private rate of return; *FT* is a binary variable equaling 1 if the project was funded as a Fast Track project and 0 otherwise; *NE* is a binary variable equaling 1 if the company conducting the research is located in a northeastern state, and 0 otherwise; and *W* is a binary variable equaling 1 if the company conducting the research is located in the west (California in fact), and 0 otherwise. The least-squares results are reported in Table 5.

TABLE 5 Regression Results from Eq. (6) ($N = 44$)

	Dependent Variable Gap		
Variable	Coefficient	*t* statistic	$p > \|t\|$
constant	0.811	10.2	0.000
FT	0.0752	0.86	0.395
NE	–0.475	–4.78	0.000
W	–0.168	–1.62	0.112

$R^2 = 0.38$
F level = 8.19, $p > F = 0.0002$

	Expected Gap as Predicted by the Model	
Variable	Fast Track	Non-Fast Track
NE	$\exists_0 + \exists_1 + \exists_2 = 0.411$	$\exists_0 + \exists_2 = 0.336$
SE	$\exists_0 + \exists_1 = 0.886$	$\exists_0 = 0.811$
W	$\exists_0 + \exists_1 + \exists_3 = 0.718$	$\exists_0 + \exists_3 = 0.643$

What is clear from the estimation of Eq. (6), after controlling for region, is that there is not a statistically significant difference between the gap of Fast Track and non-Fast Track projects, meaning that in this sample the expected spillover benefits from Fast Track projects are equal to those of non-Fast Track projects. Shown in Table 6 are the mean values of the gap by region, and clearly there are regional differences.

Although there is not a statistical difference in the gap between Fast Track and non-Fast Track projects, there are differences, as would be expected, between the component measures of *Gap*—the lower-bound estimate of the social rate of return and the estimate of the private rate of return without SBIR funding support. A priori, we expected higher prospective estimated social rates of return among Fast Track projects because these are the projects that have attracted outside investors at an early state in the research in anticipation that they were projects that were closer to commercialization. This same reasoning implies a priori that these companies also would have a greater private rate of return without SBIR funding support. As shown in Table 7, these differences are born out in the data.

TABLE 6 Descriptive statistics related to the gap variable ($N = 44$)

Variable *Gap*	Mean (%)
Northeast ($n = 17$)	36
Southeast ($n = 13$)	83
West ($n = 14$)	66

TABLE 7 Mean Values Related to the Components of the *Gap* Variable ($N = 44$)

Variable	Fast Track (n = 14)	Non-Fast Track (n = 30)
Gap	0.63	0.58
Lower-bound social rate of return	0.93	0.80
Private rate of return without SBIR funding	0.30	0.22

CONCLUDING OBSERVATIONS

Two points need to be emphasized as caveats to our conclusion that, as measured by our sample of 44 projects, SBIR is funding socially desirable projects. First, the social rates of return estimated for the SBIR projects are very conservative, lower-bound estimates because they do not include consumer surplus in the benefit stream. Second, some might be skeptical about the SBIR awardees' earnest belief that without SBIR funding the projects would not have been undertaken or at least would not have been undertaken to the same extent or with the same speed. With the SBIR program in place, the pursuit of SBIR funding probably would be a path of least resistance. However, if the research would have occurred without the public funding, the estimated upper bound and hence the average of the upper and lower bounds for the expected private returns would be too low, and the actual lower bounds for the social rates of return would be even higher than we have estimated. Further, the gap would remain, although that would not in itself necessarily justify the public funding of the projects.

REFERENCES

Audretsch, David B., Albert N. Link, and John T. Scott. 1999. "Analysis of the National Academy of Sciences' Survey of Small Business Innovation Research Awardees," this volume.

Baron, Jonathan. 1998. *DoD SBIR/STTR Program Manager. Comments at the Methodology Workshop on the Assessment of Current SBIR Program Initiatives,* Washington, D.C., October.

Bator, Francis. 1958. "The anatomy of market failure," *Quarterly Journal of Economics.* pp. 351–379.

Bush, Vannevar. 1946. *Science—the Endless Frontier.* Republished in 1960 by U.S. National Science Foundation, Washington, D.C.

Clinton, William Jefferson. 1994. *Economic Report of the President.* Washington, D.C.: U.S. Government Printing Office.

Jaffe, Adam B. 1998. "The importance of 'spillovers' in the policy mission of the Advanced Technology Program," *Journal of Technology Transfer* (Summer).

Lerner, Joshua. 1999. "Public Venture Capital: Rationale and Evaluations, in National Academy of Sciences," *The Small Business Innovation Research Program: Challenges and Opportunities.* Washington, D.C.: National Academy Press. pp. 115-128.

Link, Albert N. 1998. *Public/Private Partnerships as a Tool to Support Industrial R&D: Experiences in the United States, Final report prepared for the Working Group on Technology and Innovation Policy,* Division of Science and Technology, Paris: Organization for Economic Cooperation and Development, January.

Link, Albert N. 1999. "An Assessment of the Small Business Innovation Research Fast Track Program in Southeastern States," this volume.

Link, Albert N., and John T. Scott. 1998a. "Assessing the infrastructural needs of a technology-based service sector: A new approach to technology policy planning," *STI Review* 22:171–207.

Link, Albert N., and John T. Scott. 1998b. *Public Accountability: Evaluating Technology-Based Institutions.* Norwell, MA: Kluwer Academic.

OECD (Organization for Economic Cooperation and Development). 1998. *Technology, Productivity and Job Creation: Toward Best Policy Practice.* Paris: OECD.

Office of the President. 1990. *U.S. Technology Policy.* Washington, DC: Executive Office of the President.

OMB (Office of Management and Budget). 1996. Economic analysis of federal regulations under Executive Order 12866. Mimeo.

Scott, John T. 1998. :Financing and leveraging public/private partnerships: The hurdle-lowering auction," *STI Review* 23:67–84.

Scott, John T. 1999. "An Assessment of the Small Business Innovation Research Program in New England: Fast-Track Compared with non-Fast Track Projects," this volume.

Tassey, Gregory. 1997. *The Economics of R&D Policy.* Westport, CT: Quorum Books.

Statistical Analysis of the National Academy of Sciences Survey of Small Business Innovation Research Awardees: Analyzing the Influence of the Fast Track Program

David B. Audretsch, Albert N. Link,** and John T. Scott****
Indiana University, University of North Carolina at Greensboro,** and Dartmouth College****

EXECUTIVE SUMMARY

This paper summarizes the findings from a statistical analysis of the survey data collected by Peter Cahill for the National Academy of Sciences, under the sponsorship of the Department of Defense's Small Business Innovation Research program. Also, the findings from the statistical analysis are related to the findings from several regional case-based studies that compare, along a number of dimensions, Fast Track Phase II projects to non-Fast Track projects.

The primary conclusions from the statistical analysis of the survey data are:

- Fast Track projects have greater expected sales (commercialization) than do non-Fast Track projects,
- Fast Track projects experience a shorter funding gap between Phase I and Phase II awards than do non-Fast Track projects, and
- Fast Track projects have greater employment growth than do non-Fast Track projects.

These findings are extremely robust. In addition, they complement the case-based analyses by other researchers.

INTRODUCTION

Responding to a paucity of evidence about the impact of the Small Business Innovation Research (SBIR) program, recent studies document the fact that the SBIR does make a number of positive economic contributions. Most notably, research has found that

- the growth rates of SBIR firms exceed those of comparable small firms not receiving SBIR support (Lerner and Kegler, 1999),
- the social returns from SBIR-funded projects exceed the private returns (Link and Scott, 1999), and
- the SBIR program influences the entrepreneurial behavior of scientists by changing their career paths and inducing them to commercialize (Audretsch, Weigand, and Weigand, 1999).

Although these studies conclude that the SBIR program makes a positive contribution to the commercialization of knowledge, they are somewhat limited in their scope of coverage.

One of the more controversial aspects of the SBIR program was the introduction of the Fast Track Initiative in 1996. Under this initiative, firms winning Fast Track designation have priority for the funding of the Phase II award because additional outside funding is committed to the research. It is conjectured that this Fast Track option bestows at least three main advantages to firms. First, it provides a mechanism for avoiding, or at least reducing, the funding gap that often occurs between Phase I and Phase II research. The significance and impact of this funding gap is made clear in the case studies by Audretsch et al. (1999), Feldman (1999), Link (1999), and Scott (1999). Audretsch et al. (1999) report examples related to funding-gap problems in their Indiana-based case studies. For example, the cofounders of a new startup developing genetically based rats experienced a funding gap between Phase I and Phase II research. The turnover of key personnel that resulted from the lapse of funding forced the company to incur retraining because of key personnel turnover. One purpose of the Fast Track is to assist small firms in avoiding such redundant cost burdens.

Second, the Fast Track program may reduce complications arising during the normal review process. For example, Audretsch et al. (1999) report in their case studies that Anthony Hubbard, founder of Endotech, Inc., noted from his non-Fast Track experience that

> [t]here was no continuity of reviewers between our Phase I and Phase II proposals. It was like the Phase II review board ignored our Phase I results and overlooked that we had met our Phase I goals.

A third possible gain from the Fast Track program comes through certification. As Kegler and Lerner (1999) point out, there is a growing body of empirical research that suggests that new, technology-based firms are burdened with asymmetric information between them and external financing institutions. The certification involved in the Fast Track process may encourage third-party financing by alleviating information asymmetries.

In responding to the lack of evidence about the impact of the Fast Track Initiative, the U.S. Department of Defense (DoD) requested that the National Academy of Sciences (NAS) review its SBIR Fast Track program to determine, to the extent possible,

- if the Fast Track Initiative encourages more rapid commercialization of research results through the acquisition of private investment capital, and
- if Fast Track projects progress more rapidly than do the standard SBIR awards.

To accomplish this, the NAS undertook a multifaceted research strategy that included both a broad-based mail survey to a representative sample of SBIR awardees and focused regional case studies of firms taken from that sample. An overview of the mail survey is presented by Cahill (1999). The focused regional case studies are described by Audretsch et al. (1999), Cramer (1999), Link (1999), and Scott (1999).

We expand on the Cahill (1999) analysis in one very important aspect. We present a number of statistical analyses that examine the relationship between the Fast Track program and several different performance output variables. The analyses described herein control statistically for the relationship between other factors and performance output. The remainder of the paper is outlined as follows: In the second section, the rationale for public funding of small technology-based firms is explained. In the third section, selected characteristics of the Cahill survey sample are presented. In the fourth section, the statistical models that we considered are presented, and the relevant variables are defined. In the fifth section, we present our results and our interpretation of them. We conclude with a brief summary of our findings.

THE FINANCE GAP CONFRONTING SMALL TECHNOLOGY-BASED FIRMS

Are small technology-based firms merely replicas of their larger counterparts? This is an important question because an affirmative answer would indicate that there is no reason to expect them to be financed differently than large firms. A large literature in economics has addressed this question and provided a resolute answer. Small technology-based firms are, in fact, markedly different from large enterprises in a number of key dimensions that have important implications for public policy, in general, and the SBIR, in particular.

The first important difference is that large firms have a proven track record based on success. New firms have no such proven track record, and many SBIR firms are relatively new firms. Having a track record is crucial because efforts to commercialize new technologies are characterized by a greater degree of uncertainty, knowledge asymmetries, and transactions costs than other types of economic activity. The greater degree of uncertainty is the result of not knowing how the new technology-based product can be made, and if there is a viable commercial market for the product (Arrow, 1962). The high degree of uncertainty makes a track record of success dealing with such uncertainly very important to outside investors.[1]

An additional complication stems from asymmetries in knowledge about the project and its prospects for success between the firm and external financiers. The firm may have a better—or at least different—understanding of how the technology can be produced and commercialized than do external investors, who generally do not have the same technological background, experience with the technology, or degree of specialization in the technology. This gap in technological competence and understanding results in high costs of transacting information about the technology, possible products, and potential commercial applications. For external investors to understand and evaluate accurately the prospects of the project, they need to invest in technological competence and experience, which can be prohibitively costly.

These knowledge asymmetries are compounded by the prohibitive cost of transacting the knowledge about the project to external parties. Not only do these asymmetries exist but, because of the specialized knowledge required for path-breaking technologies, it becomes prohibitively expensive for external financiers to learn enough to evaluate the project, or even hire experts who can.

Thus, external financiers are confronted with an inability to evaluate accurately a proposed new technology-based product. This is true for larger, established firms as well as for new small firms. However, there is an important difference that tilts the decision to provide finance toward large enterprises. Larger enterprises have established a proven track record, whereas the new small firms have not. External financiers may be uncertain about the outcome of the proposed project and even unable to evaluate accurately its technological and commercial prospects, but they can be certain about the past performance of the established large firm. This often is not the case for new enterprises, with no proven track record. As one of the founders of Genetic Models, Joe Pesek, learned (Audretsch et al., 1999) after unsuccessfully trying to obtain funding from traditional financial institutions: "Nobody is going to finance a firm with no assets, no product, and no track record." The founders of Genetic Models, like many of the other SBIR firms, have no commercial track record because they had been involved in scientific research at the university prior to starting their firm.

SBIR provides the needed certification to reduce the uncertainty confronting external financiers. According to Paul J. Hall, President and CEO of Integrated Biotechnology:

[1]As Hebert and Link (1989, p. 47) argue in their synthetic definition of entrepreneurship, "An entrepreneur is someone who specializes in taking responsibility for and making judgmental decisions that affect the location, form, and the use of goods, resources, or institutions."

> Even if you only received a $1 award, the fact remains that you have people extensively reviewing your product with impeccable credentials. Receiving an award really boosts your reputation and your credibility. The greatest benefit from receiving an award is the sense of comfort it provides for investors when our small firm is trying to raise capital.

It is well known that the rate of success of innovative activity is low. Many, if not most, innovative projects do not succeed. However, studies have documented that because a specific project did not result in a viable commercial product does not necessarily mean that the knowledge and experience generated by that project have no economic value. In fact, most successful new products are the result of previous attempts at either the same product or a related but different one that failed. For large firms, much of the knowledge and experience resulting from failed projects is then applied to successive projects, enabling the firm and its investors to capture the economic value from the learning process inherent in risky innovative activity. However, a failed project for a small firm typically means that the firm will go out of business. The economics literature has documented as a virtual "stylized fact" (Geroski, 1995, Caves, 1998) that the failure rate of small firms is systematically higher than that for larger enterprises. Thus, the scientists from these small firms going out of business will take their knowledge and experience to other firms, often other small firms. Although this experience and knowledge may result in a commercial success for a different firm, the investors in the original failed firm are unable to realize any of the financial gains from the original investments. However, although the high rate of failure by technology-based small firms may deter private investors, the public concern is that the science be commercialized. This is the direct result of the externality of knowledge and experience created in innovative efforts by small firms that ultimately fail. The gap between the valuable and useful knowledge with a potential commercial value created in small firms and the ability of private investors to earn a return on that knowledge results in an underinvestment in technology-based small firms.

This underinvestment in technology-based small firms is particularly pronounced in regions where there is a deficiency of technology-based small firms. Several important studies (Link and Rees, 1990; Feldman, 1994a,b) have documented how external sources of scientific knowledge are much more critical to small firms than to their larger counterparts. This means that the success of new technology-based firms is highly conditional upon the existence of other small technology-based firms in the same geographic region. This second type of externality associated with small technology-based firms results in a high propensity for these firms to cluster within tightly concentrated geographic regions (Audretsch and Feldman, 1996, Audretsch and Stephan, 1996). An important and valuable function of the SBIR is to induce the start-up of a critical mass of new technology-based firms, which can trigger the start-up of subsequent new firms. In examining the decision to start a new biotechnology firm, both Feldman

(1999) and Audretsch and Stephan (1996) show the importance of accessibility to similar small biotechnology firms.

Private investors will underinvest in small technology-based firms in regions where a cluster of technology-based firms is lacking. However, the externality generated by creating such a cluster of small technology-based firms, which will then make it profitable for private investors to finance subsequent start-ups, provides a clear mandate for public support to compensate for the finance gap existing between large and small firms.

OVERVIEW OF THE FAST TRACK SURVEY

As described in Cahill (1999), a mail survey was sent in early 1999 to a representative sample of companies that had received an SBIR Phase II award since 1992. This sample of 379 projects consisted of all 48 Fast Track projects funded since the inception of the program in 1996, all 127 BMDO co-investment projects funded between 1992 and 1996, a matched control group for Fast Track and BMDO projects, and an additional 29 projects for population adjustments. A total of 232 surveys were returned partially or totally completed. That responding sample was defined as the parent sample for this statistical study.

The statistical models described in Section III, and the relevant findings discussed in Section IV, relate to a subset of the parent sample of 232 projects. That subset contains information related to 112 projects. A number of surveys were returned partially completed; when information relevant to the analyses of this paper was missing the project was deleted from consideration.

In terms of Fast Track projects, which are the focus of this paper, 12.7 percent of the initial sample of 379 projects were Fast Track projects, 18.1 percent of the 232 returned surveys represented Fast Track projects, and 16.1 percent of the 112 projects considered herein are Fast Track projects.

THE STATISTICAL MODEL

The fundamental model considered in this study is described, in the most general terms, by Eq. (1). Performance output (*OUTPUT*) associated with an active research project is assumed to be functionally related to the experience of the research company (*EXP*), the associated strategy adopted by company (*STRATEGY*), the technical characteristics of the project (*TECH*), and whether the project was funded as a Fast Track (*FT*) project or not. We represent this behavioral model as

$$OUTPUT = f\,(EXP,\ STRATEGY,\ TECH,\ FT) \tag{1}$$

Five alternative performance output measures are considered:

1. actual sales (in dollars) realized to date resulting from the technology developed during the Phase II project (*ActSales*);
2. sales (in dollars) expected from the technology developed during the Phase II project between now and the end of 2001 (*ExpSales*);
3. actual sales (in dollars) realized to date from the technology developed during the Phase II project plus sales expected between now (the time of the survey) and the end of 2001 (*ActExpSales*);
4. duration of the funding gap (in months) between the completion of Phase I and the beginning of Phase II for the project (*DurGap*); and
5. number of employees hired as a result of the technology developed during the Phase II project (*Employ*).[2]

The precedence for these five performance output measures comes from evidence, both quantitative and anecdotal, collected by Audretsch et al. (1999), Link (1999), and Scott (1999) while conducting case studies of SBIR awardees.

The experience of each research company in the subset of 112 projects was characterized along five dimensions, each dimension hypothesized to have an independent influence on performance output. The variables that characterize these dimensions are

- age of the business (in years) defined in terms of its founding date (*AgeBus*);
- experience of the business founder(s) measured dichotomously in terms of whether the founder(s) most recently came from another private company, or not (*ExpFounder*)[3];
- size of the business defined in terms of its total revenues during the previous fiscal year (*Revenues*)[4];
- research experience of the company as measured by the number of previous Phase II awards that it has received (*PhaseII*);
- research stage of the company as measured dichotomously in terms of whether the Phase II award has been completed (*Complete*).[5]

Company strategy was characterized in only one dimension:

- marketing plans of the company as measured dichotomously if the company has a marketing (e.g., commercialization) plan under way or completed, or not (*Market*).[6]

---

[2]A company could report a fractional unit of an employee's time.

[3]*ExpFounder* equals 1 if the founder's most recent employment was in another private company and 0 otherwise (e.g., with a college or university or the government).

[4]On the Cahill survey, respondents noted company revenue as the range of total revenue (less than \$100,000; \$100,000 to \$499,999, \$500,000 to \$999,999, \$1,000,000 to \$4,999,999, \$5,000,000 to \$19,999,999, and more than \$20,000,000). The variable *Revenues* is defined as the midpoint of each stated range, with the lower bound defined as \$50,000 and the upper bound defined as \$25,000,000.

[5]*Complete* equals 1 if the Phase II research is completed, and 0 otherwise.

[6]*Market* equals 1 if the company has under way or has completed a marketing plan, and 0 otherwise.

Two technical characteristics of each project were considered:

- length of time (in years) that the Phase II award has been active, measured as the time between the year that the award started and 1999 (*Active*)[7];
- primary technology area of the research, defined in terms of the Small Business Administration's (SBA) general technology areas.[8]

Finally, the process for funding each award was defined by one variable:

- funding process was measured dichotomously in terms of whether the award was a Fast Track (*FT*) Phase II, or not.[9]

Table 1 reports the mean value for each of the variables described above.[10]

STATISTICAL FINDINGS

Variations of Eq. (1), for each of the five alternative performance output variables defined above, were estimated using ordinary least-squares analysis. Each of the explanatory variables discussed earlier was included in each of the five estimated equations. The estimated regression results are in the second column of Tables 2 through 6.

The findings in column 2 of Tables 2 through 4 can be discussed as a group because each has as a dependent variable a sales or expected sales measure. Thus, as a group, these three specifications relate to commercialization activity. An inspection of the regression results in column 2 of Tables 2 through 4 suggests that, *ceteris paribus*:

- The survey respondents associated with Fast Track projects have a greater expectation of future sales than those associated with non-Fast Track projects.

---

[7]All projects considered in the sample of 112 are active projects even if the Phase II research is complete.

[8]On the basis of a reading of the technical abstracts of each award, SBA assigned several detailed technology codes to each project. Based on the assumption that the first code identified by SBA is the dominant technology, a more aggregate technology area was assigned to each project. SBA defines seven broad technology areas: Computer, Information Processing, Analysis (*Computer*); Electronics (*Electronics*); Materials (*Materials*); Mechanical Performance of Vehicles, Weapons, Facilities (*Mechanical*); Energy Conversion and Use (*Energy*); Environment and Natural Resources (*Environment*); Life Sciences (*LifeScience*).

[9]*FT* equals 1 if the award was a Fast Track Phase II award, and 0 otherwise.

[10]Two constructed variables are discussed. These variables are a probability of response to the survey variable (*ProbResponse*) and an associated hazard rate (*HazardRate*). These two related variables were constructed using all of the Cahill sample to estimate a model of the probability of response. The two variables were available for only 109 of the 112 observations. Observations were lost to either a perfect prediction in the probit model that generated the probability of response, or to missing observations that were used as controls in the probit model. The results from the underlying probit model are not reported herein, but are available upon request from the authors.

TABLE 1 Mean Values for Variables Used in the Statistical Analysis

Variable	Mean
ActSales	$175,021
ExpSales	$6,299,554
ActExpSales	$6,474,574
DurGap	7.58 months
Employ	2.38 individuals
AgeBus	11.13 years
ExpFounder	0.7857 (78.57% from business)
Revenues	$5,547,321
PhaseII	6.01 awards
Complete	0.3661 (36.61% completed)
Market	0.5893 (58.93% with operational market plan)
Active	2.82 years
Electronics	0.5000 (50.00% in electronics technology)
Computer	0.2143 (21.43% in computer technology)
Materials	0.0714 (7.14% in materials)
Mechanical	0.0496 (4.46% in mechanical performance of vehicles, weapons, and facilities)
Energy	0.0982 (9.82% in energy conversion and use)
Environment	0.0446 (4.46% in environment and natural resources)
LifeScience	0.0268 (2.68% in life sciences)
FT	0.1607 (16.07% Fast Track)
ProbResponse	0.659 (65.9% probability of response)
HazardRate	1.168 (conditional density)

- The survey respondents associated with Fast Track projects, in contrast, report lower current sales than those associated with non-Fast Track projects, although the estimated regression coefficient is only marginally significant in a statistical sense. This finding is not unexpected because Fast Track projects from awards in 1996 are just now being completed.
- Companies with a marketing strategy in place have realized, and expect to realize in the near future, greater sales than companies that do not have one.
- Older, more established companies, as measured by the age of the company, seem to have somewhat dampened expectations of future sales than do younger companies. This finding may reflect greater reasonableness in forecasting expectations.

Regarding duration of the funding gap as a performance output measure, the results reported in column 2 of Table 5 suggest that, *ceteris paribus*:

- Fast Track projects are associated with a shorter funding gap, compared to non-Fast Track projects, as expected, given the focus of the Fast Track Initiative.

TABLE 2 Estimated Regression Results: Dependent Variable = *ActSales*

(1) Variable	(2) Estimated Coefficient	t statistic	(3) Estimated Coefficient Probability of Response	t statistic	(4) Estimated Coefficient Hazard rate for Response	t statistic
Intercept	41,094.9	(0.175)	553,777.5	(1.298)	253,384.2	(0.732)
AgeBus	–11,518.1	(–1.048)	–11,879.3	(–1.071)	–11,784.8	(–1.055)
ExpFounder	–47,862.1	(–0.350)	–35,194.0	(–0.249)	–30,692.7	(–0.215)
Revenues	0.0236	(2.485)	0.0189	(1.882)	0.0207	(2.053)
PhaseII	–2,979.0	(–0.533)	–1,973.2	(–0.346)	–2700.2	(–0.473)
Complete	263,649.4	(1.840)	201,098.6	(1.335)	232,003.6	(1.551)
Market	231,439.1	(1.943)	258,994.1	(2.110)	250,487.2	(2.029)
Active	–42,668.4	(–0.667)	–70,092.2	(–0.998)	–56,503.3	(–0.802)
Computer	327,922.9	(2.297)	298,169.2	(2.036)	313,263.7	(2.129)
Materials	284,034.9	(1.251)	299,487.7	(1.262)	300,828.3	(1.257)
Mechanical	–14,2017.4	(–0.545)	–92,018.0	(–0.347)	–117,836.2	(–0.443)
Energy	114,603.2	(0.611)	75,228.9	(0.394)	83,641.0	(0.434)
Environment	290,793.3	(1.054)	256,384.1	(0.918)	265,770.8	(0.945)
LifeScience	–8,726.6	(–0.026)	–48,978.1	(–0.145)	–49,581.1	(–0.146)
FT	–210,619.4	(–1.276)	–147,079.3	(–0.851)	–166,812.4	(–0.954)
ProbResponse	—		–624,951.6	(–1.550)	—	
HazardRate	—		—		–142,912.1	(–0.991)
R^2	0.224		0.244		0.233	
F–level	2.00		2.00		1.88	

- Older, more established companies, as measured by the age of the company, seem to experience slightly longer funding gaps than do the younger companies, although the estimated regression coefficient is not significant in a statistical sense. One possible explanation for the positive coefficient is that there is less urgency among older companies to obtain gap funding because they have other sources of funds to rely upon; another possible explanation is that diseconomies of research and production scope associated with the relatively older companies decrease the ability of these companies to maintain an administrative schedule once new research begins.

Finally, regarding employment growth during the Phase II projects, the results reported in Table 6 suggest that, *ceteris paribus*:

- Employment growth is greater in Fast Track projects than in non-Fast Track projects.
- Company founders who have a business background, compared to an academic or public-sector background, expand staffing slower, perhaps reflecting prior experience or lessons learned.

TABLE 3 Estimated Regression Results: Dependent Variable = *ExpSales*

(1) Variable	(2) Estimated Coefficient	*t* statistic	(3) Estimated Coefficient Probability of Response	*t* statistic	(4) Estimated Coefficient Hazard rate for Response	*t* statistic
Intercept	7,018,013.0	(1.193)	2.63E+07	(2.493)	2.08E+07	(2.46)
AgeBus	–575,499.8	(–2.088)	–594,588.8	(–2.169)	–593,407.7	(–2.17)
ExpFounder	–2,467,573.0	(–0.719)	–2,531,579.0	(–0.723)	–2,237,995.0	(–0.640)
Revenues	–0.063	(–0.264)	–0.231	(–0.931)	–0.2319	(–0.939)
PhaseII	91,310.1	(0.651)	133,096.2	(0.944)	118,248.7	(0.846)
Complete	241,319.1	(0.067)	–1,875,286.0	(–0.503)	–111,378.0	(–0.304)
Market	6,712,359.0	(2.246)	7,451,051.0	(2.456)	7,375,226.0	(2.44)
Active	248,885.0	(0.155)	–873,956.6	(–0.503)	–869,935.8	(–0.504)
Computer	841,505.3	(0.235)	–417,943.6	(–0.115)	–277,023.8	(–0.077)
Materials	302,986.1	(0.053)	664,196.8	(0.113)	457,868.9	(0.078)
Mechanical	–3,242,383.0	(–0.497)	–1,475,863.0	(–0.225)	–1,780,125.0	(–0.273)
Energy	6,488,823.0	(1.380)	5,079,322.0	(1.077)	4,830,751.0	(1.024)
Environment	1.54E+07	(2.229)	1.43E+07	(2.075)	1.42E+07	(2.067)
LifeScience	–7,648,164.0	(–0.915)	–8,879,436.0	(–1.065)	–9,498,163.0	(–1.140)
FT	9,317,159.0	(2.252)	1.15E+07	(2.702)	1.18E+07	(2.764)
ProbResponse	—		–2.22E+07	(–2.225)	—	
HazardRate	—		—		–8,204,561.0	(–2.325)
R^2	0.266		0.305		0.308	
F–level	2.52		2.72		2.76	

- Companies with a marketing strategy in place report greater Phase II employment growth than those without a market strategy.[11]
- Older, more established companies, as measured by the age of the company, seem to expand their staff slower than do the younger companies, although the estimated regression coefficient is only marginally significant in a statistical sense. As with the experience of the company founders, this may reflect prior experience or lessons learned.

Estimation of our five basic models yields essentially the same results, as discussed earlier, whether or not we control for the possibility that the error in the estimating equation is correlated with the binary variable denoting Fast Track status, *FT*. There are two reasons that we might expect such a correlation, and if the correlation exists, our previously discussed estimates of the relationship between Fast Track status and performance output would be biased.

First, if the probability of responding to the Cahill survey is associated with performance output, and if that response effect in the error term is correlated with

[11]This finding should be interpreted with caution because the survey does not distinguish the category of employee growth being measured and hence it is possible that part of the reported growth included individuals in marketing.

TABLE 4 Estimated Regression Results: Dependent Variable = *ActExpSales*

(1) Variable	(2) Estimated Coefficient	t statistic	(3) Estimated Coefficient: Probability of Response	t statistic	(4) Estimated Coefficient: Hazard rate for Response	t statistic
Intercept	7,059,108	(1.190)	2.68E+07	(2.529)	2.11E+07	(2.472)
AgeBus	–587,017.9	(–2.113)	–606,468.1	(–2.197)	–605,192.5	(–2.197)
ExpFounder	–2,515,435.0	(–0.727)	–2,566,773.0	(–0.728)	–2,268,688.0	(–0.644)
Revenues	–0.0391	(–0.163)	–0.2123	(–0.849)	–0.2112	(–0.849)
PhaseII	88,331.1	(0.625)	131,123.0	(0.924)	115,548.5	(0.820)
Complete	504,968.5	(0.139)	–1,674,188.0	(–0.446)	–881,776.4	(–0.239)
Market	6,943,852.0	(2.305)	7,710,045.0	(2.524)	7,625,713.0	(2.505)
Active	206,216.6	(0.127)	–944,048.8	(–0.540)	–926,439.1	(–0.533)
Computer	1,169,428.0	(0.324)	–119,774.4	(–0.033)	36,240.0	(0.010)
Materials	587,021.1	(0.102)	963,684.6	(0.163)	758,697.2	(0.129)
Mechanical	–3,384,400.0	(–0.514)	–1,567,881.0	(–0.238)	–1,897,961.0	(–0.289)
Energy	6,603,426.0	(1.393)	5,154,551.0	(1.085)	4,914,392.0	(1.034)
Environment	1.57E+07	(2.253)	1.46E+07	(2.098)	1.45E+07	(2.09)
LifeScience	–7,656,890.0	(–0.908)	–8,928,414.0	(–1.064)	–9,547,744.0	(–1.137)
FT	9,106,539.0	(2.184)	1.14E+07	(2.649)	1.17E+07	(2.705)
ProbResponse	—		–2.28E+07	(–2.272)	—	
HazardRate	—		—		–8,347,473.0	(–2.348)
R^2	0.266		0.306		0.309	
F–level	2.51		2.74		2.77	

the variables included in our model, our estimates could be biased. To control for that possibility, we used the entire Cahill sample and estimated a probit model of response and constructed the probability of response and an associated hazard rate for each surveyed company. In our estimations, we controlled for the possibility that the probability of response affects our previously discussed results. We added in turn to our equations the probability density of response and the hazard rate for response. The specifications allowed us to see if there was something important in the error that was correlated with either the probability of response or the counterfactual conditional probability of response, counterfactually assuming nonresponse. In sum, we asked across the observations in the sample whether the variance in probability of response has an effect on the dependent variable, and whether controlling for that matters for the other estimates. Further, using the hazard rate, we ask across the observations whether the variance in the counterfactual hazard for response has an effect and whether the control affects the other estimates. Our results regarding the association of Fast Track projects with performance outputs are essentially unchanged when either the probability of response (results are shown in each table of regression results in column 3) or the hazard rate for response (results are shown in each table of regression results in column 4) is added to the specifications.

TABLE 5 Estimated Regression Results: Dependent Variable = *DurGap*

(1) Variable	(2) Estimated Coefficient	*t* statistic	(3) Estimated Coefficient: Probability of Response	*t* statistic	(4) Estimated Coefficient: Hazard rate for Response	*t* statistic
Intercept	6.8103	(2.918)	6.4712	(1.543)	7.7186	(2.288)
AgeBus	0.1213	(1.109)	0.1165	(1.068)	0.116	(1.065)
ExpFounder	–0.6879	(–0.506)	–0.7913	(–0.568)	–0.7570	(–0.544)
Revenues	1.26E–08	(0.133)	2.33E–08	(0.236)	7.52E–09	(0.077)
PhaseII	–0.0131	(–0.236)	–0.0162	(–0.289)	–0.0136	(–0.245)
Complete	1.1487	(0.806)	0.6838	(0.462)	0.6059	(0.416)
Market	0.5510	(0.465)	0.3477	(0.288)	0.4026	(0.335)
Active	–0.0154	(–0.024)	0.2084	(0.302)	0.0915	(0.133)
Computer	–2.7073	(–1.907)	–2.5066	(–1.74)	–2.6024	(–1.816)
Materials	5.7556	(2.549)	6.9308	(2.97)	6.8676	(2.946)
Mechanical	–2.9899	(–1.154)	–2.8411	(–1.091)	–2.693	(–1.039)
Energy	3.7107	(1.989)	3.7802	(2.014)	3.645	(1.941)
Environment	–1.4416	(–0.525)	–1.1827	(–0.431)	–1.285	(–0.469)
LifeScience	0.3557	(0.107)	0.3762	(0.113)	0.2268	(0.068)
FT	–4.5941	(–2.799)	–4.2573	(–2.506)	–4.0109	(–2.353)
ProbResponse	—		–0.1179	(–0.030)	—	
HazardRate	—		—		–0.810	(–0.577)
R^2	0.289		0.312		0.315	
F–level	2.81		2.82		2.85	

Second, the error in the estimating equations could be correlated with the binary variable denoting Fast Track status because Fast Track status can be modeled as an endogenous variable, with Fast Track status having as its ultimate cause better expected commercial performance. Thus, while we expect Fast Track status to help propel the commercial success of the so-funded SBIR project, it is also the case that projects with better prospects for commercial success will attract outside funding and thus will be placed on a Fast Track status. We recognize that both directions for causality—from Fast Track to expected performance and from expected performance to Fast Track—exist. That causality issue *in itself* is not necessarily a problem for estimating the relationship between Fast Track and performance output. However, we might expect the associated problem of the error in our estimating equations being correlated with the binary Fast Track variable. To control for that possible simultaneity bias, we estimated each of our five basic equations with instrumental variables (results not shown here), letting the binary Fast Track variable and the response variable be endogenous in addition to the left-hand-side variable for each equation.[12] With both the survey

[12]These results are also available upon request from the authors.

TABLE 6 Estimated Regression Results: Dependent Variable = *Employ*

(1) Variable	(2) Estimated Coefficient	t statistic	(3) Estimated Coefficient Probability of Response	t statistic	(4) Estimated Coefficient Hazard rate for Response	t statistic
Intercept	3.0047	(2.897)	6.1047	(3.276)	4.9009	(3.252)
AgeBus	–0.0737	(–1.517)	–0.0762	(–1.572)	–0.0758	(–1.559)
ExpFounder	–1.4127	(–2.336)	–1.3659	(–2.208)	–1.3238	(–2.130)
Revenues	5.62E–08	(1.342)	2.87E–08	(0.655)	3.22E–08	(0.733)
PhaseII	–0.0025	(–0.102)	0.0037	(0.150)	0.0006	(0.025)
Complete	0.4001	(0.632)	0.0238	(0.036)	0.1705	(0.262)
Market	1.1675	(2.216)	1.310	(2.444)	1.2848	(2.39)
Active	–0.1308	(–0.462)	–0.2964	(–0.965)	–0.2693	(–0.878)
Computer	0.0327	(0.052)	–0.1464	(–0.229)	–0.1009	(–0.158)
Materials	0.5082	(0.506)	0.6195	(0.598)	0.5988	(0.575)
Mechanical	–0.8445	(–0.734)	–0.5383	(–0.465)	–0.6232	(–0.538)
Energy	0.2280	(0.275)	–0.0007	(–0.001)	–0.0124	(–0.015)
Environment	4.3522	(3.568)	4.165	(3.415)	4.174	(3.408)
LifeScience	–0.4954	(–0.336)	–0.719	(–0.488)	–0.7902	(–0.533)
FT	2.0884	(2.863)	2.4804	(3.287)	2.4757	(3.251)
ProbResponse	—		–3.7334	(–2.119)	—	
HazardRate	—		—		–1.2083	(–1.925)
R^2	0.357		0.387		0.382	
F–level	3.85		3.92		3.84	

data and the associated project-specific data collected by DoD, a number of instruments were identified—variables highly correlated with Fast Track status, yet arguably not affected by the error in expected performance—to use in the estimations. For example, experience with private investors prior to the SBIR award, the number of founders with a business background, and the agency making the SBIR award were among the instruments available. Our results regarding the relationship between Fast Track status and performance output are essentially the same qualitatively in the instrumental variable specifications.

CONCLUSIONS

Although the empirical results provide compelling evidence that the SBIR generally has promoted the commercialization of scientific knowledge, detailed case studies have revealed several impediments hindering the effectiveness of the program. Small technology-based firms often have trouble financing the gap between Phase I and Phase II awards. This funding gap is particularly troublesome because it can cause the loss of investments in scientific training for key personnel. Although the Fast Track Initiative was initiated to overcome this impediment, until now there has been no systematic evaluation of the program.

The empirical results presented in this paper clearly support the claim that the Fast Track Initiative has measurable impacts on the performance output of Phase II award recipients. The evidence is overwhelming, when other factors are controlled for, that respondents associated with Fast Track projects report greater expectations of sales associated with the Phase II research, shorter funding gaps between Phase I and Phase II awards, and more employment growth. Thus, the Fast Track program improves the ability of SBIR firms to transform science into commercially successful products.

In addition, the empirical results presented here are consistent with a basic finding reported by Link and Scott (1999) based on a very different methodology. In the present paper, we have shown that Fast Track projects are associated with better commercial performance, where that performance is estimated directly by the SBIR respondents through survey responses. The paper by Link and Scott also finds that Fast Track projects are associated with better commercial performance, but in contrast to the present paper, a key measure of such performance is derived indirectly. In particular, instead of expected sales that the respondents have estimated on the Cahill survey, Link and Scott asked the respondents during extensive interviews a set of questions about investments and the duration of those investments, and then derived the expected profit stream from an economic model of investment behavior. Using their indirect method for estimating expected profits from the SBIR investments, Link and Scott find that the Fast Track projects are associated with greater rates of return on investment. Both our paper and the Link and Scott paper find that the Fast Track projects as a group have greater commercial potential than the non-Fast Track projects.

Further, the basic finding of our paper—the association of Fast Track projects with better performance regarding expected sales and employment growth, and a more rapid movement from Phase I to Phase II development—remains when we control for the possibility that the failure to respond by some recipients of the survey could affect the results of our estimations. Additionally, the basic finding of the paper remains when we control for the possibility that there is a simultaneous-equations effect because Fast Track status itself is affected by expected performance. The basic result—the association of Fast Track with better performance—survives econometrically both the hazard-rate response bias scrutiny, and the instrumental-variables simultaneity bias scrutiny.

REFERENCES

Arrow, Kenneth J. 1962. "Economic welfare and the allocation of resources for invention," pp. 609–625 in *The Rate and Direction of Inventive Activity: Economic and Social Factors.* Princeton, N.J.: Princeton University Press.

Audretsch, David B., and Maryann P. Feldman. 1996. "R&D spillovers and the geography of innovation and production," *American Economic Review* 86(3):630–640.

Audretsch, David B., and Paula E. Stephan. 1996. "Company-scientist locational links: The case of biotechnology," *American Economic Review* 86(3):641–642.

Audretsch, David B., Juergen Weigand, and Claudia Weigand. 1999. "Does the Small Business Innovation Research program foster entrepreneurial behavior? Evidence from Indiana," this volume.

Cahill, Peter. 1999. "Fast Track: Is It Speeding Commercialization of Department of Defense Small Business Innovation Research Projects?" this volume.

Caves, Richard E. 1998. "Industrial organization and new findings on the turnover and mobility of firms," *Journal of Economic Literature* 36(4):1947–1982.

Cramer, Reid. 1999. "Patterns of firm participation in the Small Business Innovation Research Program in Southwestern and Mountain States," this volume.

Feldman, Maryann P. 1994a. "Knowledge complementarity and innovation," *Small Business Economics* 6(5):363–372.

Feldman, Maryann P. 1994b. *The Geography of Knowledge.* Boston: Kluwer Academic.

Feldman, Maryann P. 1999. "Role of the Department of Defense in Building Biotech Expertise," this volume.

Geroski, Paul A. 1995. "What do we know about entry?" *International Journal of Industrial Organization* 13(4):421–440.

Hebert, Robert F., and Albert N. Link. 1989. "In search of the meaning of entrepreneurship," *Small Business Economics* 1(1):39–49.

Lerner, Josh. 1999. "The government as venture capitalist: The long-run effects of the SBIR program," *Journal of Business.* July, v. 72 3, pp. 285-297.

Josh Lerner, and Kegler, Colin. 1999. "Evaluating the Small Business Innovation Research Program: A Literature Review," this volume.

Link, Albert N. 1999. "An Assessment of the Small Business Innovation Research Fast Track Program in Southeastern States," this volume.

Link, Albert N., and John Rees. 1990. "Firm size, university based research and the returns to R&D," *Small Business Economics*, 2(1):25–32.

Link, Albert N., and John T. Scott. 1999. "Estimates of the Social Returns to Small Business Innovation Research Projects," this volume.

Scott, John T. 1999. "An Assessment of the Small Business Innovation Research Program in New England: Fast-Track Compared with Non-Fast Track Projects," this volume.

Evaluating the Small Business Innovation Research Program: A Literature Review

Josh Lerner and Colin Kegler***
Harvard Business School and Harvard University***

EXECUTIVE SUMMARY

Despite the proliferation of public efforts to finance small high-technology firms in recent years, there has been relatively little assessment of these programs' economic impacts. This article first explores the underlying challenges that the financing of young firms poses, the ways that specialized financial intermediaries address them, and the rationales for and problems faced by public efforts to finance these companies. The final two sections review earlier efforts to assess these programs, and discuss the proposed evaluation of the Small Business Innovation Research program.

INTRODUCTION

The federal government has played an active role in financing new firms, particularly in high-technology industries, since the Soviet Union's launch of the Sputnik satellite. In recent years, European and Asian nations and many U.S. states have adopted similar initiatives. Although these programs' precise structures have differed, the efforts have been predicated on two shared assumptions: (1) that the private sector provides insufficient capital to new firms, and (2) that the government either can identify investments that ultimately will yield high social and/or private returns or can encourage effective financial intermediaries.[1] In contrast to many forms of government intervention designed to boost economic growth, such as privatization programs, these claims have received little scrutiny by economists.

The neglect of these questions is unfortunate. Although the sums of money involved are modest relative to public expenditures on defense procurement or retiree benefits, these programs are very substantial when compared to contemporaneous private investments in new firms. Several examples, documented by Gompers and Lerner (1998b), underscore this point:

- The Small Business Investment Company (SBIC) program led to the provision of more than $3 billion to young firms between 1958 and 1969, more than three times the total private venture capital investment during these years (Noone and Rubel, 1970).
- In 1995, the sum of the equity financing provided through and guaranteed by public small business financing programs was $2.4 billion, more than 60 percent of the amount disbursed by traditional venture funds in that year. Perhaps more significant, the bulk of the public funds went to early-stage firms, which in the past decade had accounted for only about 30 percent of the disbursements by independent venture capital funds (Venture Economics, 1996).
- Some of America's most dynamic technology companies received support through the SBIC and the Small Business Innovation Research (SBIR) programs while the companies were still privately held entities, including Apple Computer, Chiron, Compaq, and Intel.
- Public venture capital programs also have had a significant impact overseas: For example, Germany has created about 800 federal and state government financing programs for new firms over the past two decades, which provide the bulk of the financing for technology-intensive start-ups (OECD, 1995).

[1]It is striking to note the similar emphasis on these rationales in, for instance, the statement of Senator John Sparkman (1958) upon passage of the Small Business Investment Act and the recent testimony of Dr. Mary Good (1995), Under Secretary for Technology at the U.S. Department of Commerce. The rationales for such programs are discussed in depth in a report from the U.S. Congressional Budget Office (1985).

Government programs in this arena have been divided between those efforts that directly fund entrepreneurial firms and those that encourage or subsidize the development of outside investors.

Although these efforts have proliferated, a consensus as to how to evaluate these programs remains elusive. The gap between the approaches employed by academics and practitioners is substantial. Furthermore, there is a lack of consensus among economists as to what the proper approaches are.

This article provides an overview of the motivations for these efforts to encourage individual investors. In the second section, the underlying challenges that the financing of young growth firms poses are discussed, as well as the ways that specialized financial intermediaries address them. The rationales for and common problems of public programs are explored in the third section. The fourth section discusses earlier efforts to assess these efforts, and the challenges that they have faced.

VENTURE CAPITALISTS AND THE FINANCING CHALLENGE

The initial reaction of a financial economist to the argument that the government needs to invest in young firms is likely to be skepticism. A lengthy literature has highlighted the role of financial intermediaries in alleviating moral hazard and information asymmetries. Young high-technology firms often are characterized by considerable uncertainty and informational asymmetries, which permit opportunistic behavior by entrepreneurs. Why one would want to encourage public officials instead of specialized financial intermediaries (venture capital organizations) as a source of capital in this setting is not immediately obvious.

The Challenge of Financing Young High-Technology Firms

To briefly review the types of conflicts that can emerge in these settings, Jensen and Meckling (1976) demonstrate that agency conflicts between managers and investors can affect the willingness of both debt and equity holders to provide capital. If the firm raises equity from outside investors, the manager has an incentive to engage in wasteful expenditures (e.g., lavish offices) because he does not bear their entire cost. Similarly, if the firm raises debt, the manager may increase risk to undesirable levels. Because providers of capital recognize these problems, outside investors demand a higher rate of return than would be the case if the funds were internally generated.

Even if the manager is motivated to maximize shareholder value, informational asymmetries may make raising external capital more expensive or even preclude it entirely. Myers and Majluf (1984) and Greenwald et al. (1984) demonstrate that equity offerings of firms may be associated with a "lemons" problem (first identified by Akerlof [1970]). If the manager is better informed about the investment opportunities of his or her firm than the investors and acts in the interest of current shareholders, then the manager issues new shares only when the company's stock is overvalued. Indeed, numerous studies have documented that stock prices decline upon the announcement of equity issues, largely because of the negative signal sent to the market.

These information problems also have been shown to exist in debt markets. Stiglitz and Weiss (1981) show that if banks find it difficult to discriminate among companies, raising interest rates can have perverse selection effects. In particular, the high interest rates discourage all but the highest-risk borrowers, so the quality of the loan pool declines markedly. To address this problem, banks may restrict the amount of lending rather than increasing interest rates.

These problems in the debt and equity markets are a consequence of the information gaps between the entrepreneurs and the investors. If the information asymmetries could be eliminated, financing constraints would disappear. Financial economists argue that specialized financial intermediaries can address these problems. By intensively scrutinizing firms before providing capital and then monitoring them afterward, they can alleviate some of the information gaps and reduce capital constraints.

Responses by Venture Capitalists

The financial intermediary that specializes in funding young high-technology firms is the venture capital organization. The first modern venture capital firm, American Research and Development (ARD), was formed in 1946 by Massachusetts Institute of Technology President Karl Compton, Harvard Business School Professor Georges F. Doriot, and local business leaders. A small group of venture capitalists made high-risk investments in emerging companies that were formed to commercialize technology developed for World War II. The success of the investments ranged widely: Almost half of ARD's profits during its 26-year existence as an independent entity came from its $70,000 investment in Digital Equipment Corporation (DEC) in 1957, which grew in value to $355 million. Because institutional investors were reluctant to invest, ARD was structured as a publicly traded closed-end fund and marketed mostly to individuals (Liles, 1977). The few other venture organizations begun in the decade after ARD's formation also were structured as closed-end funds.

The first venture capital limited partnership, Draper, Gaither, and Anderson, was formed in 1958. Imitators soon followed, but limited partnerships accounted for a minority of the venture pool during the 1960s and 1970s. Most venture organizations raised money either through closed-end funds or SBICs, federally guaranteed risk capital pools that proliferated during the 1960s. Although investor demand for SBICs in the late 1960s and early 1970s was strong, incentive problems ultimately led to the collapse of the sector. The annual flow of money into venture capital during its first three decades never exceeded a few hundred million dollars and usually was substantially less.

The activity in the venture industry increased dramatically in the late 1970s and early 1980s. Industry observers attributed much of the shift to the U.S. Department of Labor's clarification of the Employee Retirement Income Security Act's (ERISA) "prudent man" rule in 1979. Prior to that year, ERISA regulations limited pension funds from investing substantial amounts of money in venture capital or other high-risk asset classes. The Department of Labor's clarification of the rule explicitly allowed pension managers to invest in high-risk assets, including venture capital. In 1978, when $424 million was invested in new venture capital funds, individuals accounted for the largest share (32 percent). Pension funds supplied just 15 percent. Eight years later, when more than $4 billion was raised, pension funds accounted for more than half of all contributions. (These annual commitments represent pledges of capital to venture funds raised in a given year. This money typically is invested over three to five years, starting in the year the fund is formed.)

The subsequent years saw both very good and trying times for venture capitalists. On the one hand, venture capitalists had backed during the 1980s and 1990s many of the most successful high-technology companies, including Apple Computer, Cisco Systems, Genentech, Netscape, and Sun Microsystems. A substantial number of service firms (including Staples, Starbucks, and TCBY) also received venture financing. At the same time, commitments to the venture capital industry were very uneven. The annual flow of money into venture funds increased by a factor of 10 during the early 1980s, peaking at just under 6 billion 1996 dollars. From 1987 through 1991, however, fund-raising steadily declined. Over the past five years, the pattern has been reversed; 1997 represented a record fund-raising year, in which nearly $10 billion was raised by venture capitalists. This process of rapid growth and decline has created a great deal of instability in the industry.

To address the information problems that preclude other investors in small high-technology firms, the partners at venture capital organizations employ a variety of mechanisms. First, business plans are intensively scrutinized: Of those firms that submit business plans to venture capital organizations, historically fewer than 1 percent have been funded (Fenn et al., 1995). The decision to invest frequently is made conditional on the identification of a syndication partner who agrees that this is an attractive investment (Lerner, 1994). In exchange for their capital, the venture capital investors demand preferred stock with numerous restrictive covenants and representation on the board of directors.

Once the decision to invest is made, the venture capitalists frequently disburse funds in stages. Managers of these venture-backed firms are forced to return repeatedly to their financiers for additional capital in order to ensure that the money is not squandered on unprofitable projects. In addition, venture capitalists intensively monitor managers, often contacting firms on a daily basis and holding monthly board meetings during which extensive reviews of every aspect of the firm are conducted. (Various aspects of the oversight role played by venture capitalists are documented by Gompers and Lerner [1999]; the theoretical literature is reviewed by Barry [1994].)

Note that, even with these many mechanisms, the most likely primary outcome of a venture-backed investment is failure or, at best, modest success. Gompers (1995) documents that, out of a sample of 794 venture capital investments made over three decades, only 22.5 percent ultimately succeeded in going public, the avenue through which venture capitalists typically exit their successful investments. (A Venture Economics [1988] study finds that a $1 investment in a firm that goes public provides an average cash return to venture capitalists of $1.95 in excess of the initial investment, with an average holding period of 4.2 years. The next best alternative, a similar investment in an acquired firm, yields a cash return of only 40 cents over a 3.7-year mean holding period.) Similar results emerge from Huntsman and Hoban's (1980) analysis of the returns from 110 investments by three venture capital organizations. About one in six investments was a complete loss, while 45 percent were either losses or simply broke even. The elimination of the top-performing 9 percent of the investments was sufficient to turn a 19 percent gross rate of return into a negative return.

In short, the environment in which venture organizations operate is extremely difficult. It is the mechanisms bundled with the venture capitalists' funds that are critical in ensuring a satisfactory return. These circumstances have led to venture capital organizations emerging as the dominant form of equity financing for privately held technology-intensive businesses.[2]

RATIONALES FOR AND PROBLEMS OF PUBLIC VENTURE CAPITAL PROGRAMS

At the same time, there are reasons to believe that despite the presence of venture capital funds, there still might be a role for public venture capital programs. In this section, we assess these claims. We highlight two arguments: that public venture capital programs may play an important role by certifying firms to outside investors, and that these programs may encourage technological spillovers. We then highlight two classes of problems that can affect these programs.

The Certification Hypothesis

A growing body of empirical research suggests that new firms, especially technology-intensive ones, may receive insufficient capital because of the infor-

[2]Although evidence regarding the financing of these firms is imprecise, Freear and Wetzel's (1990) survey suggests that venture capital accounts for about two-thirds of the external equity financing raised by privately held technology-intensive businesses from private-sector sources.

mation problems discussed in the preceding section.[3] If public venture capital awards could certify that firms are of high quality, then these information problems could be overcome and investors could confidently invest in these firms.

As discussed earlier, venture capitalists specialize in financing these types of firms. They address these information problems through a variety of mechanisms. Many of the studies that document capital-raising problems examine firms during the 1970s and early 1980s, when the venture capital pool was relatively modest in size. Since the pool of venture capital funds has grown dramatically in recent years (Gompers and Lerner, 1996, 1998a), even if small high-technology firms had numerous value-creating projects that they could not finance in the past, one might argue that it is not clear that this problem remains today.

A response to this argument emphasizes the limitations of the venture capital industry. Venture capitalists back only a tiny fraction of the technology-oriented businesses begun each year. In 1996, a record year for venture disbursements, 628 companies received venture financing for the first time (VentureOne, 1997); to put this in perspective, the Small Business Administration estimates that in recent years close to one million businesses have been started annually. Furthermore, these funds have been very concentrated: 49 percent of venture funding in 1996 went to companies based in either California or Massachusetts, and 82 percent went to firms specializing in information technology and the life sciences (VentureOne, 1997).

It is not clear, however, what lessons to draw from these funding patterns. Concentrating investments in such a manner may well be an appropriate response to the nature of opportunities. Consider, for instance, the geographic concentration of awards. Recent models of economic growth—building on earlier works by economic geographers—have emphasized powerful reasons why successful high-technology firms may be very concentrated. The literature highlights several factors that lead similar firms to cluster in particular regions, including knowledge spillovers, specialized labor markets, and the presence of critical intermediate goods producers.[4] Case studies of the development of high-technology regions (e.g., Saxenian [1994]) have emphasized the importance of such intermediaries as venture capitalists, lawyers, and accountants in facilitating this clustering.

A related argument for public investments is that the structure of venture investments may make them inappropriate for many young firms. Venture funds tend to make quite substantial investments, even in young firms; the mean venture investment in a start-up or early-stage business between 1961 and 1992 (ex-

[3]The literature on capital constraints (reviewed by Hubbard [1998]) documents that an inability to obtain external financing limits many forms of business investment. Particularly relevant are works by Hall (1992), Hao and Jaffe (1993), and Himmelberg and Petersen (1994). These show that capital constraints appear to limit research and development (R&D) expenditures, especially in smaller firms.

[4]The theoretical rationales for such effects are summarized by Krugman (1991).

pressed in 1996 dollars) was $2.0 million (Gompers, 1995). The substantial size of these investments may be partially a consequence of the demands of institutional investors. The typical venture organization raises a fund (structured as a limited partnership) every few years. Because investments in partnerships are often time-consuming to negotiate and monitor, institutions (limited partners) prefer making relatively large investments in venture funds, typically $10 million or more. Furthermore, governance and regulatory considerations lead institutions to limit the share of any fund that any one limited partner holds.[5] As a consequence, venture organizations typically raise substantial funds of $100 million or more. Because each firm in the venture capitalist's portfolio must be closely scrutinized, the typical venture capitalist is typically responsible for no more than a dozen investments. Consequently, venture organizations are unwilling to invest in very young firms that require only small capital infusions.[6]

This problem may be increasing in severity with the growth of the venture industry, as discussed earlier. As the number of dollars per venture fund and dollars per venture partner has grown, so too has the size of venture investments. For instance, the mean financing round for a start-up firm has climbed (in 1996 dollars) from $1.6 million in 1991 to $3.2 million in 1996 (VentureOne, 1997).

Again, it is not clear what lessons to draw from these financing patterns. Venture capitalists may have eschewed small investments because they were simply not profitable, because of either the high costs associated with these transactions or the poor prospects of the thinly capitalized firms.[7] Encouraging public investments in small firms may be counter-productive and socially wasteful if the financial returns are unsatisfactory and the companies financed are not viable.

Support for these claims is found in recent work on the long-run performance of initial public offerings (IPOs). Brav and Gompers (1997) show that IPOs that had previously received equity financing from venture capitalists outperform other offerings. These findings underscore concerns about policies that seek to encourage public investments in companies that are rejected by professional investors.

Furthermore, it appears that there were in 1997 a number of financial innovations to address the needs of early-stage entrepreneurs. These included the creation of incubators and "entrepreneur-in-residence" programs by established ven-

---

[5]The structure of venture partnerships is discussed at length by Gompers and Lerner (1996, 1998a).

[6]There are two primary reasons that venture funds do not simply hire more partners if they raise additional capital. First, the supply of venture capitalists is quite inelastic. The effective oversight of young companies requires highly specialized skills that can only be developed with years of experience. A second important factor is the economics of venture partnerships. The typical venture fund receives a substantial share of its compensation from the annual fee, which is typically between 2 percent and 3 percent of the capital under management. This motivates venture organizations to increase the capital that each partner manages.

[7]For a theoretical discussion of why poorly capitalized firms are less likely to be successful, see Bolton and Scharfstein (1990).

ture organizations such as Mayfield and Mohr Davidow. Other examples are innovative efforts to direct the resources of individual investors to small venture capital funds (an example is Next Generation Partners, a "fund-of-funds" for wealthy families developed by FLAG Venture Partners). Finally, some institutional investors are displaying an increased willingness to provide capital to first-time and seed venture funds. Thus, market forces may be addressing whatever problem has existed.

The Presence of R&D Spillovers

A second rationale emerges from the literature on R&D spillovers. Public finance theory emphasizes that subsidies are an appropriate response in the case of activities that generate positive externalities. Such investments as R&D expenditures and pollution control equipment purchases may have positive spillovers that help other firms or society as a whole. Because the firms making the investments are unlikely to capture all the benefits, public subsidies may be appropriate.

An extensive literature (reviewed by Griliches [1992] and Jaffe [1996]) has documented the presence of R&D spillovers. These spillovers take several forms. For instance, the rents associated with innovations may accrue to competitors who rapidly introduce imitations, developers of complementary products, or to the consumers of these products. Whatever the mechanism of the spillover, however, the consequence is the same: The firm invests below the social optimum in R&D.

After reviewing a wide variety of studies, Griliches estimates that the gap between the private and social rates of return is substantial: The gap is probably equal to between 50 percent and 100 percent of the private rate of return. Although few studies have examined how these gaps vary with firm characteristics, a number of case-based analyses (Jewkes, 1958; Mansfield et al., 1977) suggest that spillover problems are particularly severe among small firms. These organizations may be particularly unlikely to effectively defend their intellectual property positions or to extract most of the rents in the product market.

Distortions in the Award Process

Even if these problems are substantial, however, the government may not be able to address them dispassionately. An extensive political economy and public finance literature has emphasized the possible distortion that may result from government subsidies as particular interest groups or politicians seek to direct subsidies in a manner that benefits themselves. As articulated by Olson (1965) and Stigler (1971), and formally modeled in works such as those of Peltzman (1976) and Becker (1983), the theory of regulatory capture suggests that direct and indirect subsidies will be captured by parties whose joint political activity is not too difficult to arrange (i.e., when "free-riding" by coalition members is not too large a problem).

These distortions may manifest themselves in several ways. One possibility (discussed, for instance, by Eisinger [1988]), is that firms may seek transfer payments that directly increase their profits. Politicians may acquiesce in such transfers in the case of companies that are politically connected. Alternatively, past "public venture capital" recipients may develop relationships with evaluators and managers that aid in the selection process.

A more subtle distortion is discussed by Cohen and Noll (1991) and Wallsten (1996): Officials may seek to select firms based on their likely success, and fund them regardless of whether the government funds are needed. In this case, they can claim credit for the firms' ultimate success even if the marginal contribution of the public funds was very low.

Inappropriate Program Design

Even if "political capture" is not a problem, the programs' effectiveness may be impaired because of poor design. In this section, we examine two classes of problems that affect the design of public venture capital programs. First, certain company characteristics—attributes that may not be adequately considered in the award selection process—appear to be highly correlated with a company's ability to achieve its research and commercialization goals. Second, the structure of the financing may not match the needs of the entrepreneurial firm.

The first of these problems has been highlighted in a variety of case study evidence, including Gompers and Lerner's case studies of the Advanced Technology Program (1998b).[8] Our field research indicates that a prevalent characteristic among underachieving companies is the existence of research grants from numerous government sources. Because a lack of results can easily be attributed to the high-risk nature of technology development, many of these companies can avoid accountability indefinitely. As a result, some of these government grant-oriented research organizations are able to drift from one federal contract to the next.

Adding to the problem is the fact that companies with substantial government grant experience appear to have several advantages over other firms when applying for future public awards. Past grants, regardless of project outcomes, help a company to gain legitimacy in a particular area of research as well as to acquire the equipment and personnel needed to do future work. There is also a tendency for some government programs to try to "piggyback" on other government programs, hoping to leverage the impact of their grant dollars. In addition,

---

[8]These findings are also corroborated by surveys of SBIR awardees by the U.S. General Accounting Office and the Lerner (1996) study discussed later.

firms gain considerable insight into the grant application process with each proposal they submit. Because of all of these factors, these firms frequently have a greater chance of being awarded future government grants than other firms. The end result can be a stream of government funding being awarded to companies that consistently underachieve. In fact, we have encountered examples in which awardees frequently advise first-time applicants on how to write and structure award proposals.

Another telltale characteristic of underachieving firms was the existence of factors outside the scope of "public venture capital" projects that undermined their ability to successfully complete and later commercialize technology. Legal troubles, for instance, can divert substantial amounts of human and financial resources away from a company's R&D projects. For early-stage firms, legal problems may even cause dramatic changes in the size and structure of the company. And when a firm is ready to commercialize its technology, the liability concerns associated with pending legal battles will often drastically impair the company's ability to attract venture capital investment dollars.

The existence of resource-draining auxiliary research projects also may undermine a company's performance. One company in the sample of Advanced Technology Program awardees, for instance, was involved in a project that was only distantly related to the company's core technology. Although the public funds were not used to fund this auxiliary project, it appeared that a substantial amount of the company's time, energy, and capital was diverted toward this tangential research. This, in turn, diluted the company's focus on its publicly funded research project, and thus slowed the development of its core technology.[9] The existence of unrelated R&D projects, especially for smaller companies, can cause a company's resources to be spread too thin.

For early-stage companies, additional limiting factors frequently involve managers who lack experience in running small companies. Although some of these managers may have accumulated business experience as consultants or as members of large organizations, the successful operation of early-stage companies can demand very different management skills. It thus comes as no surprise that when venture capitalists sink substantial funds in a company, they often place their own hand-picked manager at the helm—typically an individual who has already been successful in managing an early-stage company in a similar industry. Because much of the skills needed for managing start-up companies comes through experience, the existence of managers who do not have this background can significantly undermine a company's ability to carry out its commercialization plans.

---

[9]Part of the problem in this instance is the lack of corporate discipline. If a venture capital firm had invested in this company, it likely would have provided this discipline by closely monitoring the company, and limiting the company's R&D activities to areas that are directly related to its core technology.

In a broader context, each of these performance-undermining factors emphasizes the need for program managers to critically evaluate whether a particular company is, in fact, a viable vehicle for actually accomplishing its commercialization goals. This goes far beyond a simple assessment of the feasibility of a business plan. In fact, many of these potentially limiting factors will not even be discussed in a company's proposal. It is tempting, of course, to attribute the failures resulting from such factors to the high-risk nature of the technology. However, to a large extent, companies exhibiting a high potential for underachievement could be more thoroughly weeded out by placing a greater emphasis on these factors during the selection process. The R&D project itself may be high risk, but the risks of turning the technology into a product should be minimized. Regardless of how innovative or enabling a technology may be—or how well a business plan is constructed—if these undermining factors are substantial, a company will be hard pressed to overcome such roadblocks.

A second example of inappropriate program design involves the structure of the financing provided. Before considering the example of the SBIR program, it is worth highlighting the typical attributes of the early-stage, technology-driven firms that are the typical recipients of public venture capital funds. First, a great deal of uncertainty always accompanies these types of firms. Because such companies are in their formative stages and have little or no track record, it is extremely difficult for their managers to predict the optimal magnitude and duration of R&D expenditures at the onset of a project. Regardless of the time and energy devoted to such forecasts, it is likely that initial estimates will have to be revised over time. Second, R&D and management resources are typically very scarce in these firms. If a project is delayed because of a lack of financing, talented researchers and managers are unlikely to be left "on hold"; rather, they are likely to be drawn into other projects. Finally, the firms have a tremendous need to enter the marketplace rapidly. Rapid market entry is critical in the technology-driven industries targeted by public venture capital programs, particularly for small firms without the large marketing budgets that major corporations enjoy. The primary strength of small firms is their ability to get products into the market quickly. Through such early entry, the small firm may be able to build up a defensible market position, even against larger competitors. This may be accomplished through the establishment of a dominant industry standard, the creation of "network externalities" that encourage later adapters to choose the same product as that selected by early users, or product improvements gleaned from early interactions with customers.

Several public venture capital programs have structured their financing in ways that appear to be at odds with these conditions. With respect to SBIR, although it is a multiagency program, the structure of the awards is constrained to be similar across agencies. Promising proposals are granted Phase I awards (originally no more than \$50,000, today \$100,000 or smaller), which are intended to allow firms to conduct research to determine the feasibility of their ideas. Approximately one-half of the Phase I awardees are then selected for the more substantial Phase II grants. Phase II awards of at most $750,000 (originally, one-half million dollars) are designed to support two years of development work.

This structure was designed to provide many of the same benefits as a staged venture capital financing. Information generated in the first phase of the project would be useful in assessing the application for a Phase II award. In theory, the program managers would be able to make small investments in a wide variety of projects, providing the bulk of the SBIR funds to the most promising of these projects. Survey findings, however, suggest that firms have serious concerns about the delays between Phase I and Phase II awards (GAO, 1987, 1992). Firms believed that the long delays—sometimes two years or more—between the original application and the receipt of the Phase II funds often had a detrimental effect on their ability to commercialize technologies. Because of the characteristics of high-technology firms discussed earlier, these delays often made it difficult for the firm to sustain its innovative effort or to commercialize its findings.

THE CHALLENGE OF EVALUATION

As public venture capital programs have increased in number, policy makers and economists increasingly are grappling with the question of how to assess these programs. Not only do substantial divisions exist between the approaches employed by academics and practitioners, but there is little consensus within the academic community itself about the best evaluation methodologies. In this section, we review some of the most frequently encountered approaches and discuss their strengths and limitations.

The approaches most frequently employed by practitioners have the virtue of being relatively straightforward to implement and communicate. One approach—utilized by many agencies when examining their SBIR programs—has been to highlight successful firms.[10] Another approach has been to survey firms that have been funded under the SBIR program, asking such questions as whether the technologies funded were ever commercialized, the extent to which their development would have occurred without the public award, and how firms assessed their experiences with the program more generally.[11]

These approaches have important limitations. First, many awardees may have a stake in the programs that have funded them, and consequently feel inclined to give favorable answers (i.e., that they have received benefits from the program and that commercialization would not have taken place without the awards). This may be a particular problem in the case of the SBIR initiative

[10]In the context of the SBIR program, see SBA (1994), and many agency publications.

[11]Examples of evaluations of the SBIR program include Myers, Stern, and Rorke (1983), Price, Waterhouse (1985), and U.S. General Accounting Office (1987, 1989, 1992).

because many small high-technology company executives have organized to lobby for its renewal. Second, in other cases, the results may be biased the other way: Firms may be unwilling to acknowledge that they received important benefits from participating in public programs, lest they attract unwelcome attention. This is especially likely to be a problem in the life sciences, because periodic press and congressional investigations have highlighted "give-aways" of research funded by the National Institutes of Health (NIH) to biotechnology and pharmaceutical companies. Third, in many cases, it may simply be very difficult to identify the marginal contribution of a public venture capital award, which may be one of many sources of financing that a firm employed to develop a given technology. Finally, as argued by Wallsten (1996), these evaluation criteria may have a distorting effect on which firms are selected for participation in these programs, leading to an emphasis on "safe" firms that would have succeeded anyway.

The approaches employed by academics have important limitations as well. The most common approach is to examine in a regression framework the marginal impact of public funding on private research spending. Studies of federal technology programs by academic economists, beginning with Levy and Terleckyj (1983), have tended to focus on the short-run effects of these efforts. In particular, they often ask whether federal funds substitute for or stimulate private R&D spending. In another application, Irwin and Klenow (1996) show that semiconductor manufacturers substantially reduced their own R&D spending while participating in the SEMATECH consortium. In theory, these frameworks should be applicable to the assessment of public venture capital programs.

An analysis along these lines is undertaken by Wallsten (1996). He examines whether the SBIR program managers may select firms with too high a probability of success. In keeping with the earlier literature on "crowding out," he seeks to distinguish between *marginal funding* (i.e., cases in which investments in firms would yield a high social return, yet are commercially unprofitable) and *inframarginal funding* (i.e., companies that would still be successful in the absence of federal assistance). Wallsten's analysis concludes that the SBIR program is more inclined to fund inframarginal projects rather than support firms on the margins of commercial profitability. Every dollar of SBIR funding awards is likely to lead to a reduction of about one dollar of private research spending by the awardee firm. To remain true to its original purpose, Wallsten therefore recommends a restructuring of SBIR policy to fund marginal firms whose commercial success is less certain.

However valuable a framework it may be when examining the macroeconomic impact of public expenditures, it is less clear that this econometric approach is appropriate when assessing public efforts to assist small high-technology firms. In many cases, small high-technology firms are organized around one key scientist or engineer and his research laboratory or product development team. It may not be possible to accelerate the project's progress by "scaling up" the project through the addition of researchers or technicians. It may well be rational for a firm not to increase its rate of spending, but rather to use the funds to prolong the time before it needs to seek additional capital. To interpret such a short-run reduction in other research spending as a negative signal is very problematic.

A second academic approach is to examine the long-run impact of participation in public venture capital programs on the growth of the firms themselves, relative to a matched set of firms. In this way, it is possible to assess whether either superior firms were selected for the program or participation in the program was associated with ultimate success, although disentangling the two effects, as discussed later, is challenging. In the context of the SBIR program, Lerner (1996) analyzes the growth of 1,435 SBIR awardees and matching firms over a 10-year period and documents that the awardees appear to have superior employment growth.

This approach also has some important limitations. Most fundamentally, policy makers should seek to maximize social, not private, returns. If the growth of the SBIR awardees is merely at the expense of their rivals, the impact of the program on public welfare is likely to be minimal. Second, even the measures of private benefits that can be employed are imperfect. Ideally, the increase in firm value would be measured. Unfortunately, over 98 percent of the firms were privately held at the time of their first SBIR award. Consequently, assessing the valuation and profitability of these awards is very difficult. Thus, Lerner's examination is confined to two measures that are only imperfectly correlated with firm value, employment and sales. Finally, it is difficult to disentangle whether the superior performance of the awardees is due to the selection of better firms or the positive impact of the awards.[12]

[12]Lerner (1996) tries to address this issue in a supplemental analysis using the following argument: Firms whose key assets are intangible intellectual property are much harder for outside investors to evaluate using traditional financial measures. If SBIR awards are certifying firm quality to outside investors, then these signals may be particularly valuable in these industries. SBIR awards should then be more strongly associated with firm growth in high-technology industries. An alternative hypothesis is that federal officials are selecting firms likely to grow rapidly, even without public subsidies. A potential motive would be that politicians could claim credit for the firms' ultimate success, even if the marginal contribution of the public funds was very low. Though the insights of federal officials may give them a greater insight relative to that of other investors (and thus make a signal more valuable), it is by no means certain that it is easier to select successful firms in these industries. Empirical studies suggest that predicting success is much more difficult in high-technology industries. This suggests the reverse pattern: SBIR awards should be more correlated with firm growth in low-technology industries. Consistent with the certification hypothesis, he finds that the relationship between SBIR awards and growth is much stronger in high-technology industries.

REFERENCES

Akerlof, G. A. 1970. "The market for 'lemons': Qualitative uncertainty and the market mechanism," *Quarterly Journal of Economics* 84:488–500.

Barry, C. B. 1994. "New directions in research on venture capital finance," *Financial Management* 23(Autumn):3–15.

Becker, G. S. 1983. "A theory of competition among pressure groups for political influence," *Quarterly Journal of Economics* 98:371–400.

Bolton, P., and D. Scharfstein. 1990. "A theory of predation based on agency problems in financial contracting," *American Economic Review* 80:93–106.

Brav, A., and P. A. Gompers. 1997. "Myth or reality? Long-run underperformance of initial public offerings; Evidence from venture capital and nonventure capital-backed IPOs," *Journal of Finance* 52:1791–1821.

Cohen, L. R., and R. G. Noll, eds. 1991. *The Technology Pork Barrel.* Washington, D.C.: Brookings Institution.

Eisinger, P. K. 1988. *The Rise of the Entrepreneurial State: State and Local Economic Development Policy in the United States.* Madison: University of Wisconsin Press.

Fenn, G. W., N. Liang, and S. Prowse. 1995. *The Economics of the Private Equity Market.* Washington, D.C.: Board of Governors of the Federal Reserve System.

Freear, J., and W. E. Wetzel, Jr. 1990. "Who bankrolls high-tech entrepreneurs?" *Journal of Business Venturing* 5:77–89.

Gompers, P. A. 1995. "Optimal investment, monitoring, and the staging of venture capital," *Journal of Finance* 50:1461–1489.

Gompers, P. A., and J. Lerner. 1996. "The use of covenants: An empirical analysis of venture partnership agreements," *Journal of Law and Economics* 39:463–498.

Gompers, P. A., and J. Lerner. 1998a. "An analysis of compensation in the U.S. venture capital partnership," *Journal of Financial Economics.* Vol. 51, no. 1, pp. 7-51.

Gompers, P. A., and J. Lerner. 1998b. *Capital formation and investment in venture markets: A report to the National Bureau of Economic Research and the Advanced Technology Program.* Working Paper, Harvard University, Cambridge, Mass.

Gompers, P. A., and J. Lerner. 1999. *The Venture Cycle.* Cambridge, Mass.: MIT Press.

Good, M. L. 1995. *Prepared testimony before the Senate Commerce, Science, and Transportation Committee of the Subcommittee on Science, Technology, and Space,* Washington, D.C., June, 23. Photocopy, U.S. Department of Commerce.

Greenwald, B. C., J. E. Stiglitz, and A. Weiss. 1984. "Information imperfections in the capital market and macroeconomic fluctuations," *American Economic Review Papers and Proceedings* 74:194–199.

Griliches, Z. 1992. "The search for R&D spillovers," *Scandinavian Journal of Economics* 94:S29-S47.

Hall, B. H. 1992. *Investment and research and development: Does the source of financing matter?* Working Paper No. 92-194, Department of Economics, University of California at Berkeley.

Hao, K. Y., and A. B. Jaffe. 1993. "Effect of liquidity on firms' R&D spending," *Economics of Innovation and New Technology* 2:275–282.

Himmelberg, C. P., and B. C. Petersen. 1994. "R&D and internal finance: A panel study of small firms in high-tech industries," *Review of Economics and Statistics* 76:38–51.

Hubbard, R. G. 1998. "Capital-market imperfections and investment," *Journal of Economic Literature* 36:193–225.

Huntsman, B., and J. P. Hoban, Jr. 1980. "Investment in new enterprise: Some empirical observations on risk, return, and market structure," *Financial Management* 9 (Summer): 44–51.

Irwin, D. A., and P. J. Klenow. 1996. "High tech R&D subsidies: Estimating the effects of Sematech," *Journal of International Economics* 40:323–344.

Jaffe, A. B. 1996. *Economic analysis of research spillovers—Implications for the Advanced Technology Program.* Washington, D.C. National Institute of Standards and Technology.

Jensen, M. C., and W. H. Meckling. 1976. "Theory of the firm: Managerial behavior, agency costs and ownership structure," *Journal of Financial Economics* 3:305–360.

Jewkes, J., D. Sawers, and R. Stillerman. 1958. *The Sources of Invention.* New York: St. Martin's Press.

Krugman, P. R. 1991. *Geography and Trade.* Cambridge, MA: MIT Press.

Lerner, J. 1994. "The syndication of venture capital investments," *Financial Management* 23(Autumn):16–27.

Lerner, J. 1996. *The government as venture capitalist: The long-run effects of the SBIR program.* Working Paper 5753, National Bureau of Economic Research.

Lerner, Josh. 1999. "The government as venture capitalist: The long-run effects of the SBIR program," *Journal of Business.* July, v. 72 (3), pp. 285-297.

Levy, D. M., and N. Terleckyj. 1983. "Effects of government R&D on private R&D investment and productivity: A macroeconomic analysis," *Bell Journal of Economics* 14:551–561.

Liles, P. 1977. *Sustaining the Venture Capital Firm.* Cambridge, MA: Management Analysis Center.

Mansfield, E., J. Rapoport, A. Romeo, S. Wagner, and G. Beardsley. 1977. "Social and private rates of return from industrial innovations," *Quarterly Journal of Economics* 91:221–240.

Myers, S. C., and N. Majluf. 1984. "Corporate financing and investment decisions when firms have information that investors do not have," *Journal of Financial Economics* 13:187–221.

Myers, S., R.L. Stern, and M.L. Rorke, 1983, *A study of the Small Business Innovation Research program* (Lake Forest, Illinois, Mohawk Research Corporation).

Nelson, C. W. 1996. "From free money to free market," *International Journal of Technology Management* 12:748–759.

Noone, C. M., and S. M. Rubel. 1970. *SBICs: Pioneers in Organized Venture Capital.* Chicago: Capital Publishing.

OECD (Organization for Economic Cooperation and Development). 1995. *Venture Capital in OECD Countries.* Paris: OECD.

Olson, M. 1965. *The Logic of Collective Action.* Cambridge, MA: Harvard University Press.

Peltzman, S. 1976. "Towards a more general theory of regulation," *Journal of Law and Economics* 19:211–240.

Price Waterhouse. 1985. *Survey of small high-tech businesses shows Federal SBIR awards spurring job growth, commercial sales.* Washington, DC: Small Business High Technology Institute.

Saxenian, A. 1994. *Regional Advantage: Culture and Competition in Silicon Valley and Route 128.* Cambridge, MA: Harvard University Press.

SBA (Small Business Administration). 1994. *Small Business Innovation Development Act: Tenth-Year Results.* Washington, D.C.: U.S. Government Printing Office (and earlier years).

Sparkman, J., 1958. "Introduction". in U.S. Congress, Senate, Small Business Committee, Small Business Investment Act of 1958. Washington, D.C.: U.S. Government Printing Office.

Stigler, G. 1971. "The economic theory of regulation," *Bell Journal of Economics* 2:3–21.

Stiglitz, J. E., and A. Weiss. 1981. "Credit rationing in markets with incomplete information," *American Economic Review* 71:393–409.

U.S. Congressional Budget Office. 1985. *Federal Financial Support for High-Technology Industries.* Washington, D.C.: U.S. Congressional Budget Office.

U.S. General Accounting Office. 1989. *Federal research: Assessment of Small Business Innovation Research program.* Washington, DC: U.S. General Accounting Office.

U.S. General Accounting Office. 1992. *Federal Research: Small Business Innovation Research Shows Success But Can Be Strengthened.* Washington,D.C.: General Accounting Office.

U.S. General Accounting Office. 1998. *Federal Research: Observations on the Small Innovation Research Program.* Washington, D.C.: General Accounting Office.

U.S. General Accounting Office. 1987. *Federal Research: Small Business Innovation Research Participants Give Program High Marks.* Washington, D.C.: General Accounting Office.

Venture Economics. 1988. *Existing Venture Capital Investments.* Wellesley, MA: Venture Economics.

Venture Economics. 1996. "Special report: Rose-colored asset class," *Venture Capital Journal* 36(July). 32–34 (and earlier years).

VentureOne. 1997. *National Venture Capital Association 1996 Annual Report.* San Francisco: VentureOne.

Wallsten, S. J., 1996. *The Small Business Innovation Research program: Encouraging technological innovation and commercialization in small firms?* Working paper, Stanford University, Stanford, Calif.

V

ANNEX

Annex A

Research Team Biographies

Robert B. Archibald

Robert B. Archibald is a Professor of Economics at The College of William and Mary in Williamsburg, Virginia. Professor Archibald received a B.A. from the University of Arizona in 1968 and an M.S. and Ph.D. from Purdue University in 1972 and 1974, respectively. Professor Archibald finished work on his doctoral dissertation as a Research Fellow at the Brookings Institution.

After completing the requirements for his Ph.D., Professor Archibald spent two years as Research Economist at the Division of Price and Index Number Research of the Office of Prices of the Bureau of Labor Statistics. He came to the College of William and Mary in 1976. At William and Mary he has served as Chair of the Department of Economics for five years, Interim Dean of the Faculty of Arts and Sciences for one year, and in the 2000-01 academic year he will be the Interim Director of the Thomas Jefferson Program in Public Policy.

Professor Archibald teaches macroeconomics, statistics, a seminar in behavioral economics, and microeconomics for public policy analysis. He has published over 20 papers in academic journals. His published research has covered many topics in economics including macroeconomics, index number construction, the economics of small business, the economics of energy, behavioral economics, rankings of economic journals, and federal R&D policy. He is currently doing research on the economics of financial aid in higher education as well as continuing his work on federal R&D policy.

David B. Audretsch

David B. Audretsch is the Ameritech Chair of Economic Development and Director of the Institute for Development Strategies at Indiana University. He is also a Research Fellow of the Centre for Economic Policy Research (London). He was at the Wissenschaftszentrum Berlin fuer Sozialforschung in Berlin, Germany, which is a government-funded research think tank between 1984 and 1997. Between 1989 and 1991 he served as Acting Director of the Institute. In 1991 he was named the Research Professor. Professor Audretsch's research has focused on the links between entrepreneurship, government policy, innovation, economic development, and global competitiveness. He has consulted with the World Bank, the National Academy of Sciences, the Department of State, United States Federal Trade Commission, General Accounting Office, and International Trade Commission, as well as the United Nations, Commission of the European Union, the European Parliament, the Organization for Economic Cooperation and Development (OECD), and numerous private corporations, state governments, and a number of European governments. He is a member of the Advisory Board to a number of international research and policy institutes, including the Zentrum fuer Europaeisch Wirtschaftsforschung (ZEW, Centre for Economic Research), Mannheim, the Hamburgisches Welt-Wirtschafts-Archiv (HWWA, Hamburg Institute of International Economics), and the American Institute for Contemporary German Studies (AICGS), Washington, D.C. His research has been published in over 100 scholarly articles in the leading academic journals. He has published 25 books including *Innovation and Industry Evolution* with MIT Press. He is founder and editor of the premier journal on small business and economic development, *Small Business Economics: An International Journal.*

Peter J. Cahill

Peter J. Cahill is a Senior Principal Analyst and Program Manager at BRTRC, Inc. In this position he has performed extensive analysis of and provided support to the Small Business Innovation Research Program (SBIR). Following an in-depth survey and series of interviews of DoD agencies and SBIR awardees, he conducted a two-year survey and interview study of the entire federal SBIR program for the Small Business Administration. He developed and implemented a Web-based system to measure past commercialization performance of SBIR firms as a part of the award evaluation process for DoD SBIR proposals. Other recent projects have included research and analysis of a number of military systems, including bridging, mine clearing, and battle simulation models. Prior to joining BRTRC, Inc., in 1993, Mr. Cahill's U.S. Army assignments included professor of management engineering at the U.S. Military Academy at West Point, Deputy Commander of the U.S. Army Research Laboratory, and command and staff positions in construction, development, and engineering.

Reid Cramer

Dr. Cramer is a policy analyst at the Office of Management and Budget in the Executive Office of the President. Prior to working at the federal level, he worked locally in the public and non-profit sectors. He received his doctorate from the Lyndon B. Johnson School of Public Affairs, the University of Texas at Austin. During his doctoral studies, he served as a research associate for the International Workshop on Governance, a project sponsored by the Ford Foundation designed to promote international collaboration in the promotion of effective governance. In 1997, he was appointed to the City of Austin Telecommunications Commission. His research on economic development, urban planning, and affordable housing has recieved support from the U.S. Department of Housing and Urban Development and the Nonprofit Sector Research Fund of the Aspen Institute. He received his M.S. in city and regional planning from Pratt Institute in Brooklyn, New York, and his bachelor of arts degree from Wesleyan University in Middletown, Connecticut.

Maryann P. Feldman

Maryann P. Feldman is Research Scientist at the Institute for Policy Studies at Johns Hopkins University. She received her B.A. from Ohio State University and her Ph.D. from Carnegie Mellon University. She previously taught economics at Western Maryland College and Goucher College. Before coming to Hopkins, she was on the faculty at the H.J. Heinz School at Carnegie Mellon.

Maryann Feldman is the author of over 30 academic articles that have been published in such journals as the *American Economic Review, The Review of Economics and Statistics,* and *The Annals of the Association of American Geographers.* Her Ph.D. dissertation, *The Geography of Innovation,* was published in 1994 by Kluwer Academic Publishers. She is currently editing the forthcoming *Handbook of Economic Geography* for Oxford University Press. In addition, Maryann Feldman also served as book review editor of the journal, *Growth and Change: An International Journal of Urban Policy.*

Maryann Feldman has also served as a consultant to private business, various federal, state and local agencies and non-profit organizations. She has received grants from the National Science Foundation, the Andrew W. Mellon Foundation, the German Marshall Fund, the U.S. Small Business Administration and the Edison Electric Institute, among others. Forthcoming books by Dr. Feldman include *Innovation Policy for a Knowledge-Based Economy* (with A.N. Link) and *The Oxford Handbook of Economic Geography* (with G. Clark and M. Gertler).

David H. Finifter

David H. Finifter is Professor of Economics at The College of William and Mary. He is also director of the Center for Public Policy Research within the

Public Policy Program. He served as founding director of The Thomas Jefferson Program in Public Policy at William and Mary, a position he held from 1987-2000. His teaching and scholarly interests include the economics of education and public policy, human resource economics, evaluation and benefit/cost analysis, labor economics, science and technology policy, public health service delivery and finance, microeconomics applied to public policy analysis, and econometrics applied to public policy analysis. Dr. Finifter has been on the faculty at The College of William and Mary since completing his Ph.D. in economics at the University of Pittsburgh. He also holds a B.S. from Loyola College of Maryland and an M.A. in economics from the University of Pittsburgh. He has published several articles and reports in the area of evaluation of human resources and public policy on issues including federally subsidized employment and training programs, unemployment insurance policy, performance standards for employment and training programs, veterans' job training programs, and the Job Corps program. He has also published research on workplace literacy and productivity. He has co-edited two books on higher education and public policy and a special edition of the *Quarterly Review of Economics and Business* on health care policy. He has served as a consultant to several federal government agencies, including the United States Department of Labor, the Veterans Administration, NASA, Sandia National Laboratories, and the Environmental Protection Agency. During 1978-79, he served as a Staff Associate in Employment Policy at the Brookings Institution and the United States Department of Labor. During the summer of 1995, he served as a faculty summer fellow, American Society for Engineering Education (ASEE) at NASA Langley Research Center, and worked on technology transfer policy and performance measurement/metrics. His research over the past few years has emphasized work in collaboration with Dr. Robert Archibald on the Small Business Innovation Research Program including an evaluation of the SBIR Program at NASA Langley Research Center.

John B. Horrigan

John Horrigan served as Program Officer at the STEP Board from 1998 to February 2000 where he worked on several reports that are part of the "Government-Industry Partnerships for the Development of New Technologies" project, including volumes on the Advanced Technology Program, the Sandia Science and Technology Park, U.S./European Union science and technology cooperation, and government-industry partnerships in biotechnology and computing. He is presently Senior Research Specialist at the Pew Internet and American Life Project, which conducts research into how the Internet is affecting Americans' day-to-day lives. In addition to helping design surveys of Internet usage patterns at Pew, Horrigan conducts research into the social and economic impact of the Internet on communities. Horrigan has also served as a consultant to the World Resources Institute, the U.S. Department of Commerce, and Texas Perspectives of Austin, Texas. He received a B.A. in Economics and Government from the University of Virginia and a Ph.D. in Public Policy from the Lyndon B. Johnson School of Public Affairs at the University of Texas at Austin. Prior to receiving his Ph.D., Horrigan served as Press Secretary and Legislative Assistant to U.S. Representative J.J. Pickle.

Joshua Lerner

Josh Lerner is a Professor of Business Administration at Harvard Business School, with a joint appointment in the Finance and Entrepreneurial Management Units. He graduated from Yale College with a special divisional major that combined physics with the history of technology. He worked for several years on issues concerning technological innovation and public policy, at the Brookings Institution, for a public-private task force in Chicago, and on Capitol Hill. He then undertook his graduate study at Harvard's Economics Department.

Professor Lerner's research focuses on the structure of venture capital organizations and their role in transforming scientific discoveries into commercial products. Much of his research is collected in his co-authored volume, *The Venture Capital Cycle*, published by MIT Press in 1999. He also examines the effects of intellectual property protection, particularly patents, on the competitive strategies of firms in high-technology industries. He is a Research Associate in the National Bureau of Economic Research's Corporate Finance and Productivity Programs, and serves as Co-Organizer of the Innovation Policy and the Economy Group.

In the 1993-94 academic year, he introduced an elective course for second-year MBAs on private equity finance. The course materials are collected in *Venture Capital and Private Equity: A Casebook*, John Wiley & Sons, 1999. He serves as the Business School's representative on the Harvard University Patent, Trademark and Copyright Committee and as Faculty Chair of the Focused Financial Management Series, a set of targeted executive education courses on current issues in finance.

Albert N. Link

Albert N. Link is Professor of Economics at the University of North Carolina at Greensboro. He received his B.S. in mathematics from the University of Richmond in 1971 and his Ph.D. in economics from Tulane University in 1976. Prior to joining the faculty at the University of North Carolina at Greensboro in 1982, he was on the faculty at Auburn University and was scholar in residence at Syracuse University.

Professor Link's research focuses broadly on the economics of science and technology policy. His publications encompass many dimensions of that field ranging from philosophy of science to the mathematical theory of productivity growth. More specifically, he has written extensively on methods for evaluating public sector and private sector research and development, technology policies to promote economic growth, and corporate strategies to increase competitiveness.

Professor Link has served on data and evaluation advisory panels of the National Academy of Sciences, the Organization for Economic Cooperation and Development (OECD), and government agencies such as the National Science Foundation, NASA, and the National Institute of Standards and Technology. He has also consulted for numerous European Union and APEC governments on science policy and program evaluation.

Among his most recent books are *Public Accountability: Evaluating Technology-Based Institutions* (with John T. Scott), *A Generosity of Spirit: The Early History of Research Triangle Park*, and *Evaluating Public Sector Research and Development*. His scholarly papers have appeared in such journals as the *Journal of Political Economy*, *American Economic Review*, *Research Policy*, *STI Review*, and the *International Journal of Industrial Organization*. Professor Link is also the editor of the international *Journal of Technology Transfer*.

John T. Scott

John T. Scott received a Ph.D. in economics from Harvard University and holds the position of Professor of Economics at Dartmouth College. His research is in the areas of industrial organization and the economics of technological change. He has served as the President of the Industrial Organization Society and on the editorial boards of the *International Journal of Industrial Organization,* the *Review of Industrial Organization,* and the *Journal of Industrial Economics.* He has consulted in matters of technology policy for the National Institute of Standards and Technology, the National Science Foundation, the National Academy of Sciences, and the Organization for Economic Cooperation and Development, and he has served as an economist at the Board of Governors of the Federal Reserve System and at the Federal Trade Commission.

Claudia Weigand

Claudia Weigand is currently a Researcher for the Dutch Central Bank in Amsterdam and an Ameritech Research Fellow at the Institute for Development Strategies at Indiana University. She also served as an Assistant Professor of Economics at the University of Erlangen-Nürnberg from 1994-1998 where she earned her doctoral degree in economics. Her principal research interests include banking supervision and prudential regulation.

Dr. Weigand is the author of *Bank Lending and Product Market Competition,* (Hamburg: Dr. Kovac, 1998) and co-authored "Internationalization and the Spatial Structure of Markets" (with David Audretsch) in *Corporate Strategies in*

Domestic and Globalizing Markets (Hans-Eckart Scharrer, ed., Baden-Baden: Nomos Verlagsgesellschaft, 1999).

Juergan Weigand

Juergen Weigand is a Professor of Economics at the Otto Beisheim School of Management (Wissenschaftliche Hochschule für Unternehmensführung WHU) in Koblenz-Vallendar, Germany, and an Advisor to the Netherlands Bureau for Economic Policy Analysis in The Hague. He previously served as an Ameritech Research Scholar at the Institute for Development Strategies at Indiana University. His research interests include industrial organization, economics of finance and banking, corporate governance, economics of sports, and competition policy.

Dr. Weigand is the author of many articles which have been published in journals such as *Kredit und Capital* and *DIW-Vierteljahreshefte zur Wirtschaftsforschung*. He is also the author of three books including *Innovation, Competition, and the Business Cycle*, (Berlin: Dunker & Humblot, 1996). Among his forthcoming works are "Market Size, Fixed Costs and Horizontal Concentration" (with Manfred Neumann, Alexandra Gross, and Markus Münter) in the *International Journal of Industrial Organization* and "Does the Governed Corporation Perform Better? Governance Structures and the Market for Corporate Control in Germany" (with Erik Lehmann) in the *European Finance Review*. He has also made conference presentations for organizations including the American Economic Association and European Economic Association and recently was invited to make a seminar presentation before the Board of Governors of the United States Federal Reserve System.

Dr. Weigand holds both a doctoral and post-doctoral degree in economics from the University of Erlangen-Nürnberg.

Charles W. Wessner

Dr. Wessner is the Director of the Program on Technology and Competitiveness for the National Research Council's Board on Science, Technology, and Economic Policy. Dr. Wessner began his federal career with the U.S. Treasury, served overseas as an international civil servant with the OECD and as a senior officer with the U.S. Diplomatic Corps, and directed the Office of International Technology Policy in the Technology Administration of the Department of Commerce. Since joining the National Research Council, he has led several major studies working closely with the senior levels of the U.S. government, leading industrialists, and prominent academics. Recent work includes a White House-initiated study on "The Impact of Offsets on the U.S. Aerospace Industry" and a major international study on "Competition and Cooperation in National Competition for High Technology Industry" in cooperation with the HWWA in Hamburg and the IFW in Kiel, Germany.

Currently, he is directing a portfolio of activities centered around "Government-Industry Partnerships for the Development of New Technologies" and initiating work on "Measuring and Sustaining the New Economy." The Partnerships program constitutes one of the first program-based efforts to assess U.S. policy on government-industry partnerships. Recent publications include *Conflict and Cooperation in National Competition for High Technology Industry*, *Policy Issues in Aerospace Offsets*, *International Friction and Cooperation in High-Technology Development and Trade*, *Trends and Challenges in Aerospace Offsets*, *New Vistas in Transatlantic Science and Technology Cooperation*, *Industry-Laboratory Partnerships: A Review of the Sandia Science and Technology Park Initiative*, *The Advanced Technology Program: Challenges and Opportunities*, and *The Small Business Innovation Research Program: Challenges and Opportunities*. Dr. Wessner holds degrees in International Affairs from Lafayette College (Phi Beta Kappa) and the Fletcher School of Law and Diplomacy where he obtained an M.A., an M.A.L.D., and a Ph.D. as a Shell Fellow.

Robert H. Wilson

A member of the faculty of the LBJ School of Public Affairs, University of Texas at Austin, since 1979, Robert Wilson holds the Mike Hogg Professorship of Urban Policy. He teaches local and state economic development policy, technology policy, applied econometrics, public policy in Brazil, and local governance in developing countries. He was Assistant Dean at the LBJ School from 1980 through 1983 and served as the Coordinator of the Ph.D. Program in Public Policy from 1991 through 1994. Dr. Wilson has served as the Director of the Urban Issues Program, a university-wide program based in the Office of the Provost, since 1995, and Director of the Brazil Center since 2000. His most recent books include *States and the Economy: Policymaking and Decentralization,* and *Public Policy and Community: Activism and Governance in Texas.* Before coming to UT, Wilson taught urban planning at the Federal University of Pernambuco in Brazil. During the spring of 1999, Wilson held the International Philips Chair at the Getulio Vargas Foundation in Sao Paulo, Brazil. During the summer of 2000, Wilson held the Fulbright/FLAD Chair in Knowledge Management and Policy at the Instituto Tecnico Superior in Lisbon. He served as Fulbright Fellow in Belo Horizonte, Brazil and as a United States Information Agency Lecturer in Brazil and Argentina and has served as a consultant to the United Nations Development Program, Organization of American States, National Research Council, Urban Institute, Texas Legislative Education Board, and Texas Historical Commission. He holds a Ph.D. in City and Regional Planning from the University of Pennsylvania.

Annex B

Participants*

**The Small Business Innovation Research Program:
A Review of Current Research**

**May 5, 1999
Washington, D.C.**

Leslie Aitcheson
National Technology Transfer Center

Martin Apple
Council of Scientific Society Presidents

Robert Archibald
College of William and Mary

David Audretsch
Indiana University

Ken Bannister
United States Army Research Office

Jon Baron
Department of Defense

Joseph Bishop
Department of Commerce

Richard Bissell
National Research Council

William Bonvillian
Office of Senator Joseph Lieberman

Peter Cahill
BRTRC

Dennis Carroll
General Accounting Office

*Speakers are *italicized.*

Dennis Chamot
National Research Council

Connie Chang
National Institute of Standards and Technology

Charles Cleland
Department of Agriculture

William Colglazier
National Research Council

Jennifer Connell Dowling
SEMI

Paul Cooksey
Senate Committee on Small Business

Ron Cooper
United States Small Business Administration

Giovanni Coratolo
United States Chamber of Commerce

Ritchie Coryell
Industrial Innovation Programs

Reid Cramer
University of Texas at Austin

Mark Crawford
New Technology Week

Laureen Daly
Office of Senator Lieberman

K.C. Das
Council on Information Management

Arlene de Blanc
Department of Energy

Marshall Dick
Environmental Protection Agency

Damon Dozier
Senate Committee on Small Business

Lee Eiden
Department of Education

Maryann Feldman
Johns Hopkins University

David Finifter
College of William and Mary

Kevin Finneran
Issues in Science and Technology

James Gallup
Environmental Protection Agency

Scott Geesey
House Science Committee

David Goldston
Office of Congressman Boehlert

Jo Anne Goodnight
National Institutes of Health

Wanda Gozdy
National Association of Women Business Owners

Heike Grimm
German-American Center for Visiting Scholars

Jeff Grove
House Science Committee

Adi Guzdar
Foster-Miller, Inc.

Kathy Hale
General Accounting Office

Bruce Hamilton
National Science Foundation

Laura Holliday
National Research Council

John Horrigan
National Research Council

Dean Hudson
Army Research Lab

Connie Jacobs
DARPA

Robert Jaeger
National Institute on Disability and Rehabilitation

John Jankowski
National Science Foundation

James Kalshoven, Jr.
NASA Goddard Space Flight Center

E.V. Khan
Department of Encrgy

Robert Kispert
Massachusetts Technology Collaborative

Richard Koski
SEMATECH

Norman Kreisman
Department of Energy

Patrice Laget
European Commission

Ruth Lange
Department of Agriculture

Clifford Lanham
Technology Planning and Management

Sean Lilly
Technology Administration
Department of Commerce

Steven Lingle
Environmental Protection Agency

Albert Link
University of North Carolina at Greensboro

Molly Macauley
Resources for the Future

Sheila Maith
Office of Senator Kennedy

Christina Mattar
Department of Commerce

Clark McFadden
Dewey Ballantine

Yogesh Mehrotra
Materials Technology Corporation

Stephen Merrill
National Research Council

Duncan Moore
White House Office of Science and Technology Policy

Julie Moses
British Embassy

Kesh Narayanan
National Science Foundation

Allen Neece
Neece, Cator & Associates, Inc.

Carl Nelson
Carl Nelson Consulting, Inc.

Karen Pera
Department of Defense

William Peterson
National Institute on Disability and Rehabilitation

Kees Plangue
Embassy of the Netherlands

Gregory Reck
NASA Headquarters

Veronica Romanov
Small Business Exporters Association

Sally Rood
Federal Laboratory Consortium

Morrie Ruffin
Biotechnology Industry Association

Wendy Schacht
Congressional Research Service

Craig Schultz
National Research Council

John Scott
Dartmouth College

Larry Steranka
Vanderbilt University

Gregory Tassey
National Institute of Standards and Technology

Roland Tibbets
National Science Foundation, retired

James Turner
House Science Committee

Christine Villa
BRTRC

John Williams
Department of the Navy

Charles Wessner
National Research Council

Robert Wilson
University of Texas at Austin

James Woo
Interscience, Inc.

William Wulf
National Academy of Engineering

Annex C

Case Study Template

I. Description of the Firm

1) History of firm: founding date, founder, founder's background.

2) Product or service: description of the firm's main good or service.

3) Technology: description of technology produced.

4) Projected market size and major competitors.

5) What is the firm's competitive advantage?

6) Reasons for location of the firm (e.g., labor pool, nearby research facilities or science parks, markets, access to capital).

II. SBIR Project Information

7) SBIR technology: description of technology for which SBIR award was given and technology's origin.

8) Number of SBIR awards won.

9) What is the relationship to the agency mission?

III. Impact of SBIR

A. SBIR & Firm Strategy

10) Role of SBIR award in company strategy: How important was the award to the firm's current position? What alternative sources of funding were considered and/or enhanced by the SBIR award? Do they anticipate applying for further SBIR awards? If the firm received earlier SBIR awards, how did these affect the current award?

11) Was the SBIR and/or Fast Track award used as a marketing tool? That is, did the firm use the presence of the award as leverage to attract additional outside capital? Did the firm use the Fast Track policy—i.e., the opportunity for outside investors to obtain up to a 4:1 match on their investment—as leverage to attract outside capital?

B. Commercialization

12) Does a strategy exist for commercializing the product? For example, strategic alliances for production, such as a joint venture or licensing. If possible, please disclose names of firms.

13) Has the SBIR-funded technology generated any patents? Is it expected to generate patents? Have any scientific papers resulted?

13a) How was the commercialization strategy affected by participation in SBIR? By participation in Fast Track?

14) Has the firm sold a product resulting from the SBIR project? If not, when does it anticipate selling its first product? Does the firm have specific customers interested in the product? If so, who? Does the firm have anyone on its board of directors or in a management position that has built a successful company before and taken it public?

C. Financing and external partners

15) Does the SBIR awardee have external financing? If so, how much and will it identify the partner? Did the SBIR award play a role in the external partner's decision to provide funding. Where is the partner located? How was the relationship with the partner developed? Did the SBIR award play a role in securing other external investment? Does the SBIR awardee have internal financing? If so, how much?

D. Other Impacts

16) Has participation in the SBIR program generated other types of impacts, relationships, or opportunities?

IV. SBIR Program Administration

17) Did the requirements of SBIR in general and Fast Track in particular prove helpful or onerous in terms of delay or impact on external funding? For standard SBIR awards, did the delay between Phase I and Phase II increase time to market for the firm's product? Did the Fast Track award improve time to market for the firm or allow the firm to maintain continuity of effort?

18) How could the SBIR and especially the Fast Track program be improved?

19) How did the firm become aware of the SBIR program? Of the Fast Track element of SBIR?

20) Did the firm experience difficulties in preparing the SBIR application?

V. Perspectives of Third Party Investors (if relevant)

21) How did the third-party investor become aware of the project being funded by SBIR?

22) What prospects does the SBIR project hold for the third-party investor's organization?

23) Did the SBIR award influence the third-party investor's decision to invest in the project? Did the Fast Track policy (i.e., significantly higher chance of Phase II award for projects attracting outside investors) influence the decision by (a) enabling the third-party investor to leverage its investment in the company, (b) "certifying" the promise of the technology (through a government review of and implicit approval of the technology)?

VI. Cross-Cutting Research Questions:

24) Does the firm think that the DoD policy of giving a higher chance of Phase II award to companies that attract outside investors (per Fast

Track) is (a) a useful way of focusing the SBIR program on companies with strong commercialization capabilities; (b) good public policy?) What factors influence a firm's decision to participate in Fast Track?

25) What factors inhibit a firm's participation in Fast Track?

26) What benefits do firms expect to gain from a Fast Track award not available through the regular Phase II process?

27) Does the Fast Track award affect the performance of firms? How (e.g., in terms of research capabilities or commercialization prospects)?

28) What specific effects does the presence of a third-party investor have on performance?

29) Are differential impacts of Fast Track observed by region, DoD funding agency, or firm characteristics?

30) What do participating firms see as strengths and weaknesses of the Fast Track program?

Annex D

Description of Department of Defense SBIR Fast Track Program*

Since October 1995 DoD's SBIR and STTR programs have featured a "Fast Track" process for SBIR/STTR projects that attract outside investors who will match Phase II funding, in cash, contingent on the project's selection for Phase II award. Projects that obtain such outside investments and thereby qualify for the Fast Track will (subject to qualifications described in the solicitation):

- Receive interim funding of $30,000 to $50,000 between phases I and II;
- Be evaluated for Phase II award under a separate, expedited process; and
- Be selected for Phase II award provided they meet or exceed a threshold of "technically sufficient" and have substantially met their Phase I technical goals.

Consistent with DoD policy, this process should prevent any significant gaps in funding between phases I and II for Fast Track projects, and result in a significantly higher percentage of Fast Track projects obtaining Phase II award than non-Fast Track projects.

Thus far, over 90 percent of projects qualifying for the Fast Track have received interim funding and been selected for Phase II award. As of April 2000, 156 projects are on the Fast Track and, under these projects, $110 million in DoD SBIR funds has directly leveraged at least $50 million in matching funds from outside investors.

* http://www.acq.osd.mil/sadbu/sbir/overview.html

Under the Fast Track, a small company can offer an investor the opportunity to obtain a match of between one and four dollars in DoD SBIR (or STTR) funds for every dollar the investor puts in.

To qualify for the Fast Track, small companies and their outside investors must follow the procedures detailed in *Section 4.5* of the SBIR solicitation. The most important of these procedures are summarized as follows.

First, toward the end of a small company's Phase I SBIR (or STTR) project, the company and its investor submit a Fast Track application. In the Fast Track application, the company and investor:

- State that the investor will match both interim and phase II SBIR (or STTR) funding, in cash, contingent on the company's selection for Phase II award. The matching rates needed to qualify for the Fast Track are as follows:
- *For small companies that have never before received a phase II SBIR or STTR award* from DoD or any other federal agency, the matching rate is 25 cents for every SBIR (or STTR) dollar. (For example, if such a company receives interim and Phase II SBIR funding that totals $750,000, it must obtain matching funds from the investor of $187,500.)
- *For all other companies,* the matching rate is 1 dollar for every SBIR (or STTR) dollar. (For example, if such a company receives interim and Phase II SBIR funding that totals $750,000, it must obtain matching funds from the investor of $750,000.)

The matching funds may pay for additional R&D on the company's SBIR (or STTR) project or, alternatively, they may pay for other activities (e.g., marketing) that further the development and/or commercialization of the technology.

- Certify that the outside funding qualifies as a "Fast Track investment," and the investor qualifies as an "outside investor," as defined in *DoD Fast Track Guidance.* Outside investors may include such entities as another company, a venture capital firm, an individual "angel" investor, a non-SBIR, non-STTR government program; they do not include the owners of the small business, their family members, and/or affiliates of the small business.

Second, DoD will notify each Fast Track company, no later than 10 weeks after the end of Phase I, whether it has been selected for Phase II award. Once notified, the company and investor must certify, within 45 days, that the entire amount of the matching funds from the outside investor has been transferred to the company.

Annex E

Bibliography*

Acs, Z. and D. Audretsch. 1991. *Innovation and Small Firms*. Cambridge, MA: MIT Press.

Alic, John A., and Lewis Branscomb, Harvey Brooks, Ashton B. Carter, and Gerald L. Epstein. 1992. *Beyond Spinoff: Military and Commercial Technologies in a Changing World.* Boston: Harvard Business School Press.

Arrow, Kenneth. 1962. "Economic welfare and the allocation of resources for invention." Pp. 609-625 in *The Rate and Direction of Inventive Activity: Economic and Social Factors.* Princeton, NJ: Princeton University Press.

Arrow, Kenneth. 1973. "The theory of discrimination." Pp. 3-31 in *Discrimination In Labor Markets,* Orley Ashenfelter and Albert Rees, eds. Princeton, NJ: Princeton University Press.

Audretsch, David B. 1995. *Innovation and Industry Evolution*. Cambridge, MA: MIT Press.

Audretsch, David B., and Maryann P. Feldman. 1996. "R&D spillovers and the geography of innovation and production." *American Economic Review* 86(3):630-640.

Audretsch, David B., and Paula E. Stephan. 1996. "Company-scientist locational links: The case of biotechnology." *American Economic Review* 86(3):641-642.

Audretsch, D. and R. Thurik. 1999. *Innovation, Industry Evolution, and Employment.* Cambridge, MA: MIT Press.

Barfield, C. and W. Schambra, eds., 1986. *The Politics of Industrial Policy.* Washington, DC: American Enterprise Institute for Public Policy Research.

Baron, Jonathan. 1998. "DoD SBIR/STTR Program Manager." Comments at the Methodology Workshop on the Assessment of Current SBIR Program Initiatives, Washington, DC, October.

Barry, C.B. 1994. "New directions in research on venture capital finance." *Financial Management* 23 (Autumn):3-15.

Bator, Francis. 1958. "The anatomy of market failure." *Quarterly Journal of Economics* 72:351-379.

Bingham, R. 1998. *Industrial Policy American Style: From Hamilton to HDTV*. New York: M.E. Sharpe.

Birch, D. 1981. "Who Creates Jobs." *The Public Interest* 65 (Fall):3-14.

* Includes selected references from the papers as well as a variety of references of general interest.

Branscomb, L. and J. Keller, 1998. *Investing in Innovation: Creating a Research and Innovation Policy*. Cambridge, MA: MIT Press.

Brav, A. and P. A. Gompers, 1997. "Myth or reality?: Long-run underperformance of initial public offerings; Evidence from venture capital and nonventure capital-backed IPOs." *Journal of Finance* 52:1791-1821.

Brown, G. and Turner J., 1999. "Reworking the Federal Role in Small Business Research." *Issues in Science and Technology* XV(4, Summer).

Bush, Vannevar. 1946. *Science—the Endless Frontier*. Republished in 1960 by U.S. National Science Foundation, Washington, D.C.

Caves, Richard E. 1998. "Industrial organization and new findings on the turnover and mobility of firms." *Journal of Economic Literature* 36(4):1947-1982.

Clinton, William Jefferson. 1994. *Economic Report of the President*. Washington, DC: U.S. Government Printing Office.

Coburn, C. and Bergland, D., 1995. *Partnerships: A Compendium of State and Federal Cooperative Technology Programs*. Columbus, OH: Battelle.

Cohen, L. R. and R. G. Noll. 1991. *The Technology Pork Barrel*. Washington, DC: The Brookings Institution.

Council of Economic Advisers, 1995. *Supporting Research and Development to Promote Economic Growth: The Federal Government's Role*. Washington, DC.

Davis, S.J., J. Haltiwanger, and S. Schuh. 1994. "Small Business and Job Creation: Dissecting the Myth and Reassessing the Facts," *Business Economics* 29(3):113-22.

Eisinger, P. K., 1988. *The Rise of the Entrepreneurial State: State and Local Economic Development Policy in the United State*. Madison, WI: University of Wisconsin Press.

Feldman, Maryann P. 1994a. "Knowledge complementarity and innovation." *Small Business Economics* 6(5):363-372.

Feldman, Maryann P. 1994b. *The Geography of Knowledge*. Boston: Kluwer Academic.

Fenn, G. W., N. Liang, and S. Prowse, 1995. *The Economics of the Private Equity Market*. Washington, DC: Board of Governors of the Federal Reserve System.

Flamm, K., 1988. *Creating the Computer*. Washington, DC: The Brookings Institution.

Freear, J., and W. E. Wetzel, Jr. 1990. "Who bankrolls high-tech entrepreneurs?" *Journal of Business Venturing* 5:77-89.

Freeman, Chris and Luc Soete. 1997. *The Economics of Industrial Innovation*. Cambridge, MA: MIT Press.

Galbraith, J. K. 1957. *The New Industrial State*. Boston: Houghton Mifflin.

Geroski, Paul A. 1995. "What do we know about entry?" *International Journal of Industrial Organization* 13(4):421-440.

Gompers, P. A., 1995. "Optimal investment, monitoring, and the staging of venture capital." *Journal of Finance* 50:1461-1489.

Gompers, P. A. and J. Lerner. 1996. "The use of covenants: An empirical analysis of venture partnership agreements." *Journal of Law and Economics* 39:463-498.

Gompers, P. A. and J. Lerner. 1999. "An analysis of compensation in the U.S. venture capital partnership." *Journal of Financial Economics* 51(1):3-7.

Gompers, P. A. and J. Lerner. 1998. "What drives venture capital fund-raising?" Unpublished working paper, Harvard University.

Gompers, P. A. and J. Lerner. 1998b. "Capital formation and investment in venture markets: A report to the NBER and the Advanced Technology Program." Unpublished working paper, Harvard University.

Gompers, P. A. and J. Lerner. 1999. *The Venture Cycle*. Cambridge, MA: MIT Press.

Good, M. L. 1995. Prepared testimony before the Senate Commerce, Science, and Transportation Committee, Subcommittee on Science, Technology, and Space (Photocopy, U.S. Department of Commerce).

Graham, O. L. 1992. *Losing Time: The Industrial Policy Debate*. Cambridge, MA: Harvard University Press.

Greenwald, B. C., J. E. Stiglitz, and A. Weiss, 1984. "Information imperfections in the capital market and macroeconomic fluctuations." *American Economic Review Papers and Proceedings* 74:194-199.

Griliches, Z. 1990. *The Search for R&D Spillovers*. Cambridge, MA: Harvard University Press.

Hall, Bronwyn H. 1992. "Investment and research and development: Does the source of financing matter?" Working Paper No. 92-194, Department of Economics, University of California at Berkeley.

Hall, Bronwyn H. 1993. "Industrial research during the 1980s: Did the rate of return fall? " Brookings Papers: *Microeconomics* 2:289-343.

Hamberg, Dan, 1963. "Invention in the industrial research laboratory." *Journal of Political Economy* (April):95-115.

Hao, K. Y. and A. B. Jaffe, 1993. "Effect of liquidity on firms' R&D spending." *Economics of Innovation and New Technology* 2:275-282.

Hebert, Robert F. and Albert N. Link. 1989. "In search of the meaning of entrepreneurship." *Small Business Economics* 1(1):39-49.

Himmelberg, C. P. and B. C. Petersen, 1994. "R&D and internal finance: A panel study of small firms in high-tech industries." *Review of Economics and Statistics* 76:38-51.

Hubbard, R. G. 1998. "Capital-market imperfections and investment." *Journal of Economic Literature* 36:193-225.

Huntsman, B. and J. P. Hoban, Jr. 1980. "Investment in new enterprise: Some empirical observations on risk, return, and market structure." *Financial Management* 9 (Summer): 44-51.

Jaffe, A. B. 1996, "Economic analysis of research spillovers—Implications for the Advanced Technology Program." (Washington, Advanced Technology Program, National Institute of Standards and Technology, U.S. Department of Commerce).

Jaffe, A. B. 1998. "The importance of 'spillovers' in the policy mission of the Advanced Technology Program." *Journal of Technology Transfer* (Summer).

Jewkes, J., D. Sawers, and R. Stillerman. 1958. *The Sources of Invention*. New York: St. Martin's Press.

Kleinman, D. L.. 1995. *Politics on the Endless Frontier: Postwar Research Policy in the United States*. Durham, NC: Duke University Press.

Krugman, P. 1990. *Rethinking International Trade*. Cambridge, MA: MIT Press.

Krugman, P. 1991. *Geography and Trade*. Cambridge, MA: MIT Press.

Lebow, I. 1995. *Information Highways and Byways: From the Telegraph to the 21st Century*. New York: Institute of Electrical and Electronic Engineering.

Lerner, J. 1994. "The syndication of venture capital investments." *Financial Management* 23 (Autumn):16-27.

Lerner, J. 1995. "Venture capital and the oversight of private firms." *Journal of Finance* 50:301-318.

Lerner, J. 1996. "The government as venture capitalist: The long-run effects of the SBIR program" (Working paper no. 5753, National Bureau of Economic Research).

Lerner, J. 1998. "Angel financing and public policy: An overview" *Journal of Banking and Finance* 22(6-8):773-784.

Lerner, J. 1999. "The government as venture capitalist: The long-run effects of the SBIR program." *Journal of Business* 72(3):285-297.

Lerner, J. 1999. "Public venture capital: Rationales and evaluation," in *The SBIR Program: Challenges and Opportunities*. Washington, D.C.: National Academy Press.

Levy, D. M. and N. Terleckyk, 1983. "Effects of government R&D on private R&D investment and productivity: A macroeconomic analysis" *Bell Journal of Economics* 14:551-561.

Liles, P. 1977. *Sustaining the Venture Capital Firm*. Cambridge, MA: Management Analysis Center.

Link, Albert N. and John T. Scott. 1998. *Overcoming Market Failure: A Case Study of the ATP Focused Program on Technologies for the Integration of Manufacturing Applications (TIMA).* Draft final report submitted to the Advanced Technology Program. Gaithersburg, MD: National Institute of Technology. October.

Link, Albert N. 1998. "Public/Private Partnerships as a Tool to Support Industrial R&D: Experiences in the United States." Paper prepared for the working group on Innovation and Technology Policy of the OECD Committee for Science and Technology Policy, Paris.

Link, Albert N. and John T. Scott. 1998a. "Assessing the infrastructural needs of a technology-based service sector: A new approach to technology policy planning." *STI Review* 22: 171-207.

Link, Albert N. and John T. Scott. 1998b. *Public Accountability: Evaluating Technology-Based Institutions.* Norwell, Mass.: Kluwer Academic.

Link, Albert N. and John Rees. 1990. "Firm size, university based research and the returns to R&D." *Small Business Economics* 2(1):25-32.

Malone, T. 1995. *The Microprocessor: A Biography.* Hamburg, Germany: Springer Verlag/Telos.

Mansfield, E., J. Rapoport, A. Romeo, S. Wagner, and G. Beardsley, 1977. "Social and private rates of return from industrial innovations" *Quarterly Journal of Economics* 91: 221-240.

Mansfield, E. 1985. "How Fast Does New Industrial Technology Leak Out?" *Journal of Industrial Economics.* 34(2).

Mansfield, E. 1996. *Estimating Social and Private Returns from Innovations Based on the Advanced Technology Program: Problems and Opportunities.* Unpublished report.

McCraw, T. 1986. Mercantilism and the market: antecedents of American industrial policy, in Barfield, C. and Schambra, W., eds., *The Politics of Industrial Policy.* Washington, DC: American Enterprise Institute for Public Policy Research.

Mervis, Jeffrey D. 1996. A $1 Billion "'Tax' on R&D Funds." *Science* 272:942-944.

Mowery, D. ed. (1999) *U.S. Industry in 2000: Studies in Competitive Performance.* Washington, DC: National Academy Press.

Mowery, D. 1998. "Collaborative R&D: how effective is it," *Issues in Science and Technology* Fall, pp. 37-44.

Mowery, D. and N. Rosenberg. 1998. *Paths of Innovation: Technological Change in 20th Century America.* New York: Cambridge University Press.

Mowery, D. and N. Rosenberg. 1989. *Technology and the Pursuit of Economic Growth.* New York: Cambridge University Press.

Myers, S., R. L. Stern, and M. L. Rorke. 1983. *A Study of the Small Business Innovation Research Program.* Lake Forest, Illinois: Mohawk Research Corporation.

Myers, S. C. and N. Majluf, 1984. "Corporate financing and investment decisions when firms have information that investors do not have," *Journal of Financial Economics* 13:187-221.

National Research Council, 1996. *Conflict and Cooperation in National Competition for High-Technology Industry.* Washington, DC: National Academy Press.

National Research Council, 1997. *Review of the Research Program of the Partnership for a New Generation of Vehicles: Third Report.* Washington, DC: National Academy Press.

National Research Council, 1999. *Funding A Revolution: Government Support for Computing Research.* Washington DC: National Academies Press.

Nelson, R. R. 1982. *Government and Technological Progress.* New York: Pergamon.

Nelson, R. R. 1986. "Institutions supporting technical advances in industry." *American Economic Review, Papers and Proceedings* 76(2): 188.

Nelson, R. R., ed. 1993. *National Innovation System: A Comparative Study.* New York: Oxford University Press.

Office of the President. 1990. *U.S. Technology Policy.* Washington, DC: Executive Office of the President.

Office of Management and Budget. 1996. *Economic analysis of federal regulations under Executive Order 12866.* Mimeo.

Organization for Economic Cooperation and Development. 1982. *Innovation in Small and Medium Firms*. Paris: OECD.

Organization for Economic Cooperation and Development, 1997. *Small Business Job Creation and Growth: Facts, Obstacles, and Best Practices*. Paris: OECD.

Organization for Economic Cooperation and Development. 1998. *Technology, Productivity and Job Creation: Toward Best Policy Practice*. Paris: OECD.

Organisation for Economic Co-operation and Development, 1995. *Venture capital in OECD countries*. Paris: OECD.

Perret, G. 1989. *A Country Made by War: From the Revolution to Vietnam—The Story of America's Rise to Power*. New York: Random House.

Powell, Walter W. and Peter Brantley. 1992. "Competitive cooperation in biotechnology: Learning through networks?" Pp. 366-394 in: N. Nohria and R. G. Eccles (eds.), *Networks and Organizations: Structure, Form and Action*. Boston: Harvard Business School Press.

Price Waterhouse. 1985. *Survey of small high-tech businesses shows Federal SBIR awards spurring job growth, commercial sales*. Washington, DC: Small Business High Technology Institute.

Roberts, Edward B. 1968. "Entrepreneurship and technology." *Research Management* (July):249-266.

Romer, P. 1990. "Endogenous technological change." *Journal of Political Economy* 98:71-102.

Rosenbloom, R. and Spencer, W. 1996. *Engines of Innovation: U.S. Industrial Research at the End of an Era*. Boston: Harvard Business Press.

Rubenstein, A. H. 1958. *Problems Financing New Research-Based Enterprises in New England*. Boston Mass.: Federal Reserve Bank.

Sahlman, W. A. 1990. "The structure and governance of venture capital organizations." *Journal of Financial Economics* 27:473-521.

Saxenian, Annalee. 1994. Regional advantage: Culture and competition in Silicon Valley and Route 128. Cambridge, MA: Harvard University Press.

Scherer, F. M. 1970. *Industrial Market Structure and Economic Performance*. New York: Rand McNally College Publishing.

Small Business Administration. 1994. *Small Business Innovation Development Act: Tenth-Year Results*. Washington, DC: U.S. Government Printing Office (and earlier years).

Small Business Administration. 1992. *Results of Three-Year Commercialization Study of the SBIR program*. Washington, DC: U.S. Government Printing Office.

Stiglitz, J. E. and A. Weiss. 1981. "Credit rationing in markets with incomplete information." *American Economic Review* 71:393-409.

Schumpeter, J. 1950. *Capitalism, Socialism, and Democracy*. New York: Harper and Row.

Scott, John T. 1998. "Financing and leveraging public/private partnerships: The hurdle-lowering auction." *STI Review* 23:67-84.

Stiglitz, J. E. and A. Weiss. 1981. "Credit rationing in markets with incomplete information." *American Economic Review* 71:393-409.

Stowsky, J. 1996. "Politics and Policy: The Technology Reinvestment Program and the Dilemmas of Dual Use." Mimeo: University of California.

U.S. Congress, House. 1992. *SBIR and Commercialization: Hearing Before the* Subcommittee on Technology and Competitiveness of the House Committee on Science, Space, and Technology, on the Small Business Innovation Research [SBIR] *Program* (testimony of James A. Block, President of Creare Inc.), pp. 356-361.

U.S. Congress, Senate Committee on Small Business. 1981. Small Business Research Act of 1981. S.R. 194, 97th Congress.

U.S. Congressional Budget Office. 1985. *Federal financial support for high-technology industries*. Washington, DC: U.S. Congressional Budget Office.

U.S. General Accounting Office. 1987. *Federal research: Small Business Innovation Research participants give program high marks*. Washington, DC: U.S. General Accounting Office.

U.S. General Accounting Office. 1989. *Federal research: Assessment of Small Business Innovation Research program.* Washington, DC: U.S. General Accounting Office.

U.S. General Accounting Office. 1992. *Federal research: Small Business Innovation Research shows success but can be strengthened.* Washington, DC: U.S. General Accounting Office.

U.S. General Accounting Office. 1997. *Federal Research: DoD's Small Business Innovation Research Program.* RCED-97-122, Washington, DC: U.S. General Accounting Office.

U.S. General Accounting Office. 1999. *Federal Research: Evaluations of Small Business Innovation Research Can Be Strengthened.* RCED-99-198, Washington, DC: U.S. General Accounting Office.

U. S. General Accounting Office. 1998. *Federal Research: Observations on the Small Business Innovation Research Program.* RCED-98-132. Washington, DC: GAO.

U.S. General Accounting Office. 1992. *Small Business Innovation Research Program Shows Success But Can Be Strengthened.* RCED-92-32. Washington, DC: GAO.

U.S. Senate Committee on Small Business. 1981. Senate Report 97-194. *Small Business Research Act of 1981.* September 25, 1981. Washington, DC: U.S. Government Printing Office.

U.S. Senate Committee on Small Business. 1999. Senate Report 106-330, *Small Business Innovation Research (SBIR) Program.* August 4, 1999. Washington, DC: U.S. Government Printing Office.

U.S. Small Business Administration. 1994. *Small Business Innovation Development Act: Tenth-year results.* Washington, DC: U.S. Government Printing Office.

Venture Economics. 1988. *Exiting venture capital investments.* Wellesley, MA: Venture Economics.

Venture Economics. 1996. "Special Report: Rose-colored asset class." *Venture Capital Journal* 36 (July) 32-34 (and earlier years).

VentureOne. 1997. National Venture Capital Association 1996 annual report. San Francisco: VentureOne.

Wallsten, S. J. 1996. The Small Business Innovation Research Program: Encouraging Technological Innovation and Commercialization in Small Firms. (Unpublished working paper, Stanford University).

Wessner, Charles W., ed. 1999. *Industry-Laboratory Partnerships: A Review of the Sandia Science and Technology Park Initiative.* Washington, DC: National Academy Press.

Wessner, Charles W., ed. 1999. *The Small Business Innovation Research Program: Challenges and Opportunities.* Washington, DC: National Academy Press.

Wessner, Charles W., ed. 1999. *The Advanced Technology Program: Challenges and Opportunities.* Washington, DC: National Academy Press, 1999.

Wessner, Charles W., ed. 1999. *New Vistas in Transatlantic Science and Technology Cooperation.* Washington, DC: National Academy Press, 1999.

Tassey, Gregory. 1997. *The Economics of R&D Policy.* Westport, CT: Quorum Books.

Tirman, John. 1984. *The Militarization of High Technology.* Cambridge, MA: Ballinger.

Tyson, Laura, Tea Petrin, and Halsey Rogers. 1994. "Promoting entrepreneurship in Eastern Europe." *Small Business Economics* 6:165-184.